·BASIC·
BIOTECHNOLOGY

Edited by

JOHN BU'LOCK

Weizmann Microbial Laboratory
Department of Chemistry
University of Manchester
Manchester M13 9PL
England

BJORN KRISTIANSEN

Department of Bioscience and Biotechnology
Applied Microbiology Division
University of Strathclyde
Glasgow G1 1XW
Scotland

1987

ACADEMIC PRESS
Harcourt Brace Jovanovich, Publishers
London   Orlando   San Diego   New York   Austin
Boston   Sydney   Tokyo   Toronto

ACADEMIC PRESS INC. (LONDON) LTD.
24/28 Oval Road,
London NW1

United States Edition published by
ACADEMIC PRESS INC.
Orlando, Florida 32887

British Library Cataloguing in Publication Data
Basic Biotechnology.
1. Biotechnology
I. Bu'Lock, J.D.   II. Kristiansen, Bjorn
660'.6   TP248.2

ISBN 0–12–140752–7
ISBN 0–12–140753–5 Pbk

Printed in Great Britain by Galliard (Printers) Ltd, Great Yarmouth

Contributors

K. Allermann   Institute of Plant Physiology, University of Copenhagen, Ø Farimagsgade 2A, DK-1353 Copenhagen K, Denmark

H. W. Blanch   Department of Chemical Engineering, University of California, Berkeley, CA 94720, USA

J. W. Brewer   Biotechnology Division, Miles Laboratories Inc, Elkhart, IN 46515, USA

J. D. Bu'Lock   Microbial Chemistry Laboratory, University of Manchester, Manchester M13 9PL, UK

K. Corbett   Beecham Pharmaceuticals, UK Division, Worthing, West Sussex, BN14 8QH, UK

R. P. Elander   Industrial Division, Bristol-Meyers Co, PO Box 4755, Syracuse, NY 13201-4755, USA

M. W. Fowler   Wolfson Institute of Biotechnology, Sheffield University, Sheffield S10 2 TN, UK

W. E. Goldstein   Research and Development Group, Miles Laboratories Inc, Elkhart, IN 46515, USA

C. F. Gölker   Bayer AG, Verfahrensentwicklung Biochemie, Friedrich-Ebert-Strasse 217-333, D-5600 Wuppertal 1, West Germany

D. L. Hawkes   Department of Mechanical and Production Engineering, The Polytechnic of Wales, Pontypridd CF37 1DL, UK

F. R. Hawkes   Department of Science, The Polytechnic of Wales, Pontypridd CF37 1DL, UK

E. A. Jackman   Jackman's Economic and Technical Services Ltd, 244 Highlands Blvd, Leigh-on-Sea S59 3QZ, UK

B. Kristiansen   Department of Bioscience & Biotechnology, Applied Microbiology Division, University of Strathclyde, Royal College Building, 204 George Street, Glasgow G1 1XW, UK

J. L. Meers   Sturge Biochemicals Ltd, Denison Road, Selby SO8 8EF, UK

P. E. Milsom   Sturge Biochemicals Ltd, Denison Road, Selby YO8 8EH, UK

M. Moo-Young   Department of Chemical Engineering, University of Waterloo, Waterloo, Ontario N2L 3G1, Canada

K. Murray   Department of Molecular Biology, University of Edinburgh, King's Building, Mayfield Road, Edinburgh EH9 3JR, UK

J. Olsen   Department of General Microbiology, University of Copenhagen, Solvgarde 83 H, DK-1307 Copenhagen K, Denmark

G. W. Pace   Intra-Optics Laboratories Pty Ltd, PO Box 37, St Ives, NSW 2075, Australia

L. B. Quesnel   Department of Bacteriology and Virology, University of Manchester, Manchester M13 9PT, UK

C. Ratledge   Department of Biochemistry, University of Hull, Hull HU6 7RX, UK

J. A. Roels   Gist Brocades, Research and Development, Wateringseweg 1, PO Box 1, 2600 MA Delft, The Netherlands

G. Schmidt-Kastner   Bayer AG, Verfahrensentwicklung Biochemie, Friedrich-Ebert-Strasse 217-333, D-5600 Wuppertal 1, West Germany

C. G. Sinclair   Department of Chemical Engineering, University of Manchester Institute of Science and Technology, PO Box 88, Manchester M60 1QD, UK

R. E. Spier   Department of Microbiology, University of Surrey, Stag Hill, Guildford GU2 5XH, UK

C. Vezina   Institut Armand-Frappier, University of Quebec, 531 boulevarde des Prairies, Ville de Laval, Quebec H7N 4Z3, Canada

Preface

Jourdain: Par ma foi! il y a plus de quarante ans que je dis de la prose que j'en suisse rien, et je vois suis le plus obligé du monde de m'avoir appris çela.

Molière, Le Bourgeois Gentilhomme II.iv

And many a lakke-of-Dover hastow sold
That hath been twyes hoot and twyes cold

Chaucer, The Cook's Tale—Prologue

Why, when there seems to be such a lot of it about, do we need yet another book about biotechnology—and in particular one that claims to deal with *basic* biotechnology? There are real biotechnologists around, who were already doing the same sort of thing they are doing today well before the term became so fashionable; equally there are people who have elected to call what they are doing by this trendy new term because it sells well. So we have headed this preface with quotations that seem to define those two categories; more tediously, we shall follow the worthy but dull definition promulgated by the European Federation of Biotechnology and define our subject as the practical application of biological organisms, or their subcellular components, to manufacturing and service industries and to environmental management.

Above all, through our contributing authors and as editors, we have tried to present biotechnology as something which is already going on. Whereas for some enthusiasts biotechnology has mainly been a matter of promise—and promises—we have tried to introduce it as a productive technology which has real significance now, and even in the not-so-recent past, as well as potentiality for the future, and which can therefore be introduced to students and new researchers in a 'hard-core' way which is not at all superficial. Perhaps for each new generation of entrants into biotechnology it is the future potentialities that provide the motivation, but it must be the past and present realities that provide the machinery through which that drive will work.

v

In one sense at least, the beginnings of modern biotechnology are exemplified by Weizmann's development of a *practical* acetone–butanol fermentation in 1913–15; prior to this, various 'natural' microbial and enzymic processes had been first used, then interpreted, and—increasingly after Pasteur—controlled and even manipulated, but Weizmann's process was the first *new* productive application of biocatalysis on an industrial scale. Contemporarily, on the other side of the globe and slower to reach industrial significance, the Japanese had been developing both the applications and the production of enzymes. Also contemporary, and by coincidence coming from the same laboratory as Weizmann in Manchester, were the first conscious applications of some (quite crude) microbiology to the large-scale process of sewage treatment, hitherto simply a matter of rather simplistic civil engineering; it is a matter for real regret that some of the aspects of biotechnology arising from this last approach have had to be omitted here because of our over-riding need to present a thorough and self-contained account in quite a small compass.

We have divided our book into two major sections, the first dealing with basic principles and the second with a range of illustrative examples of practical biotechnology and some of its problems. We think that this is the order in which the book itself should be used, but its contents were actually determined by working in reverse order; we started from a range of actual examples of productive biotechnology, by no means exhaustive but hopefully representative, to be dealt with by experts of different kinds, each approaching their own topic in their own way. From these we worked back to define the kinds of basic understanding that would be needed, either to comprehend or to further advance those real processes. As a result our selection of 'basic principles' covers a wide range. However, we believe that no more than a superficial account of biotechnology can be given without taking account of the essential interdisciplinarity of the subject.

It follows that the typical scientist or engineer, and in particular the newly graduated scientist or engineer for whom this book is primarily intended, will find some of the basic principles quite familiar, while others fall quite outside his or her past experience. No matter: even among what is familiar, it will be useful to emphasize what aspects of the known discipline are the most relevant.

The best alternative to superficiality is therefore to try to cope comprehensively with all the relevant (and even partly-relevant) disciplines, and then to append comprehensive 'state-of-the-art' accounts of the equally diverse actual and potential applications. Unfortunately this leads to the assembling of multi-volume compendia, written by experts but edited by committees, meritorious but massive works to be found only in correspondingly well-endowed libraries. Knowing from experience that the task of editing such miscellanies into a coherent account would be beyond our capabilities, we have tried to do something more modest, which we hope will be of real use for a rather larger number of readers.

Sometimes we have been quite ruthless with our authors, despite their individual distinction, and so for the final results of our temerity we must pay by accepting full responsibility for all the errors, inconsistencies, omissions, misplaced emphasis, and plain wrongheadedness that the careful reader will undoubtedly find; *miserere nos*.

For some of the omissions we can make pleas in mitigation. For example, though much of our emphasis is on biotechnology's uses of microorganisms, we have excluded any descriptive microbiology as such, just as we have excluded hydrodynamics, differential calculus, protein chemistry and patent law though all these have their uses and their relevance for biotechnology. It would be possible (and indeed interesting) to compile an account of 'The Organisms of Biotechnology', but the result would probably be neither good biotechnology nor good microbiology. The biotechnologist will need the skills of a microbiologist (or a mathematician or a patent attorney), either personally or as part of the team, for example to understand the requirements of the organism actually being used or to consider what others might fit the application as well or better. Screening, selection, and strain improvement will be the major 'interface' for this, and so we have specifically covered these aspects; perhaps unfortunately, we have not been able to deal with the other interface which is the microbiologists' understanding of the morphology, habit and general life-style of organisms. This is only partly for reasons of space; our present understanding and exploitation of this aspect of biotechnology is actually rather limited. Maybe our omission will stimulate more work in this, as in other under-represented aspects. Thus for microbiology, as for biochemistry and indeed for physics and engineering principles, we have tried simply to indicate what the biotechnologist needs to know in order to ask the microbiologist, or the biochemist or engineer, the right questions—and perhaps to help the microbiologist, biochemist or engineer to understand what they are being asked, and why it matters!

Indeed, if our compilation helps to draw attention to the areas of basic science and engineering in the contributory disciplines that are most relevant to biotechnology, and to promote more research (and more support) for those areas, it will be serving a valuable purpose. In keeping with a narrowly utilitarian age, the fashion for ruthlessly target-oriented research is rapidly leading to a situation in which empirical applications, based on superficial extensions of old understanding, have outstripped the basic knowledge that is needed. Today's shiny edifice of biotechnology is in serious danger of collapsing into inadequate foundations, and the numerous national and international commissions and committees that exist to 'promote' biotechnology need to have this brought rather urgently to their attention.

This, then, is our plea in mitigation to those who will use our book to make their entry into biotechnology. The enthusiasts who are already there will have different objections; there is little here for flavour-of-the-month fans. By describing a broad selection of process biotechnology as it is, rather than speculating about what it might become, we hope our readers will emerge better-equipped to understand not only what they are doing today, but what they might do tomorrow.

Biotechnology is—or it can be—clean technology, green technology, and human-scale technology. It lends itself more readily to improving the human condition than to terminating it. It will become our major technology if mankind has any future, and today we are only seeing its very beginnings. But even in today's world it can also make good brass-nosed economics—if that is all you care about.

John D. Bu'Lock
Pune, India; November 1986

Contents

Part I | Fundamentals and Principles

1 | Introduction to Basic Biotechnology

J. D. BU'LOCK

1.1 What is biotechnology?

Biotechnology is the controlled and deliberate application of simple biological agents—living or dead cells or cell components—in technically useful operations, either of productive manufacture or as service operations. This is the sense in which the term is used here, as being useful to denote collectively a whole series of working techniques and disciplines that are actually in operation, and which are found to share common basic principles with which this book attempts to deal.

The word biotechnology is sometimes used in a much narrower sense, for the pursuit of manipulative genetics and molecular biology in hopefully practical directions; this is to mistake the fashionable part for the serviceable whole, and owes more to the philosophy of the advertising agency than to that of the factory.

Alternatively the word biotechnology can be interpreted in a very broad sense, to embrace all the operations of applied biology from agriculture to cookery. Undoubtedly our technology of dealing with complex organisms —including our own species—is being advanced by modern biology, and our understanding of the many traditional processes in which biological agents are used in less controlled or deliberate ways is being improved, but these broader disciplines are more usefully pursued in their own right (which does not mean that the biotechnologist will not find scope for useful contributions there).

Thus, for example, fermentation biotechnology as we know it today originates not from the ancient domestic discoveries of wine or sauerkraut, and not even from their observational understanding as first approached

3

in the nineteenth century, but from the first applications of selected microbial agents to specific process requirements. Examples would be the defined encouragement and recuperation of adapted biomass for the activated sludge process, or the selection and large-scale propagation of specific *Clostridia* for the production of acetone and butanol; both these developments occurred at Manchester some seventy years ago, but arose from such entirely different background circumstances that it is only now that we can see them as parts of a unified technology.

We can also attempt to define biotechnology by what it does, while hoping to avoid becoming out-of-date too quickly by considering what it might do. In directly productive industry, biotechnology is wholly involved in the production of microbial biomass for animal feed (and, foreseeably, human food), of some commodity chemicals such as citric acid, glutamic acid and some other amino acids, and of some speciality chemicals, notably antibiotics and certain vitamins. In competition with petrochemical technology it can produce bulk products such as ethanol, acetone/butanol, acetic acid, etc., and in competition with whole-organism exploitation it can be used to make special plant substances and animal cell products, the latter either directly or in transformed microbial cells to give antigens, antibodies, or various diagnostic or therapeutic agents.

Biotechnology can provide agriculture with a variety of useful agents, from soil inoculants to veterinary products, with foreseen extensions into aqua- and mari-culture. It is beginning to supplement traditional genetic methods for the development of new or improved plant and animal strains for conventional agriculture to use. It provides food industries with key agents like starter cultures and enzymes, and increasingly it adds to food process understanding and techniques. In the service industries biotechnology has a major role in both aqueous and solids waste treatment, waste valorization, and water purification.

Thus the pragmatic definition of biotechnology is a large one, which clearly changes with time and is certain to expand in new directions we have not foreseen.

1.2   Biotechnology operations

Operationally we can distinguish five main aspects in any biotechnological process, which in most cases will also correspond to stages in its development. The overall picture is summarized in Table 1.1, which also enables us to point out the main contributory disciplines of science and engineering to each aspect. At the same time this presentation is effectively a summary of the ground to be covered in this book; each individual chapter will

Table 1.1

Microbiology and cell biology		Process aspects	
Systematic	⎧	Culture choice	
Genetics	⎨ ⎧	Mass culture	⎫
Physiology	⎩ ⎨	Cell responses	⎬ Process engineering
	⎧ ⎩	Process operation	⎭
Chemistry	⎩	Product recovery	

be found to deal either with individual aspects as distinguished here, or with their overall combination in specific illustrative contexts. Emphasis on the contributory disciplines is not out of place, since these are quite fundamental to any understanding of biotechnology, and while the biotechnologist will not pretend to have mastered all of them, one must at least familiarize oneself with their basic principles, the language they use, the concepts they have developed, and the uses to which they can be put. It is our view that this familiarization should in general extend at least as far as the corresponding topics have been covered here, but always on the basis of a more thorough grounding in at least one of the basic disciplines.

The five aspects are as follows (Sections 1.2.1–1.2.5):

1.2.1 Culture choice: selecting, breeding, or otherwise creating the most suitable starter organism, or cell population

This can involve discovering and selecting the most nearly suitable strain from amongst the enormous variety of natural species of microorganisms, and then 'improving' its heritable characteristics. Such selection usually requires general biological understanding, in knowing where to look and what kind of organism to look for, combined with chemical and biochemical skills in finding how best to detect what one is looking for. Other situations can involve selecting the most suitable mixed population, or selecting a parent line of plant or animal cells with further selection amongst their progeny. A dramatically different alternative set of possibilities is opened up by the techniques which allow deliberate construction of the most suitable cell type by genetic manipulation from parents which can afford the desired hybrid characteristics.

This aspect of biotechnology consequently requires major inputs from systematic microbiology and microbial ecology, microbial and cell physiology, and both classical and molecular genetics.

1.2.2. Mass culture

For biotechnological applications it is essential to be able to conserve the chosen organism for as long as it is needed and then, at will, to multiply it up to a useful scale, which can be very large. Where the biomass itself is *per se* the desired product these requirements are seen most clearly, but a requirement for some degree of mass culture is central to all biotechnology processes.

Microbial or cell physiology is again essential, but now it must be coupled with process engineering of particular kinds, aimed at providing by macro-operations the micro-conditions that are optimal for building up the required biomass.

1.2.3 Cell responses: eliciting desired activities

In the most general case the product or active agent, for which the cells are being cultivated, will only be produced (or manifested or released) most abundantly under rather specific conditions. In general these conditions will not be the same as those for most abundant multiplication of the biomass (Section 1.2.2). Indeed, the ability to exploit the flexible expression of cell characteristics in response to external conditions is a major resource for biotechnology, just as the need to understand those responses and their limitations is a major restriction.

The necessary basic knowledge comes from small-scale experimentation and is again an aspect of microbial or cell physiology, but process engineering aspects are also very relevant in securing the optimum micro-conditions on the macro-scale.

1.2.4 Process operation

It will already be apparent that a biotechnology process does not, in general, reduce to a single operating step. The satisfactory execution of all the required steps, fully optimized for safety, reproducibility, control, and efficiency, at all stages and scales of operation, is for the most part a matter of process engineering design, applied with a full understanding of the relevant biological, chemical, and socioeconomic factors.

In many respects this is one of the least researched and most difficult aspects of biotechnology, if only because the problems must be solved afresh for each new process and even for each process improvement; on the other hand all biotechnology research depends upon this stage for its practical realization, and is only successful insofar as it has been implemented.

1.2.5 Product recovery

Any productive process is only usefully completed to the extent that its products are actually recovered in a useful form; unfortunately this quite obvious fact is very easily overlooked in laboratory-based researches. The problem is particularly acute for biotechnology, however, because of the 'inconvenient' nature of many biotechnology products and the manner in which they are first presented, especially but not exclusively their frequent dilution with large volumes of process water. Not only is the efficiency of product recovery directly reflected in product costs (more directly so than any other factor), but in modern society effective and environmentally-acceptable modes of by-product recovery (including process water and waste heat) are also called for.

The contributory disciplines here are mainly aspects of chemistry and chemical engineering, but not necessarily well-known or popular aspects of either.

1.3 What cells can do

In practice, biotechnology differs from other kinds of applied biology (like medicine or agriculture) in that the biological material is handled at cell level—either populations of individual cells, or cell components, or aggregates comprising relatively few, basically identical, cells.

When such cells constitute the organism as such, we are of course speaking of microbiology. Indeed, it is the ability to handle cells derived from more complex multicellular organisms using the techniques developed by microbiologists that brings such materials within the scope of biotechnology as we know it. Consequently in considering what potentially useful activities we can expect to be able to find for biotechnology, we can use micro-organisms as our most general examples.

A microbial cell constitutes the whole organism and must be able to carry out all the activities that characterize it, have enabled it to evolve, and continue to enable it to survive. All the necessary mechanisms for this are encoded in the cell's DNA even when they are not being actively expressed. Elementary accounts often seem to ignore this, by concentrating on the cell's ability to replicate itself (growth); however, even simple replicatory growth involves reproducing all the other capabilities of the organism. For biotechnologists these other capabilities are equally important; above all they are involved in the reasons for using one kind of cell rather than any other.

For its immediate survival the cell requires an ability to maintain and repair a full complement of cell constituents, and for replicatory growth it further requires the ability to increase each of them in a balanced way. Thus the cell must contain active and interactive systems for the balanced synthesis of all the cell components from whatever materials are environmentally accessible, as well as for some degradative processes to facilitate repair and adjustment of the mechanisms. Unless the environment can provide the cell machinery with energy directly—as in a photosynthetic organism—the cell also requires mechanisms for obtaining chemical energy to drive the synthetic processes, usually by carrying out energy-yielding reactions between other environmentally available materials.

This environmental dependence has many consequences, not least in determining the whole life-style of the microorganisms that occupy particular ecological situations. For example, organisms which in nature find their necessary materials in nutritionally rich substrates, like plant surfaces, can adopt a sedentary mode of life and fairly passive modes of obtaining raw materials—yeasts are typical. Solid substrates which are nutritionally poor, like soil particles and their adherent water films, require either active movement, as in protozoa and motile bacteria, or directed growth which constantly brings access to new substrate, as in filamentous fungi and streptomycetes. Effective access to polymeric substrates (typically polysaccharides and proteins, most abundantly in the tissues of higher organisms) requires the cells to excrete degradative enzymes into their environment, and again this is very characteristic of fungi and streptomycetes; such a mode of life is particularly fruitful if it converts a poor but abundant substrate into an accessible one. In contrast are the organisms which exploit very dilute substrates, as in most natural aqueous environments. Here extensive enzyme excretion would be very wasteful, and to obtain access to the maximum volume of water it is best to colonize some solid surface over which the water flows; hence the success of film-forming organisms in such situations.

Microbial biochemistry deals with the details and interconnections between the energy-yielding and energy-using processes. As we have seen, both are dependent on the environment, and much of microbial physiology is concerned with the interaction between environmental conditions and biochemical processes. Biotechnology can exploit either the energy-yielding processes, as in anaerobic fermentations, or the synthetic processes, as is most apparent in microbial protein production, or the immediate interface between cells and environment, as in the production of extracellular enzymes.

The responsiveness of microbial cells is such that within a certain range of environmental conditions their physiological responses enable them to continue the pattern of activity outlined above. Thus we are concerned both with biochemical mechanisms and their regulation, topics which are explored in some detail in later chapters. Some of the regulatory responses are immediate and can be effected without much change in the cell machinery. Others require more deep-seated changes in the superficial parts of that machinery, and the capacity for making such changes is built into the genetic blueprint.

The study of such adaptive responses, involving selectivity in using the total genetic capability in response to environmental changes over a wider range, is of major importance. At the molecular level the cells can display many new and chararacteristic activities which were not elicited under other conditions, and some of these are of great practical importance for biotechnology. At the cell level its whole appearance and range of functions may change (differentiation), often in such a way as to conserve all the genetically-coded capabilities while manifesting very few of them in any active way (e.g. in spore formation).

The ability to carry out such adaptations is in fact essential for long-term survival of the microbial line. For example, when bacilli are plentifully provided with easily-used substrates they proliferate rapidly, but the genes coding for extracellular degradative enzymes are not expressed; if they were, synthetic activity would be diverted from the immediate task of cell replication. When the local environment no longer provides such an easy living, the activation of such genes becomes useful, and the newly-released enzymes may well make further environmental resources available so that some further proliferation is possible. However when even these less accessible resources become limiting, the cells undergo a drastically modified version of cell division in which one of the daughter cells is consumed by the other in order to form a resistant spore which will survive through extremely adverse conditions and will only be re-activated when conditions improve. Again, such a programme would be very disadvantageous if it could not be suspended when the environment was favourable.

Both the short-term and longer-term responses of microbial cells to their environment confer very considerable flexibility on microbial activities and functions, and much of biotechnology involves the exploitation of that flexibility in controlled and useful ways. For instance, the production of lytic enzymes by bacilli, as an aspect of the sequence of events outlined above, is the basis of major industrial processes.

An even longer-term response involves adjustments in the genetic material itself, and there are good grounds for believing that this too is in some way genetically coded, by allowing a limited amount of flexibility to be 'carried' and by providing mechanisms for recombination of genetic elements from which new and advantageous combinations can emerge. The ability to do this must be limited—otherwise the line could not survive—

but it must nevertheless be realizable—otherwise the line could not have evolved in the first place; evolutionary instability is a necessary correlate of existing diversity.

Little is known about the mechanisms which allow 'silent' variations to be carried, though it is notable that in most microorganisms there are multiple copies of many genes which would facilitate this. More is known about the mechanisms for recombining genetic information. Simple microorganisms with only one chromosome, like bacteria, can only effect such recombination through exchanging chromosome fragments, either directly or through the mediation of smaller independent entities as in bacteriophage and plasmid vectors. More complex organisms like fungi, which have several chromosomes, can use the same mechanisms but in addition have highly developed systems for exchanging chromosomes in sexual reproduction. Again, it is obviously necessary to make all such exchanges relatively rare events as compared with the frequency of normal cell replication, and even as compared with the cycle of cell proliferation followed by differentiation to resistant propagules. Thus fungi have evolved very complex life cycles, with proliferation, sporulation, and sexual reproduction providing progressively less frequented patterns of activity. There has been little opportunity to enlarge upon this aspect in the present text, and the reader is referred to basic general microbiology for detailed examples.

Thus genetics is essentially conservative, even though most interest and attention is given to those occasions when it is not. Both gene recombination and chromosome reassortment are essential tools for the biotechnologist. It is by exploiting the intimate mechanisms of these responses that the manipulations of both classical and molecular genetics become possible. Equally it is because these mechanisms have already been at work throughout the whole of geological time that the existing diversity of microorganisms already provides us with such scope for discovering useful activities—if only we know how to look for them.

2 | Biochemistry of Growth and Metabolism

COLIN RATLEDGE

2.1 Introduction

The purpose of a microorganism is to make another microorganism. In some cases the biotechnologist, who seeks to exploit the microorganism, may wish this to happen as frequently and as quickly as possible. In other cases, where the product is not the organism itself, the biotechnologist must manipulate it in such a way that the primary goal of the microbe is diverted. As the microorganism then strives to overcome these restraints on its reproductive capacity, it produces the product which the biotechnologist desires. The growth of the organism and its various products are therefore intimately linked by virtue of its metabolism.

Metabolism is a matrix of two closely interlinked but divergent activities. *Anabolic* processes are concerned with the building up of cell materials, not only the major cell constituents (proteins, nucleic acids, lipids, carbohydrates, etc.) but also their intermediate precursors—amino acids, purine and pyrimidines, fatty acids, various sugars and sugar phosphates. Anabolic processes do not occur spontaneously; they must be driven by an energy flow, as discussed in Chapter 3, that for most microorganisms is provided by a series of 'energy-yielding' *catabolic* processes. The degradation of carbohydrates to CO_2 and water is the most common of these catabolic processes, but a far wider range of reduced carbon compounds can be utilized by microorganisms in this way. The coupling of catabolic and anabolic processes is the basis of all microbial biochemistry, and can be discussed either in terms of the overall balance (as in Chapter 3) or in terms of individual processes, as here.

In practice we can very usefully distinguish between organisms which carry out their metabolism *aerobically*, using oxygen from the air, and

11

those that are able to do this *anaerobically*, that is, without oxygen. The overall reaction of reduced carbon compounds with oxygen, to give water and CO_2, is a highly exothermic process; an aerobic organism can therefore balance a relatively smaller use of its substrates for catabolism to sustain a given level of anabolism, that is, of growth. Substrate transformations for anaerobic organisms are essentially *disproportionations*, with a relatively low 'energy yield', so that a larger proportion of the substrate has to be used catabolically to sustain a given level of anabolism.

The difference is very clearly illustrated in an organism such as yeast, which is a *facultative* anaerobe—that is, it can exist either aerobically or anaerobically. Transforming sugar at the same rate, aerobic yeast gives CO_2, water, and a relatively high yield of new yeast, whereas yeast grown anaerobically has a relatively slow growth which is now coupled to a high conversion of sugar into ethanol and CO_2.

2.2   Catabolism and energy

The necessary linkage between catabolism and anabolism depends upon making the varied catabolic processes 'drive' the synthesis of reactive re-agents, few in number, which in turn are used to 'drive' the full range

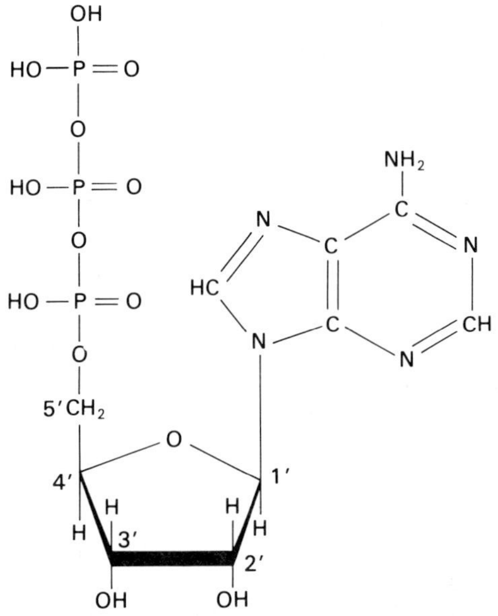

Fig. 2.1   Adenosine triphosphate (ATP).

of anabolic reactions. These key intermediates, of which the most important is adenosine triphosphate, ATP (Fig. 2.1), have what biologists term a 'high-energy bond'; in ATP, the anhydride linkage in the pyrophosphate residue. Directly or indirectly the potential exergicity of the hydrolysis of this bond is used to overcome the endergicity of bond-forming steps in anabolic syntheses. Molecules such as ATP then provide the 'energy currency' of the cell. When ATP is used in a biosynthetic reaction it generates ADP (adenosine diphosphate) or occasionally AMP (adenosine monophosphate) as the hydrolysis product:

$$A + B + ATP \rightarrow AB + ADP + P_i \ \textit{or} \ A + B + ATP/AB + AMP + PP_i$$

$$(P_i = \text{inorganic phosphate, and } PP_i = \text{inorganic pyrophosphate})$$

ADP, which still possesses a 'high-energy bond', can also be used to produce ATP by the adenylate kinase reaction:

$$ADP + ADP \rightarrow ATP + AMP$$

Phosphorylation reactions, which are very common in living cells (see Figs 2.2, 2.4 to 2.6), usually occur through the mediation of ATP:

$$-C-OH + ATP \rightarrow -C-O-\overset{\overset{\displaystyle O}{\|}}{\underset{\underset{\displaystyle OH}{|}}{P}}-OH + ADP$$

The phosphorylated product is usually more reactive (in one of several ways) than the original compound. Phosphorylation by inorganic phosphate does not occur because the equilibrium lies in the reverse direction due to the high concentration of water (55 M) in the cell:

$$-C-OH + HO-\overset{\overset{\displaystyle O}{\|}}{\underset{\underset{\displaystyle OH}{|}}{P}}-OH \rightleftarrows -C-O-\overset{\overset{\displaystyle O}{\|}}{\underset{\underset{\displaystyle OH}{|}}{P}}-OH + H_2O$$

The 'energy status' of the cell can therefore be seen as a function of the prevailing relative proportions of ATP, ADP and AMP. To give this a numerical value, the concept of *energy charge* was introduced by Daniel Atkinson, who defined the energy charge of a cell as the ratio:

$$\frac{ATP + 0.5\,ADP}{ATP + ADP + AMP}$$

A 'fully charged' cell, wherein ATP was the only adenine nucleotide, gives an energy charge value of 1.0. When, say, there are equal amounts of the three nucleotides, ATP = ADP = AMP, then the energy charge of the cell would be 0.5.

Like all conventions, the concept of energy charge has limited usefulness. No one is very sure what it means if a cell is quoted as having an energy charge of 0.7 as opposed to 0.8 or 0.6. The concept does not take into account the *absolute* amounts of the nucleotides in a cell nor does it allow for quite striking differences in the response (see Section 2.8.1.5) of individual enzymes (to which the energy charge concept was initially applied) between ATP and its magnesium complex (the form in which ATP occurs in the cell). There are unexplained differences between energy charge values found typically in bacteria, yeasts and moulds. However, the concept is helpful in following energy changes, and corresponding changes in enzyme activity, within a given type of cell, for example during growth. When a cell is growing most rapidly, the energy charge value is at its lowest; ATP is used as rapidly as it can be resynthesized. As the growth rate begins to slow at the end of growth, the proportion of ATP rises relative to those of ADP and AMP; thus the energy charge begins to rise. The maximum energy charge will be reached when the cell has ceased to grow and all the ADP and AMP have been converted into ATP.

2.3 Catabolic pathways

Though microorganisms can use a wide variety of carbon compounds for growth, we shall consider mainly the metabolism of glucose and, in view of their increasing economic importance, that of ethanol (and other C_2 compounds), hydrocarbons and fatty acids, methane and methanol.

2.3.1 Glucose and other carbohydrates

In nearly all kinds of living cells, the two most important pathways of sugar metabolism are the hexose diphosphate and hexose monophosphate pathways; they usually occur together, providing important links to anabolic processes, and their interactions are subject to key control mechanisms.

The *hexose diphosphate* pathway (often referred to as the *Embden–Meyerhof* or *glycolysis* pathway) is shown in Fig. 2.2. It converts glucose into pyruvate without loss of carbon, reducing two molecules of NAD^+ coenzyme (see Fig. 2.3a) to NADH (see Fig. 2.3b) and generating two molecules of ATP. The pyruvate formed is a source of key anabolic precursors, and in aerobic organisms it is also a substrate for oxidation. In anaerobes the pyruvate or its transformation products must also act as a re-oxidizing agent for the NADH.

The *hexose monophosphate* pathway, also known as the *pentose phosphate* pathway, is shown in Fig. 2.4. As an oxidative process it converts glucose into pentose and CO_2, reducing two molecules of $NADP^+$ (a co-

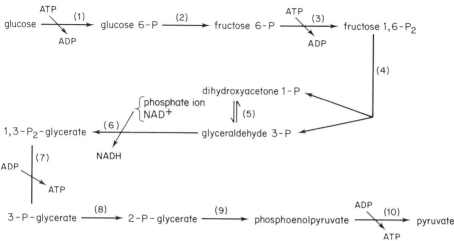

Fig. 2.2 Embden–Meyerhof glycolysis. The reactions beyond glyceraldehyde 3-phosphate, steps 6–10, will involve 2 moles of reactants and products per mole glucose utilized, so the overall reaction is

$$\text{glucose} + 2\text{NAD}^+ + 2\text{H}_2\text{PO}_4^- \rightarrow 2 \text{ pyruvate} + 2\text{NADH} + 2\text{ATP}$$

Successive enzymes are: (1) hexokinase, (2) glucose phosphate isomerase, (3) phosphofructokinase, (4) aldolase, (5) triose phosphate isomerase, (6) glyceraldehyde phosphate dehydrogenase, (7) 3-phosphoglycerate kinase, (8) phosphoglyceromutase, (9) phosphoenolpyruvate dehydratase, (10) pyruvate kinase. P represents a phosphate ester group.

enzyme related to NAD^+, see Fig. 2.3a) to NADPH. [NAD^+/NADH and NADP^+/NADPH both function by hydride transfer but have distinct roles; NADH mainly functions in energy-linked redox reactions while NADPH is mainly used for reductive steps in anabolic processes.]

Through the series of reversible interconversions shown in Fig. 2.4 the pentose phosphate is freely equilibrated with other sugar phosphates having from three to seven carbons, with various metabolic roles which depend on overall circumstances. The triose phosphates are the same as those formed in glycolysis (Fig. 2.2) and, by reversal of the cleavage step in that sequence, they can regenerate hexose as diphosphate; the tetrose phosphate is important as an anabolic precursor for aromatic amino acids, and pentose phosphate is required for the synthesis of nucleic acids (see Fig. 2.18).

In most organisms, between 66 and 80% of the glucose is metabolized via the Embden–Meyerhof pathway and the remainder by the pentose phosphate pathway. The point for controlling the proportion of carbon which flows down each pathway is normally found in the Embden–Meyerhof pathway at the phosphorylation of fructose 6-phosphate to fructose 1,6-bisphosphate, catalysed by phosphofructokinase (PFK). The molecular

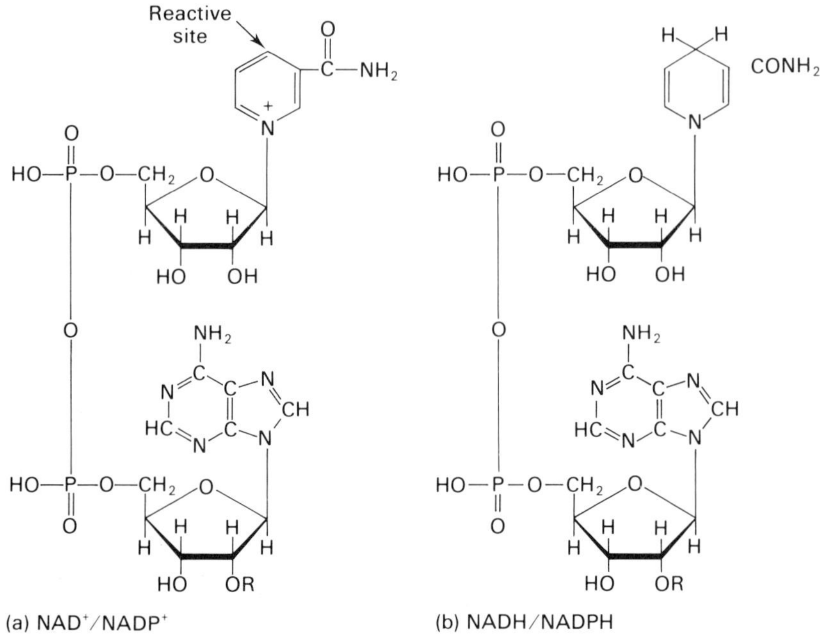

Fig. 2.3 (a) $NAD^+/NADP^+$ (oxidized). (b) $NADH/NADPH$ (reduced). In NAD^+ and $NADH$, $R = H$; in $NADP^+$ and $NADPH$, $R = PO_3^{2-}$.

constitution of this enzyme is such that its catalytic activity can be modulated according to the prevailing metabolic status of the cell: if more energy is required, activity of PFK is increased; if the cell has sufficient energy or sufficient amounts of C_3 metabolites then the activity of PFK is decreased.

This principle of enzyme control by modulation of the catalytic activity of key enzymes is widespread. Metabolic pathways must always be controlled, and for the cell to operate as efficiently as possible its whole activity must be coordinated. With respect to control of PFK this is achieved in two ways. Firstly, the enzyme is *activated*, i.e. the rate at which it can catalyse the reaction is increased, by the presence of AMP or ADP. Thus when the energy charge of the cell (see Section 2.2) is low, PFK will operate at an increased rate. Secondly the enzyme is *inhibited* by an intermediate lower down the metabolic pathway, usually either phosphoenolpyruvate or citrate. Thus if one of these is not being effectively converted to other materials the cell will not need to continue their production.

Other points where glucose metabolism can be regulated vary from organism to organism but always in such a way that the catabolic process meets the anabolic demands as closely as possible.

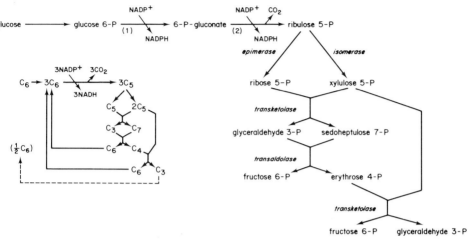

Fig. 2.4   The pentose phosphate cycle (hexose monophosphate shunt). The numbered enzymes are: (1) glucose 6-phosphate dehydrogenase, (2) phosphogluconate dehydrogenase. Inset: summary showing stoichiometry when fructose 6-phosphate is recycled to glucose 6-phosphate by an isomerase; glyceraldehyde 3-phosphate can also be recycled by reverse glycolysis (Fig. 2.2). With full recycling the pathway functions as a generator of NADPH, but the transaldolase and transketolase reactions also permit sugar interconversions which are used in other ways.

The Embden–Meyerhof pathway and the pentose phosphate cycle are not the only pathways of glucose metabolism, though they are by far the most common. A major alternative to the Embden–Meyerhof pathway is the *Entner–Doudoroff pathway*, found in several species of *Pseudomonas* and related bacteria, shown in Fig. 2.5. The enzymes of the pentose phos-

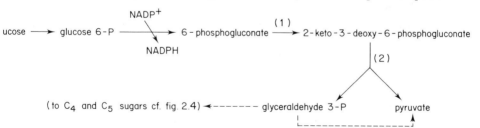

Fig. 2.5   Entner–Doudoroff pathway. This sometimes replaces glycolysis (Fig. 2.2). Numbered enzymes are: (1) phosphogluconate dehydratase, (2) a specific aldolase.

phate cycle are still required to produce the C_5 and C_4 sugars, but the flow is now reversed from the directions given in Fig. 2.4.

Another enzyme of some importance, perhaps more widespread than is generally recognized, is *phosphoketolase*. This type of enzyme (there may be more than one) acts on a C_5 or C_6 sugar phosphate to produce acetyl phosphate and either glyceraldehyde 3-phosphate or erythrose 4-phosphate (depending on whether the C_5 or the C_6 sugar is used) (see Fig. 2.6). These enzymes were originally found in the heterofermentative

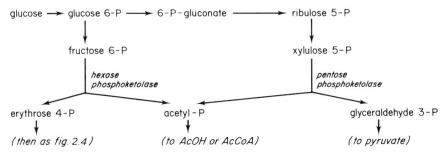

Fig. 2.6 Phosphoketolase pathways. These processes may be additional to those of Figs 2.2 and 2.4, of glucose metabolism by way of the pentose phosphoketolase pathway may replace glycolysis in some bacteria.

lactobacilli (q.v. Section 2.7.2) and acetobacter where they function in place of the Embden–Meyerhof pathway. The resulting acetyl phosphate may be converted to acetate or ethanol. More recently phosphoketolase has been found to be an induced enzyme present in most yeasts when they are grown aerobically on xylose as sole carbon source. Here, xylose is first metabolized via xylitol to xylulose and this then enters the scheme depicted in Fig. 2.6 at the level of xylulose 5-phosphate. (In bacteria growing on xylose there is an isomerase which converts xylose directly to xylulose.) Under these circumstances, the C_5-phosphoketolase does not replace the Embden–Meyerhof pathway but merely provides an efficient route for the organism to convert a pentose into C_2 and C_3 units for further metabolism. The enzyme could therefore have a wide distribution in microorganisms, not just yeasts, when they are grown on xylose or other pentoses.

2.3.2   Tricarboxylic acid cycle

The pathways so far discussed lead eventually to the production of specific C_3 or C_2 compounds, namely pyruvate or acetate, the latter as acetyl co-enzyme A which is a thioester (Fig. 2.7) with anhydride-type reactivity. The further aerobic metabolism of pyruvate and acetyl-CoA is via a cyclic process which serves two separate functions: it produces intermediates which are then used for biosynthetic reactions, and in the oxidation of compounds, eventually to CO_2 and water, it couples the oxidative reactions

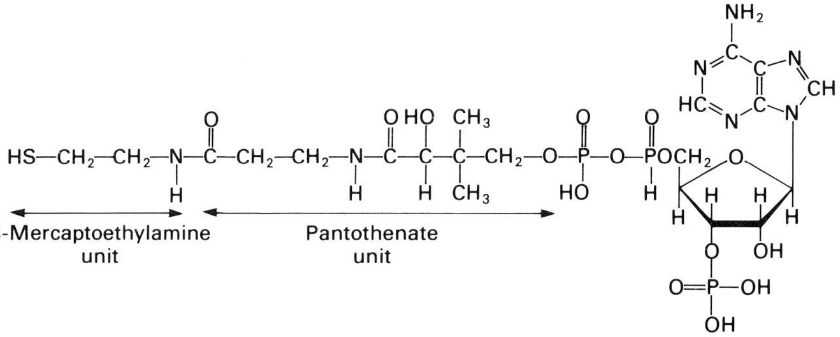

Fig. 2.7 Coenzyme A (CoA).

to the transfer of energy. This cyclic process of acetyl-CoA oxidation, ubiquitous in all aerobic cells, is termed the *tricarboxylic acid cycle* (citric acid cycle, Krebs cycle).

In eukaryotic cells the reactions of the tricarboxylic acid cycle and of energy production are carried out in the mitochondria, while in bacteria the major energy-producing enzymes are associated with the cytoplasmic membrane. As the mitochondrial process begins with the transport of pyruvate into the mitochondrion, it is convenient to include the reactions linking pyruvate to the tricarboxylic acid cycle.

Pyruvate is converted to acetyl-CoA by a multi-enzyme complex, pyruvate dehydrogenase, which catalyses the overall reaction:

$$\text{pyruvate} + \text{CoA} + \text{NAD}^+ \rightarrow \text{acetyl-CoA} + \text{Co}_2 + \text{NADH}$$

The subsequent metabolism of acetyl-CoA is through the reactions of the tricarboxylic acid cycle given in Fig. 2.8.

The functions of this cycle are:

(i) to produce intermediates which can be used in other biosynthetic pathways (see Fig. 2.18), for example:

$$2\text{-oxoglutarate} \rightarrow \text{glutamate} \rightarrow \text{proteins}$$
$$\downarrow \qquad \searrow$$
$$\text{glutamine} \qquad \text{folic acid}$$

$$\text{succinate} \rightarrow \text{porphyrins} \rightarrow \text{haems} \rightarrow \text{cytochromes}$$

$$\text{oxaloacetate} \rightarrow \text{aspartate} \rightarrow \text{proteins}$$
$$\swarrow \qquad \downarrow \qquad \searrow$$
$$\text{(lysine, methionine, threonine)}$$

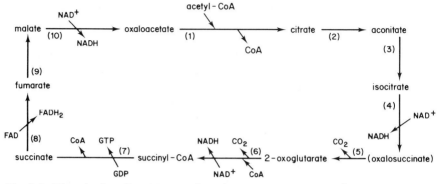

Fig. 2.8 The tricarboxylic acid cycle. (ATP/ADP may replace GTP/GDP.) The overall reaction is:

$$\text{acetyl-CoA} + 3NAD^+ + FAD + GDP$$
$$\rightarrow 2CO_2 + \text{coenzyme A} + 3NADH + FADH_2 + GTP$$

The numbered steps are catalysed by: (1) citrate synthase, (2,3) aconitase, (4,5) isocitrate dehydrogenase, (6) 2-oxogluconate dehydrogenase, (7) succinate thiokinase, (8) succinate dehydrogenase, (9) fumarase, (10) malate dehydrogenase.

The reactions leading to aspartate and glutamate are particularly important as the main pathways by which cells assimilate ammonia.

(ii) to recover energy from the oxidative reactions. The enzymes iso-citrate dehydrogenase, 2-oxoglutarate dehydrogenase, succinate dehydrogenase, and malate dehydrogenase catalyse the progressive oxidation of the intermediate with the concomitant conversion of enzyme cofactor oxidants to reductants. The coenzymes are NAD^+ and FAD which become, respectively, NADH and $FADH_2$ (cf. Fig. 2.3a), these are then reoxidized to the original coenzymes by one process of *oxidative phosphorylation* (see Section 2.5) which produces 3 moles of ATP from each mole of NADH and 2 moles of ATP from the reoxidation of $FADH_2$. Energy is also recovered in the succinate thiokinase reaction (see Fig. 2.8).

Although the cycle is apparently self-perpetuating, in that once primed by oxaloacetate it should keep working indefinitely, this cannot happen in practice. As already noted, the cycle must also furnish intermediates for biosynthetic reactions, and when any such intermediate is removed from the cycle the synthesis of oxaloacetate and the regeneration of citrate cannot occur. It is therefore necessary that additional oxaloacetate can be formed independently, and this mainly occurs by the carboxylation of pyruvate:

$$\text{pyruvate} + CO_2 + ATP \rightarrow \text{oxaloacetate} + ADP + P_i$$

This reaction is carried out by pyruvate carboxylase. However, insofar as oxaloacetate is also produced from the activity of the cycle, the carboxylation of pyruvate must be regulated so that acetyl-CoA and oxaloacetate are always produced in equal amounts. This is achieved by the pyruvate carboxylase being dependent upon acetyl-CoA as a *positive effector* (see Section 2.8.1.5), i.e one which increases its activity. The more acetyl-CoA that is present, the faster becomes the production of oxaloacetate. As oxaloacetate and acetyl-CoA are removed (to form citrate), the concentration of acetyl-CoA will fall; pyruvate carboxylase will then slow down but, as pyruvate dehydrogenase still operates as before, more acetyl-CoA will be produced. In this way not only will citric acid synthesis always continue, but the two reactions leading to the precursors of citrate will always be balanced.

There are additional controls which can regulate the activity of the cycle. Some of its enzymes are inhibited by ATP and others depend on the presence of AMP for their activity. Therefore the cycle can be regulated by the prevailing ratio of ATP to AMP, that is by the energy charge (see Section 2.2) within the cell. These control mechanisms are not universal and they need to be ascertained for individual organisms, or groups of organisms; they will not be considered in further detail, but general principles of metabolic control, as discussed for the regulation of glycolysis (p. 16), still apply.

2.3.3 Glyoxylate by-pass (growth on C_2 compounds)

If an organism grows on a C_2 compound, or on a fatty acid or hydrocarbon that is degraded primarily into C_2 units (see Section 2.3.4), the tricarboxylic acid cycle is insufficient to account for its metabolism. As shown in the previous section, any compound which is used for synthesis, and so removed from the tricarboxylic acid cycle, will, by its removal, effectively stop the regeneration of oxaloacetate. As C_2 compounds cannot be converted to pyruvate (the pyruvate dehydrogenase reaction, see p. 19, is effectively irreversible) there is no way in which oxaloacetate, or indeed any such C_4 compound, can be produced from a C_2 compound by the reactions given so far.

Acetyl-CoA can be generated directly from acetate, if this is being used as carbon source, or from a C_2 compound more reduced than acetate, i.e. acetaldehyde or ethanol:

$$C_2H_5OH \xrightarrow[\quad]{NAD^+ \quad NADH} CH_3CHO \xrightarrow[\quad]{NAD^+ \quad NADH} CH_3COO^- \xrightarrow[CoA]{ATP \quad ADP+P_i} Acetyl\text{-}CoA$$

The manner in which acetate units are converted to C_4 compounds is known as the glyoxylate by-pass (see Fig. 2.9) for which two enzymes ad-

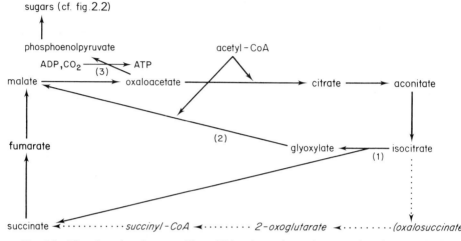

Fig. 2.9   The glyoxylate by-pass. The additional reactions, shown against the same background reactions as in Fig. 2.8, are (1) isocitrate lyase, (2) malate synthase. The scheme also shows how the bypass functions to permit sugar formation from acetyl-CoA, with the added reaction (3) phosphoenolpyruvate carboxykinase, followed by reversed glycolysis (cf. Fig. 2.2).

ditional to those of the tricarboxylic acid cycle are needed: *isocitrate lyase* and *malate synthase*. Both enzymes are 'induced' (see Section 2.8.1.3) when microorganisms are grown on C_2 compounds; the activity of the enzymes increases by some 20 to 50 times under such growth conditions. The glyoxylate by-pass does not supplant the operation of the tricarboxylic acid cycle; for example 2-oxoglutarate will still have to be produced (from isocitrate) in order to supply glutamate for protein synthesis, etc. Succinate, the other product from isocitrate lyase, will be metabolized as before to yield malate, and thence oxaloacetate. Thus through the glyoxalate cycle, the C_4 compounds can now be produced from C_2 units, and are then available for synthesis of all cell metabolites (cf. Fig. 2.18). Their conversion into sugars is detailed in Section 2.4.

2.3.4   Fatty acids and hydrocarbons

The ability to grow on a hydrocarbon is not widespread but it is found amongst bacteria, yeasts and moulds, while the ability to utilize fatty acids, or oils and fats containing them, is more common. Hydrocarbons are used as sole sources of carbon in the production of single cell protein in the

USSR (see Chapter 10) and can be used in other processes, e.g. for the production of citric acid (see Chapter 13). Fatty acids and vegetable oils are often added as cosubstrates in the manufacture of antibiotics (see Chapter 16).

For the utilization of oils and fats (triglycerides), the organism must hydrolyse the ester linkage using a lipase (intracellular or extracellular). This yields free fatty acids (3 moles) plus glycerol (1 mole); the glycerol is utilized by the Embden–Meyerhof pathway. Many microorganisms can also utilize free fatty acids. However, whether such acids are taken into the cell or actually formed there, they are extremely toxic (because of their surfactant properties) and must be converted immediately into their coenzyme A thioesters.

The thioester is suitably activated for degradation of the fatty acyl chain by the cyclic sequence of reactions depicted in Fig. 2.10. At each turn

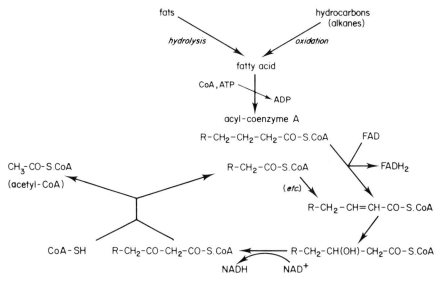

Fig. 2.10 β-Oxidation of fatty acids. With each 'turn' of the cycle acetyl-CoA is lost and a new fatty acyl-CoA with two carbon atoms fewer is formed, until the final products are 2 acetyl-CoA, or acetyl-CoA and propionyl-CoA, depending upon whether the original fatty acid has an even or an odd number of carbon atoms.

of the cycle, 1 mol of acetyl-CoA is released and the fatty acyl-CoA ester, now two carbons shorter in chain length, recommences its oxidative cycle. The sequence, known as the *β-oxidation cycle*, continues until the substrate for the final reaction is the C_4 compound, acetoacetyl-CoA, which then gives 2 moles of the acetyl-CoA. If a fatty acid has an odd number of carbon atoms, the degradation proceeds until propionyl-CoA (C_3) is reached, which is converted to pyruvate by the reverse of a sequence of reactions given in Section 2.7.3.

Microorganisms growing on *n*-alkanes usually commence with an initial attack on one of the two terminal methyl groups. The mechanism of attack by the enzyme, alkane hydroxylase, involves molecular oxygen and a re-oxidizable cofactor containing iron. The cofactor is also oxidized and regeneration of the reduced form is ultimately linked to a hydride carrier, either NADH or NADPH:

$$R.CH_3 \xrightarrow{\quad O_2 \quad} R.CH_2OH \quad (+ H_2O)$$

reduced cofactor / oxidized cofactor — FAD protein / FADH$_2$ — NAD(P)H / NAD$^+$

The fatty alcohol is then oxidized to the corresponding fatty acid in two dehydrogenase steps:

$$R.CH_2OH \xrightarrow[NAD^+ \to NADH]{} R.CHO \xrightarrow[NAD^+ \to NADH]{} R.COOH$$

Typically, all the enzymes associated with alkane degradation show broad substrate specificity and react readily with substrates from C_{10} to C_{18}. Several microorganisms can also attack either shorter or longer chains. In a few cases subterminal attack on alkanes has been reported with the production of a methyl ketone ($R.CO.CH_3$) which is eventually cleaved by further oxidation to give methyl formate and a fatty acid with two less carbon atoms than the original alkane.

Fatty acids arising from alkane degradation are usually degraded by the β-oxidation cycle (Fig. 2.10) though in some organisms there is an ω-oxidation process which produces dicarboxylic acids. These are then degraded from one end by β-oxidation. The fatty acids are also used by the cell for the direct formation of its own lipids and thus the chain length of the alkane is reflected in the chain length of the fatty acids found in the cell.

Alkanes and some branched chain hydrocarbons may also be metabolized; they are not used on a commercial scale but may be present as minor components in the feedstock. Their oxidation invariably involves conversion to a fatty acid.

2.3.5 Methane and methanol

A small number of microorganisms (both bacteria and yeasts) which are termed *methyltrophs* can use methanol as sole source of carbon; the ability to utilize methane has so far been found only in a relatively small number of bacteria which are termed *methanotrophs*. A few microorganisms can use formate as carbon source. These three compounds are metabolically linked and can be oxidized ultimately to CO_2. The mechanism of their incorporation into cell material is different from that of autotrophic CO_2 fixation.

[Utilization of CO_2 as sole carbon source is the province of photosynthetic plants and microorganisms, and of a few *chemolithotrophic* bacteria which use reactions of inorganic compounds as energy source. These organisms currently have few applications in process biotechnology. The reader who would like to know more about the pathways of autotrophic fixation of CO_2 should refer to almost any biochemistry textbook, but he should note that there are at least two distinct pathways: the Calvin cycle and a reductive carboxylic acid cycle.]

The mechanism of methane oxidation is stepwise:

$$CH_4 \rightarrow CH_3OH \rightarrow HCHO \rightarrow HCOOH \rightarrow CO_2$$

The first step is carried out by an oxygenase with NADH (or NADPH) as cofactor (compare oxidation of higher alkanes, above):

$$CH_4 + O_2 + NAD(P)H \rightarrow CH_3OH + H_2O + NAD(P)^+$$

The enzyme (a complex of three proteins) will also oxidize a variety of other compounds including various alkanes and even methanol itself.

The second reaction in the sequence is catalysed by methanol dehydrogenase, using a newly-discovered pyrroloquinoline quinone (PQQ) as cofactor:

$$CH_3OH + PQQ \rightarrow HCHO + PQQH_2$$

In some bacteria the further conversion of formaldehyde to formic acid is catalysed by the same enzyme; in others there may be a separate formaldehyde dehydrogenase with NAD as cofactor. The final step, conversion of formate to CO_2, is carried out by formate dehydrogenase and is again linked to NAD^+ reduction.

The assimilation of carbon from methane or methanol into cell material is at the level of formaldehyde, and by at least two independent routes: the *ribulose monophosphate cycle* (sometimes referred to as the Quayle cycle) and the *serine pathway*, shown in Figs 2.11 and 2.12, respectively.

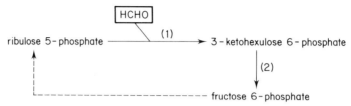

Fig. 2.11 Formaldehyde assimilation in bacteria : ribulose monophosphate cycle. Enzymes are (1) 3-ketohexulose 6-phosphate synthase, (2) phospho-3-hexulose isomerase. By the reactions of Fig. 2.4, 5/6 of the fructose-6-phosphate is recycled to regenerate the acceptor, ribulose 5-phosphate.

The ribulose monophosphate cycle (Fig. 2.11) is similar to the Calvin cycle used for the autotrophic fixation of CO_2 in using the reactions of the pentose phosphate cycle (see Fig. 2.4) to regenerate the acceptor for the incoming C_1 compound. Only two additional enzymes are needed: 3-hexulose phosphate synthase (A) and phospho-3-hexulose isomerase (B):

$$
\begin{array}{ccc}
\text{CH}_2\text{OH} & \text{CH}_2\text{OH} & \text{CH}_2\text{OH} \\
| & | & | \\
\text{C}=\text{O} & \text{HO}-\text{C}-\text{H} & \text{C}=\text{O} \\
| & | & | \\
\text{H}-\text{C}-\text{OH} \xrightarrow[\text{(A)}]{\text{HCHO}} & \text{C}=\text{O} \xrightarrow{\text{(B)}} & \text{HO}-\text{C}-\text{H} \\
| & | & | \\
\text{H}-\text{C}-\text{OH} & \text{H}-\text{C}-\text{OH} & \text{H}-\text{C}-\text{OH} \\
| & | & | \\
\text{CH}_2\text{O}.\text{PO}_3\text{H}_2 & \text{H}-\text{C}-\text{OH} & \text{H}-\text{C}-\text{OH} \\
 & | & | \\
 & \text{CH}_2\text{O}.\text{PO}_3\text{H}_2 & \text{CH}_2\text{O}.\text{PO}_3\text{H}_2
\end{array}
$$

The key enzymes of the serine pathway (Fig. 2.12) are malyl-CoA lyase, which produces acetyl-CoA and glyoxylate, and serine transhydroxymethylase, a ubiquitous enzyme which uses tetrahydrofolate (a cofactor capable of forming the necessary activated C_1 intermediate, N^{10}-formyltetrahydrofolate). The glyoxylate by-pass (see Fig. 2.9) then operates to handle the acetyl-CoA, so that the cell is effectively growing on a C_2 substrate. Isocitrate lyase is de-repressed (see p. 22) to ensure overall production of C_3 units.

In yeasts there is a further variation on the pentose phosphate cycle, by which formaldehyde reacts with xylulose 5-phosphate to produce glyceraldehyde 3-phosphate and dihydroxyacetone (see Fig. 2.13). This reaction is catalysed by a special type of transketolase enzyme (cf. Fig. 2.4). The only additional enzyme then needed to complete the cyclic assimilation of formaldehyde is a new kinase to convert dihydroxyacetone to dehydroxyacetone phosphate.

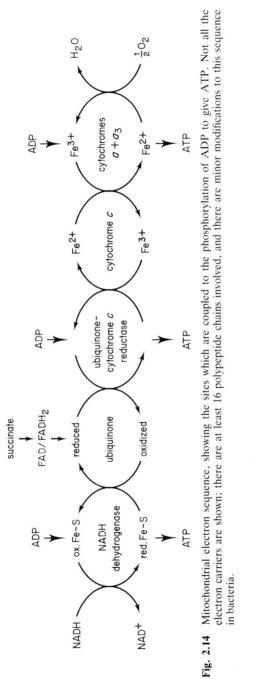

Fig. 2.14 Mitochondrial electron sequence, showing the sites which are coupled to the phosphorylation of ADP to give ATP. Not all the electron carriers are shown; there are at least 16 polypeptide chains involved, and there are minor modifications to this sequence in bacteria.

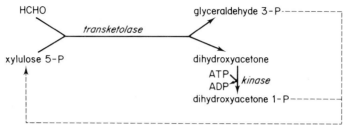

Fig. 2.13   Formaldehyde assimilation in methanol-using yeasts. The dotted line shows regeneration of the acceptor, xylulose 5-phosphate, by the reactions of Figs 2.2 and 2.4, which accounts for 5/6 of the triose phosphates formed.

2.4   Gluconeogenesis

When an organism grows on a C_2 or C_3 compound, or a material whose metabolism will produce such compounds, at or below the metabolic level of pyruvate (for example aliphatic hydrocarbons, acetate, ethanol or lactate), it is necessary for the organism to synthesize various sugars to fulfil its metabolic needs. This is termed *gluconeogenesis*. Though most of the reactions in the glycolytic pathways (see Figs 2.2 and 2.4) are reversible, those catalysed by pyruvate kinase and phosphofructokinase are not and it is necessary for the cell to circumvent these blockages.

In general, phosphoenolpyruvate cannot be formed from pyruvate, although in a few organisms there is an enzyme, phosphoenolpyruvate synthase, which can carry out the reaction:

$$\text{pyruvate} + \text{ATP} \rightarrow \text{phosphoenolpyruvate} + \text{AMP} + P_i$$

More often, oxaloacetate is used as the precursor:

$$\text{oxaloacetate} + \text{ATP} \rightarrow \text{phosphoenolpyruvate} + CO_2 + \text{ADP}$$

This reaction is catalysed by phosphoenolpyruvate carboxykinase which is the key enzyme of gluconeogenesis. The formation of oxaloacetate has already been discussed (Section 2.3.3). The irreversibility of phosphofructokinase (producing fructose 1,6-bisphosphate) is circumvented by the action of fructose bisphosphatase:

$$\text{fructose 1,6-bisphosphate} + H_2O \rightarrow \text{fructose 6-phosphate} + P_i$$

From this point hexose sugars can be formed by the reversal of glycolysis and the C_5 and C_4 sugars can now be formed via the pentose phosphate pathway (Fig. 2.4). Glucose itself is not an end-product of 'gluconeogenesis' but glucose 6-phosphate is used for the synthesis of cell wall constituents and a large variety of extracellular and storage polysaccharides.

2.5 Energy metabolism in aerobic organisms

It has already been explained how, in the metabolism of glucose (Figs 2.2 and 2.4) and in the tricarboxylic acid cycle (Fig. 2.8), oxidation of the various metabolic intermediates is linked to reduction of a limited number of cofactors (NAD^+, $NADP^+$, FAD) to the corresponding reduced forms (NADH, NADPH and $FADH_2$). The reducing power of these products is released by a complex reaction sequence which in aerobic systems is linked eventually to reduction of atmospheric O_2. During this sequence, ATP is generated from ADP and inorganic phosphate (P_i) at two or more (usually three) specific points in the electron transport chain, depending on the nature of the orginal reductant. This is shown in Fig. 2.14.

The overall reactions may be written as:

$$NADPH + 3ADP + 3P_i + \tfrac{1}{2}O_2 \rightarrow NADP^+ + 3ATP + H_2O$$

$$NADH + 3ADP + 3P_i + \tfrac{1}{2}O_2 \rightarrow NAD^+ + 3ATP + H_2O$$

$$FADH + 2ADP + 2P_i + \tfrac{1}{2}O_2 \rightarrow FAD + 2ATP + H_2O$$

The yields of ATP per mole of glucose metabolized by the Embden–Meyerhof pathway (Fig. 2.2) and from the resulting pyruvate, metabolized by the reactions of the tricarboxylic acid cycle (Fig. 2.8) are summarized in Table 2.1.

Table 2.1 ATP yields for glucose metabolism

	Moles ATP produced per mole hexose
Glycolysis (glucose to pyruvate)	
Net yield of ATP = 2 mol	2^a
NADH = 2 mol × 3	6
Pyruvate to acetyl-CoA	
NADH = 1 mol × 3 (×2 for 2 pyruvate)	6
Tricarboxylic acid cycle	
NADH = 3 mol × 3 (×2 for 2 acetyl-CoA)	18
$FADH_2$ = 1 mol × 2 (×2 for 2 acetyl-CoA)	4
ATP = 1 mol (×2 for 2 acetyl-CoA)	2
	38

[a] Under anaerobic conditions all the NADH and pyruvate must be consumed without net oxidation and these 2 moles of ATP represent the maximum attainable yield ('substrate-level phosphorylation').

The production of biologically utilizable energy, in the form of ATP, occurs in membranes—either the membrane of the mitochondria in eukaryotic organisms or the cytoplasmic membrane in bacteria. Both processes

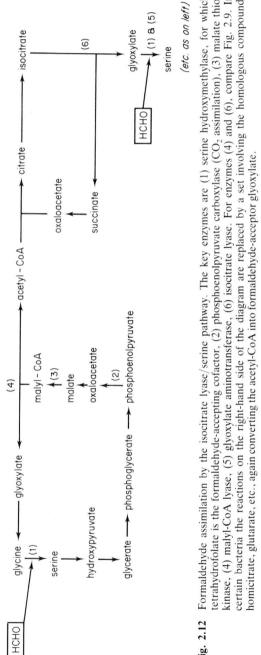

Fig. 2.12 Formaldehyde assimilation by the isocitrate lyase/serine pathway. The key enzymes are (1) serine hydroxymethylase, for which tetrahydrofolate is the formaldehyde-accepting cofactor, (2) phosphoenolpyruvate carboxylase (CO_2 assimilation), (3) malate thiokinase, (4) malyl-CoA lyase, (5) glyoxylate aminotransferase, (6) isocitrate lyase. For enzymes (4) and (6), compare Fig. 2.9. In certain bacteria the reactions on the right-hand side of the diagram are replaced by a set involving the homologous compounds homicitrate, glutarate, etc., again converting the acetyl-CoA into formaldehyde-acceptor glyoxylate.

are broadly similar, with differences of detail between individual organisms. The main components of the electron transport chain are flavoproteins, quinones and cytochromes (see Fig. 2.14) whose property is to be able to become reduced (by acceptance of hydride ions or electrons) and then oxidized, releasing electrons to the next carrier in a coupled and efficient manner. Each carrier has a different redox potential, increasing stepwise from about $-320\,mV$ for the $NADH/NAD^+$ reaction to about $+800\,mV$ for the final reaction: $\frac{1}{2}O_2/H_2O$. At certain points in the chain the difference in redox potential between two adjacent carriers is large enough to drive, in the direction of ATP synthesis, the reversible reaction: $ADP + P_i \rightleftharpoons ATP$. This activity is associated with a complex, multi-subunit, enzyme termed *ATPase*.

There are two principal views as to how the ATPase is driven. In the chemiosmotic hypothesis, developed by Mitchell over the past twenty years, it is considered that the components of the electron transport chain are spatially arranged across the membrane. This generates a pH and electrical gradient due to the passage of protons from one side to the other. Protons returning across the membrane drive the ATPase reaction into synthesis of ATP. The ATPase is so oriented that protons have access to its catalytic site from one side only. This concept is shown in its simplest form in Fig. 2.15.

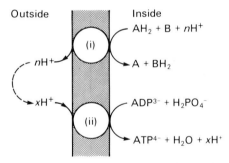

Fig. 2.15 Membrane function in coupling respiration, carried out by the electron transport carriers, with phosphorylation (energy production). Process (i) is a proton-translocating oxido-reduction, and transfers protons to the outside of the membrane; process (ii) uses the resulting proton gradient to drive the proton-translocating ATPase.

The second explanation advanced to account for ATP synthesis envisages that carriers of the electron transport chain interact with hypothetical intermediate(s) which, on becoming activated, phosphorylate ADP. These intermediates are termed *coupling factors*.

Both theories have their strengths and weaknesses. Both can be adapted to account for the effect of 'uncouplers' of oxidative phosphorylation such as rotenone, amytal, antimycin A, etc. which have the effect of stopping ATP production.

2.6   Energy production in anaerobic organisms

The process of ATP production described in Section 2.5 depends on the provision of oxygen. Some organisms can substitute nitrate, others sulphate or ferric iron, for oxygen, and if these are supplied in quantity in the medium, the organism can still produce its ATP in the absence of air, using the electron transport carriers, and thus grow anaerobically. However, if no alternative electron acceptor is supplied, or if (like the majority of bacteria) the organism lacks this facility, then the organism deprived of oxygen will be unable to produce ATP in this way. Organisms growing anaerobically must therefore achieve ATP production by coupling the reaction directly to an energy-yielding reaction; this is termed *substrate level phosphorylation*. The number of reactions in which this can happen is very limited. The free energy change of the reaction must be sufficient to drive the phosphorylation of ATP (ATP $\rightarrow$ ADP + P_i; $\Delta G^{0'}$ of -31 kJ mol^{-1}) and the most important reactions of this kind are summarized in Table 2.2.

Of these six reactions, the three last are only important in a few organisms; of the other three reactions in Table 2.2, reactions 1 and 2 involve intermediates in glycolysis (see Fig. 2.2). Reaction 3, involving acetyl phosphate, is also widespread amongst anaerobes. Acetyl phosphate is formed from acetyl-CoA by reaction with inorganic phosphate, and it is also generated by the action of phosphoketolases (p. 18).

Acetyl-CoA can be produced by the degradation of acetoacetyl-CoA (p. 23), or from pyruvate by one of three reactions: pyruvate dehydrogenase (NAD$^+$-linked)—see p. 19; pyruvate formate lyase (which catalyses the reaction pyruvate + CoA $\rightarrow$ acetyl-CoA + formate; see p. 37); or pyruvate:ferredoxin oxidoreductase, which gives the same reaction products as pyruvate dehydrogenase but involves an iron–sulphur protein, ferredoxin, rather than NAD$^+$ as oxidant (the reduced ferredoxin is oxidized back to ferredoxin by hydrogenase which releases H_2). Of these three enzymes, the latter two are sensitive to the presence of oxygen and quickly lose activity when the anaerobe containing them is exposed to air.

There is increasing evidence that electron transport phosphorylation can also occur through the mediation of fumarate reductase, which is probably important in several methane-producing bacteria, sulphate-reducing organisms, and hydrogen bacteria which ferment H_2 and CO_2. The reaction is: fumarate + 2H$^-$ + ADP $\rightarrow$ succinate + ATP. The hydride ions may be supplied from a variety of cofactors, including NADH, and in some

Table 2.2 Substrate-level phosphorylation reaction in anaerobes

Enzyme	Reaction catalysed	Occurrence
1. Phosphoglycerol kinase	1,3-bisphosphoglycerate + ADP → 3-phosphoglycerate + ATP	widespread
2. Pyruvate kinase	phosphoenolpyruvate + ADP → pyruvate + ATP	widespread
3. Acetate kinase	acetyl phosphate + ADP → acetate + ATP	widespread
4. Butyrate kinase	butyryl phosphate + ADP → butyrate + ATP	*e.g.* enterobacteria on allantoin
5. Carbamate kinase	carbamoyl phosphate + ADP → carbamate + ATP	*e.g.* clostridia on arginine
6. Formyl-tetrahydrofolate synthetase	N^{10}-formyl-H_4 folate + ADP + P_i → formate + H_4 folate + ATP	*e.g.* clostridia on xanthine

organisms such as *Escherichia coli* their generation may involve a sequence of electron carriers similar to, if not identical with, the components of the electron transport chain in aerobic organisms. Thus although O_2 is not involved the organism is able to couple various reactions so that ATP can be produced.

All anaerobes face two problems. First, lacking the coupling of NADH (or NADPH) reoxidation to ATP generation through oxidative phosphorylation, the yield of ATP per mole of substrate is much smaller than with aerobic metabolism. Second, the inability to couple the oxidation of NADH to the reduction of oxygen poses the problem of how to achieve this essential reaction, without which metabolism would rapidly come to a halt, as all the NAD+ became irreversibly converted to NADH.

Anaerobes have adopted a variety of means to achieve the reoxidation of reduced cofactors. Essentially each scheme is of the kind:

$$AH_2 \diagdown \diagup NAD^+ \diagdown \diagup BH_2$$
$$A \diagup \diagdown NADH \diagup \diagdown B$$

Here, the step $AH_2 \rightarrow A$ is part of the pathway by which substrate is being utilized by the anaerobe. Normally the substance B, required for the compensating reduction, will also be directly derived from the substrate; BH_2, once formed, need not be metabolized further. The essential metabolism of AH_2 is thus stoichiometrically linked to the compensating production of BH_2. Anaerobes must therefore accumulate reduced metabolites in order to carry out the degradation of any substrate. Moreover, since—as already noted—they derive relatively low ATP yields from substrate degradation, this accumulation of reduced metabolites is bound to be large relative to the amount of cell materials synthesized. Ways in which anaerobic metabolism is organized to achieve this are described in the following section.

2.7 Anaerobic metabolism

The choice of substrate with which to re-oxidize reductants such as NADH, NADPH, $FADH_2$, can be very wide and a corresponding variety of end-products will result. A description of pathways of anaerobic metabolism is therefore a description of what end-products are accumulated by individual organisms. Some of these, such as ethanol, have considerable commercial importance.

Even under anaerobic conditions glucose metabolism will still produce pyruvate, but only a small amount of pyruvate will be taken into the tricarboxylic acid cycle to produce intermediates for the biosynthesis of essential cell material. The tricarboxylic acid cycle reactions then function only to provide these intermediates and not to generate energy. Often the cycle is not fully operational and in particular 2-oxoglutarate dehydrogenase (see Fig. 2.8) may not operate. The 'cycle' is therefore a 'horseshoe' in which oxaloacetate is channelled to succinate and citrate is converted to 2-oxoglutarate.

2.7.1   Ethanol fermentation

In yeasts such as *Saccharomyces cerevisiae* the re-oxidant is acetaldehyde; most of the pyruvate generated from glucose is converted into ethanol:

$$\text{pyruvate} \longrightarrow \begin{array}{c} \text{2-hydroxyethyl-} \\ \text{thiamine} \\ \text{pyrophosphate} \end{array} \longrightarrow \text{acetaldehyde} \underset{\text{NADH} \quad \text{NAD}^+}{\overset{}{\searrow}} \text{ethanol}$$

As 2 moles of pyruvate are produced from 1 mole of glucose, the production of ethanol can re-oxidize both moles of NADH produced in the triose phosphate dehydrogenase reaction (see Fig. 2.2), and the overall stoichiometry is therefore:

$$\text{glucose} + 2\text{ADP} + P_i \rightarrow 2\,\text{ethanol} + 2\text{ATP}$$

The net production of ATP provides energy for the yeast cells to grow but, as will be appreciated by comparison with Table 2.1, the yield per mole of glucose transformed is less than 5% of that which occurs under aerobic conditions.

Any glucose that is metabolized by the pentose phosphate pathway, to produce the essential C_5 and C_4 sugars, can produce no more than 1 mole of pyruvate from each mole of glucose, and only with the simultaneous generation of 2 moles of NADPH and 1 mole of NADH (see Fig. 2.4). These additional reducing equivalents (for the re-oxidation of which there is now insufficient pyruvate) must be re-oxidized by being coupled to other reactions. Prime amongst these other reactions is the formation of fatty acids, which are chemically reduced compounds whose synthesis demands a considerable input of reducing equivalents (cf. Fig. 2.18).

Ethanol is also produced by some bacteria (see Section 2.7.5), often in conjunction with other end-products, and by some moulds, and anaerobic conditions are generally needed for maximum ethanol production. If the producing organism is also capable of aerobic growth, as is *S. cerevisiae*, then when oxygen is introduced the accumulated ethanol will often be taken up by cells and utilized, by way of acetic acid, as a growth substrate.

Aspects of ethanol production are dealt with at length in Chapter 11.

2.7.2 Lactic acid fermentations

Fermentations producing lactic acid are second only to alcohol fermentations, both historically and in their importance to the food industry.

Heterofermentative lactic acid bacteria produce a variety of reduced compounds besides lactate, and do not have the key glycolytic enzyme, aldolase (see Fig. 2.2); instead they use phosphoketolase (p. 18) which produces acetyl phosphate. Under anaerobic conditions this is converted both into ethanol, which regenerates NAD^+, and into acetate by a reaction which can generate ATP (see Table 2.2). The other product of phosphoketolase is glyceraldehyde 3-phosphate, which is converted to pyruvate by the usual glycolytic sequence, and thence to lactate by the action of lactate dehydrogenase:

$$pyruvate + NADH \rightarrow lactate + NAD^+$$

This reaction is also used by the homolactic acid bacteria; these organisms do not possess phosphoketolase and consequently lactate is virtually the sole end-product. Some lactobacilli produce D-lactate, others L-, and some a mixture of the two forms due to differences in the various lactate dehydrogenases.

2.7.3 Propionic acid fermentation

Propionibacteria, found for example in Gruyère cheese, convert pyruvate to propionate in a series of reactions with methylmalonyl-CoA as the key intermediate (Fig. 2.16). This is used as the immediate source of propionate in a unique transcarboxylation reaction:

The methylmalonyl-CoA is formed by an internal transcarboxylation from succinyl-CoA, and can thus be regenerated from oxaloacetate (via malate, fumarate, and succinate) while 2 moles of NADH are oxidized to NAD^+. In some *Clostridium* species propionate is produced directly from pyruvate via lactate and acrylate; again, this conversion achieves the re-oxidation of 2 moles of NADH.

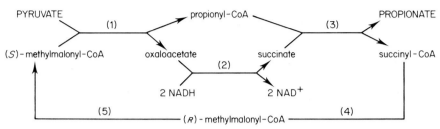

Fig. 2.16   Indirect reduction of pyruvate to propionate in propionobacteria. Enzymes are: (1) transcarboxylase, (2) as in Fig. 2.8, through malate and fumarate, (3) CoA transferase, (4) methylmalonyl-CoA mutase, (5) racemase.

2.7.4   Butanediol fermentation

In *Aerobacter* (now *Klebsiella*) spp., *Serratia* and some *Bacillus* spp., 2 moles of pyruvate undergo condensation (again, the reactive intermediate is hydroxyethyl-TPP, see p. 35) and eventually produce 2,3-butanediol:

$$2 \text{ pyruvate} \rightarrow \alpha\text{-acetolactate} + CO_2$$
$$\downarrow$$
$$\text{acetylmethylcarbinol} + CO_2$$
$$\downarrow$$
$$2,3\text{-butanediol}$$

Only the final step is linked to the oxidation of NADH, so that the yield of NAD^+ is only 0.5 mole per mole of pyruvate; these organisms also convert pyruvate to other products including lactate and formate (see below).

2.7.5   Formic acid fermentations

In various enterobacteria, pyruvate is converted partly to lactate and partly to acetyl-CoA + formate. The latter reaction is termed the *phosphoroclastic split*. The formate may accumulate in small amounts but is usually converted to CO_2 and H_2 by formate hydrogen lyase. The 'advantage' of this route from pyruvate to acetyl-CoA is that it does *not* produce NADH (compare pyruvate dehydrogenase) and so avoids the necessity of a re-oxidation reaction. Acetyl-CoA can be converted to acetaldehyde by an aldehyde dehydrogenase:

$$\text{acetyl-CoA} + NADH \rightarrow \text{acetaldehyde} + NAD^+$$

Reduction of the acetaldehyde to ethanol can occur with further oxidation of NADH. Note that this route to ethanol is not the same as that which occurs in yeast (Section 2.7.1).

2.7.6 Butyric acid fermentation

Historically, the production of butanol, acetone and propan-2-ol is the oldest of the 'deliberate' fermentation processes, i.e. ones developed from established principles using single strains of known organisms. This family of end-products from glucose metabolism are formed by members of the *Clostridium* group of bacteria, according to the scheme shown in Fig. 2.17.

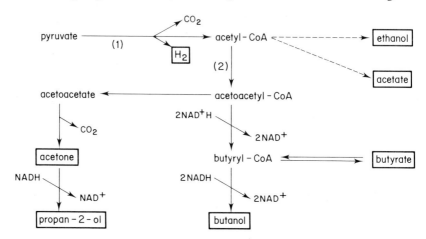

Fig. 2.17 Major end-products (boxed) from pyruvate in *Clostridium* species. Enzyme (1) is pyruvate-ferredoxin oxidoreductase plus hydrogenase; the latter enzyme also interconverts $(NAD^+ + H_2)$ and NADH. Enzyme (2) is thiolase; other CoA transfers are omitted. The relative amounts of acetone, propan-2-ol, butryrate and butanol vary between strains and according to the total $NADH/NAD^+$ flux. Ethanol and acetate are normally minor products.

There are a number of variations: some clostridia produce butyrate, acetate, CO_2 and H_2 whilst others produce mainly acetone rather than propan-2-ol. The proportions of the end-products varies according to the chosen species and strain and according to the cultural conditions.

2.7.7 Miscellaneous

There are a number of other reduced products that may be produced by microorganisms growing anaerobically, including sugar alcohols, succinic acid and trimethylene glycol. Glycerol can be formed, and indeed was for a time produced commercially, by adding bisulphite to yeast growing anaerobically. The bisulphite forms a complex with acetaldehyde, so that ethanol can no longer be produced, and the organism, seeking an alternative means to re-oxidize NADH, reduces triose phosphate to glycerol.

2.8 Biosynthesis and growth

The microbial cell is able to reproduce itself from the very simplest of nutrients. The number of pathways which a cell must use to accomplish this is enormous: a bacterial cell probably contains well over 1000 different enzymes, and a eukaryotic cell may contain twice as many. Cell macromolecules of all kinds (proteins, nucleic acids, polysaccharides, etc.) are built up from about 100 different monomer units. A general outline of the pathways of biosynthesis of these various monomers (amino acids, purines, pyrimidines, fatty acids, sugars, etc.) is given as Fig. 2.18. Pathways of biosynthesis, as can be seen, are interrelated and all rely on the maintenance of 'pools' of the necessary intermediates. Unfortunately it is not possible here to provide even outline particulars of these many pathways; their study is a major part of general biochemistry for which numerous textbooks exist. Insofar as particular pathways are especially relevant to specific fermentations, they are described in the appropriate chapters of the present book.

The cell carries out all its metabolic activities in a balanced manner so that end-products are neither over- nor under-produced; either would be disadvantageous. The microbial cell must also be able to respond to changes in its environment (temperature, pH, oxygen level, etc.) and also to take advantage of any gratuitous amounts of (say) preformed amino acids, purines or pyrimidines presented to it. This is frequently the case in natural habitats, and also where a complex nutrient (such as corn steep liquor or molasses) is used as growth substrate; such nutrients contain a plethora of carbon compounds. It is therefore possible for a cell to be able to save both carbon and energy by ceasing the synthesis of any material of which it has sufficient, and to make further economies by ceasing to synthesize any redundant enzymes. Thus there are at least two distinct ways—control of enzyme activity, and control of enzyme formation—by which a cell should be able to regulate synthesis of its various components. The same control mechanisms are also used to balance synthetic processes, even when no gratuitous materials are presented to the cell. Such general control mechanisms are described in this section.

2.8.1 Control of metabolism

2.8.1.1 Nutrient uptake
Control of cell metabolism begins by the cell regulating its uptake of nutrients. Most nutrients, apart from oxygen and a very few carbon compounds, are taken up by specific transport mechanisms so that they may be concentrated within the cell from dilute solutions outside. Such 'active'

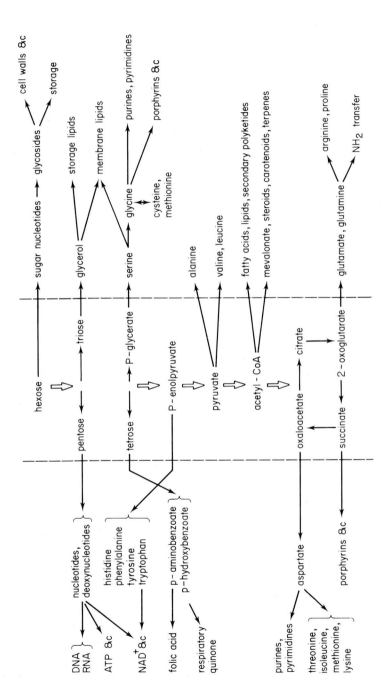

Fig. 2.18 Anabolic pathways (synthesis) and the central catabolic pathways. Only the main biosynthesis routes, and their main connections with catabolic pathways, are shown, all in highly simplified versions. Connections through 'energy' (ATP) and 'redox' (NAD⁺, NADP⁺) metabolism and through the metabolism of nitrogen, etc., are all omitted.

transport systems require an input of energy. The processes are controllable so that once the amount of nutrient taken into the cell has reached a given concentration, further unnecessary (or even detrimental) uptake can be stopped.

2.8.1.2   Compartmentalization

The second form of metabolic control is by use of compartments, or organelles, within the cell wherein separate pools of metabolites can be maintained. An obvious example is the mitochondrion of the eukaryotic cell which separates (amongst others) the tricarboxylic acid cycle reactions from reactions in the cytoplasm. Another, the peroxisome, contains enzymes for fatty acid degradation (see Fig. 2.10), whereas the somewhat similar enzymes which carry out related reactions in the reverse direction, leading to the synthesis of fatty acids, are located in the cytoplasm. Separating the two sets of enzymes prevents any common intermediates being recycled in a futile manner. Other organelles (vacuoles, the nucleus, chloroplast, etc.) are similarly used to control other reactions of the cell.

2.8.1.3   Control of enzyme synthesis

Many enzymes within a cell are present constitutively; they are there under all growth conditions. Other enzymes only 'appear' when needed; e.g. iso-citrate lyase of the glyoxylate by-pass (see Fig. 2.9) when the cell grows on a C_2 substrate. This is termed *induction* of enzyme synthesis. Conversely enzymes can 'disappear' when they are no longer required; enzymes for histidine biosynthesis stop being produced if there is sufficient gratuitous histidine available to satisfy the needs of the cell. This is termed *repression*; when the gratuitous supply of the compound has gone, the enzymes for synthesis of the material 'reappear'; their synthesis is *de-repressed*.

To understand how these controls operate it is necessary to outline the process of protein synthesis.

Proteins (including all enzymes) are synthesized by sequential addition of amino acid to amino acid by a complex of enzymes and RNA organized in the ribosome (see Fig. 2.19). The code for ensuring the right sequence of amino acids is carried by a strand of messenger RNA, which in turn is synthesized by copying from a section of DNA in the chromosome (genome) of the cell; this process is brought about by DNA-dependent RNA polymerases and is termed *transcription*. As is well known, the chromosome is composed of the double helix of DNA which is made up of a precise sequence of bases: adenine (A), cytosine (C), thymine (T) and guanine (G). The two chains are held together solely by hydrogen bonds between adjacent (A ... T) and (C ... T) base pairs. As A always pairs with T and C with G, a new strand of DNA can be made from one single

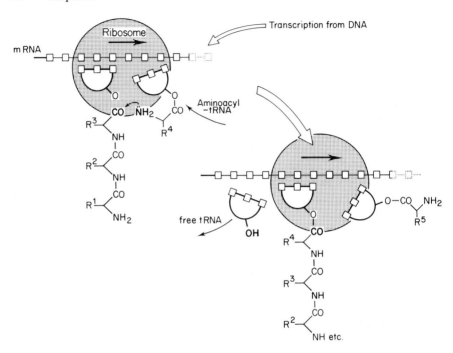

Fig. 2.19 Ribosome action (translation). Transfer RNA molecules (tRNA) complementary to successive base triplets in the messenger-RNA, and each carrying the corresponding aminoacid, transfer their aminoacyl residues in correct sequence to the growing peptide chain as the ribosome moves along the messenger RNA.

strand of the molecule, and is then *complementary* to the original. In this way DNA can be replicated and the genetic message, or code, preserved; this is shown later (Fig. 2.23). It is also from one of the strands of DNA that the messenger RNA is made (Fig. 2.19). Except that the base uracil is substituted for thymine, the RNA complements the DNA strand and similarly carries the genetic code in the sequence of its bases. Each messenger RNA is only made from a small part of the DNA, though many mRNAs can be produced by reading along the whole length of the DNA. Messenger RNAs remain single stranded.

Ribosomes become attached to the mRNA and, in the ribosome, the bases of the messenger RNA are 'read-off', that is *translated*, three at a time, to code for a particular amino acid. The three bases are termed a *codon*. Thus, for example, the codon UCA codes specifically for serine and the codon CAG for glutamine; hence when the sequence UCACAG is encountered along the mRNA attached to the ribosome this will produce seryl-glutamine. Each amino acid is enabled to 'recognize' the three bases along the mRNA to which it corresponds to by being covalently linked to a specific transfer RNA (tRNA). The aminoacyl-tRNAs are the active units used by the ribosome to produce the growing peptide chain (see Fig. 2.19).

Each individual mRNA molecule thus codes for one protein and originates from one *gene* on the chromosome (more than one protein may be needed to make some functional enzymes, for example pyruvate dehydrogenase, p. 19). Individual genes can be transcribed many times and there may be more than one copy of the gene on the chromosome. In either case, there is 'amplification' of the genetic information.

The regulation of protein synthesis through the overall mechanisms of transcription and translation is extremely complex; it differs, particularly in details, between prokaryotes and eukaryotes, and many aspects are still not elucidated. However the broad principles of the regulatory mechanisms can be illustrated from bacterial systems.

The machinery producing copies of mRNA from the DNA is controllable with respect to sections of the chromosome which code for *inducible* or *repressible* protein (see above). Such mechanisms are shown diagrammatically in Fig. 2.20. Sections of DNA termed *regulatory genes* produce (by way of the corresponding mRNA) a regulatory *repressor protein* whose function is to become attached to another, usually adjacent, gene. Binding of the protein to this *operator gene* prevents further translation occurring in the next section of DNA, that is in one or more *structural genes* which are responsible for the synthesis of mRNAs coding for enzyme proteins. If an inducer is present, this binds to the regulatory protein and prevents it binding to the operator gene. The free operator gene then permits transcription of the structural genes and the corresponding proteins can be synthesized. This is how a 'new' metabolic pathway is brought into operation. As long as the inducer molecule is present the enzymes will continue to be synthesized; if it is removed, or consumed (most frequently by the metabolic pathway it is inducing) then the enzymes will cease to be synthesized.

Repression (rather than induction) of enzyme synthesis occurs when a molecule, often the end-product of a pathway, interacts with the repressor proteins to produce a product which now blocks the operator gene. If the end-product is removed or consumed, the repressor proteins no longer bind with the operator gene, transcription of the structural genes takes place, and biosynthesis of the end-product can be resumed.

2.8.1.4 *Catabolite repression*
This type of metabolic control is an extension of the ideas already set out with respect to enzyme induction and repression, being brought about by external nutrients added to the microbial culture. The term catabolite

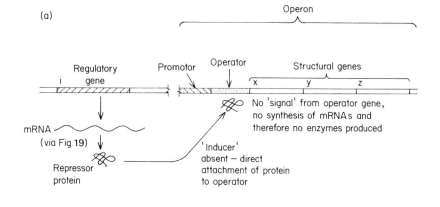

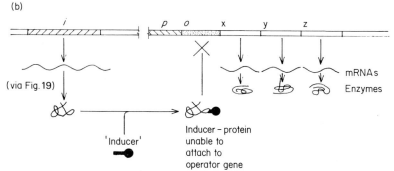

Fig. 2.20 Induction. (a) In the absence of inducer (e.g. lactose), the product of the repressor gene (*i*) combines with the operator gene (*o*) and this prevents transcription of the adjacent structural genes (*x*, *y*, *z*).

   (b) When the inducer substance is present, it combines with the repressor protein and prevents it from binding to the *o* gene; transcription of the structural genes to messenger RNAs, and translation of the mRNAs to enzymes now proceeds, and the enzymes are available for metabolism of the inducer.

   Note. In other cases, the repressor protein will only bind to the operator gene when a repressor substance is present; the enzymes are only formed when the repressor substance is removed (see text).

repression refers to several general phenomena, seen for example when a microorganism is able to select, from two or more different carbon sources simultaneously presented to it, that substrate which it 'prefers' to utilize. For example a microorganism presented with both glucose and lactose may 'ignore' the lactose until it has consumed all the glucose. Similar selection may occur for the choice of a nitrogen source if more than one is available. The advantage to the cell is that it can use the compound which involves it in the least expenditure of energy.

Detailed mechanisms of catabolite repression may well vary from organism to organism. The sequence of events shown in Fig. 2.21 is for the

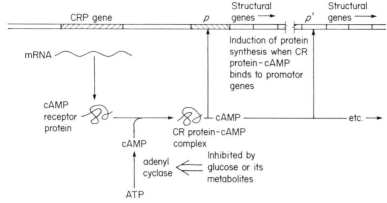

Fig. 2.21 Catabolite repression. The mechanism shown is one mediated by cyclic AMP (cAMP) in *E. coli*. One or several operons (see Fig. 2.20) are controlled simultaneously by their requirement for the product from cAMP and a catabolite repressor protein corresponding to an independent gene CRP. The eventual control in this mechanism is the effect of ATP and other nucleotides on the activities of enzymes producing or breaking-down the regulator molecule cAMP.

process as described in the bacterium *E. coli*. The key to the sequence lies in the compound cyclic AMP (cAMP; its phosphate group is attached to both the 3' and 5' hydroxyl group of the ribose moiety forming a phosphodiester—see Fig. 2.1). cAMP, formed from ATP by adenyl cyclase, interacts with a specific receptor protein (CRP = Cyclic AMP Receptor Protein) which positively promotes transcription of an operon (cf. Fig. 2.20).

Control of the pool size of cAMP is crucial for this mechanism, and is achieved by regulating the relative activities of adenyl cyclase and of the phosphodiesterase which converts cAMP to AMP. The level of cAMP relates closely to the 'energy charge' discussed in Section 2.2. Various metabolites of glucose appear to be potent inhibitors of adenyl cyclase activity (see Section 2.8.1.5b), and so long as these are present in the cell (indicating the continued availability of glucose) then the several operons under control of the cAMP–CRP complex will not be transcribed. Catabolite repression effects are also important in controlling patterns of anabolic processes, particularly in the phenomena of 'secondary biosynthesis'.

2.8.1.5 *Modification of enzyme activity*

Once an enzyme has been synthesized, its activity can be modulated by a variety of means.

(a) Post-transcriptional modifications. Some enzymes can exist in an active and a less active form, which can be interconverted by the covalent attachment of some specific group (often phosphate, sometimes AMP or UMP). The attachment is mediated by a separate enzyme, which has no other function, whose activity is in turn regulated by the presence of various metabolites (see next section). Thus the reaction catalysed by the first enzyme can be moderated by the prevailing metabolic status of the cell (see Fig. 2.22). Examples of such processes are glutamine synthetase in

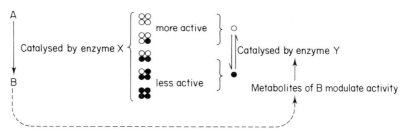

Fig. 2.22   Enzyme control by a cascade mechanism. Metabolites derived from B affect (positively or negatively) the activity of enzyme Y, which interconverts more active and less active subunits of enzyme X, which catalyses the formation of B from A.

E. coli (it is extremely important for the cell to regulate precisely the pool size of the key metabolic intermediates glutamate and glutamine which are, respectively, substrate and product of the enzyme) and glycogen phosphorylase in the mould *Neurospora crassa*.

(b) Action of effectors. Enzyme activity is very often slowed down by a build-up of the product of the reaction it is catalysing. The compound is then said to be an inhibitor, or *negative effector*, of the enzyme. This is a simple mechanism to understand: the reaction product blocks access of the substrate to the active site of enzyme, where both are able to 'fit'. However, many enzymes, particularly those at the beginning of a pathway, are similarly sensitive to the presence of compounds chemically unrelated to the reaction they are catalysing. The most common effect is known as *feedback inhibition*, in which the end-product of a pathway acts as a negative effector on the activity of an early enzyme in that pathway:

$$A \longrightarrow B \longrightarrow C \longrightarrow D \longrightarrow E$$

The effect of this inhibition is to ensure that once the end-product has been produced in sufficient quantity, no further carbon units will be channelled down the pathway; the product which is not needed is not produced. The same event will occur if the end-product is present in the growth medium taken up into the cell (see Section 2.8.1.1). Feedback inhibition often parallels repression of enzyme synthesis (see above), which is also brought about by the presence of excess end-product; it can be thought of as *fine control* rapidly brought about and readily reversible, whereas enzyme repression provides a *coarse control* which takes longer to achieve.

The general mechanism of feedback inhibition is based upon binding of the effector molecule to the enzyme at a site which is different from the active site; the effector modifies the conformation (shape) of the protein so that it is no longer so effective a catalyst for the reaction of its substrate. Such enzymes are said to be *allosterically* controlled. Examples can be found in most of the pathways of metabolism leading to the biosynthesis of amino acids, purines, pyrimidines and other monomers (see Fig. 2.17).

The process becomes more complicated where there is more than one product derived from branches in a common pathway:

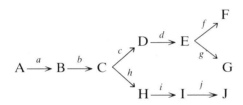

Here it is important that if one of the three end-products, F, G or J, reaches its optimum pool size, it should act to prevent more of itself being produced but at the same time not inhibit the synthesis of the other two end-products. Thus in the above diagram, assuming the three end-products were required in equal amounts, product F would be expected to inhibit reaction *f* completely, reaction *c* by 50% and reaction *a* by 33%. Thus starting from A, first B and thence C would be formed at two-thirds the usual rate; C, instead of producing twice as much D as H, would now produce equal amounts of these two compounds; all of D would now go to G, and H would go to J as before.

The manner in which this partial inhibition of enzyme activity is achieved can vary from pathway to pathway and organism to organism. A common method is for the organism to use *isoenzymes*. That is, for reaction *a* above, there will be three distinct and independent enzymes each catalysing the reaction with equal efficiency. However one isoenzyme will be sensitive to feedback inhibition by F, the second by G, and the third by H. In this way, only one isoenzyme will be inhibited if one end-product reaches its optimum pool concentration. The reaction *c* may similarly be expected to be catalysed by two isoenzymes: one sensitive to feedback inhibition by F and the other by G.

Examples of this type of control occur in the biosynthesis of the three aromatic amino acids, phenylalanine, tyrosine and tryptophan, and in the biosynthesis of threonine, methionine and lysine (see Fig. 2.18).

Feedback effects are also involved in the regulation of transport processes (see Section 2.8.1.1). However, it may not be quite accurate to extend the concept to situations where metabolites such as ATP, ADP, AMP, $NAD(P)^+$ or $NAD(P)H$ act as positive or negative effectors on a particular enzyme. For instance, several enzymes of the tricarboxylic acid (see Fig. 2.8), particularly citrate synthase, are inhibited by ATP, and as ATP is the 'end-product' of oxidative phosphorylation which is linked to the reaction of the cycle, this might be construed as a more indirect form of feedback inhibition. Irrespective of semantics, this type of control by different forms of general cofactors is quite widespread amongst the enzymes of the central pathways of metabolism.

2.8.1.6   *Degradation of enzymes*
Enzymes are not particularly stable molecules and may be quickly and irreversibly destroyed. The normal half-life is very variable; it may be as short as a few minutes, or as long as several days. Although the syntheses of enzymes can be regulated at the genetic level (see Section 2.8.1.3), once an enzyme has been synthesized it can remain functional for some time. If the environmental conditions change abruptly, it may not suffice for the synthesis of the enzyme to be 'switched off' i.e. repressed; the cell may need to inactivate the enzyme so as to avoid needless, or even perhaps deleterious, metabolic activity. Several examples are known where specific proteolytic enzymes, suitably activated, will destroy a particular enzyme. Activation is probably triggered by the presence (or absence) of a key metabolite.

2.8.2   Coordination of metabolism and growth

We have already considered how the cell is able to control the biosynthesis of its many constituent parts so that the correct amount of monomer is always synthesized as well as the appropriate number of different enzyme molecules. These control mechanisms are responses to the external environment of the cell. The cell always attempts to optimize its internal biochemistry so that it can make the most efficient use of preformed carbon and nitrogen compounds, maximizing the energy yield, minimizing energy expenditure, and growing as rapidly as it is able.

Under limiting environmental conditions, for example absence of any nitrogen source, the organism may not be able to reproduce. Under such conditions, end-products (some of which may well be desired by the biotechnologist) would be accumulated as the organism continued to metabolize the carbon available to it. The biosynthetic machinery of an organism does, and must, continue to work at all times; the only time the reactions of a cell come to equilibrium is when it is dead.

It is therefore essential for the organism to keep its cell biochemistry working and it can achieve this by a variety of means depending on the prevailing conditions: by de-repressing new anabolic enzymes it may channel the carbon substrate into any of a number of 'secondary' metabolites; it can produce large amounts of storage compounds such as poly-β-hydroxy-butyrate, lipid, glycogen and other polysaccharides. It may degrade these storage materials if placed in a 'starvation' situation where no external carbon is being supplied to the cell. What is produced is probably less important than that the cell keeps its pathways of metabolism operating.

Under 'normal' conditions, given a supply of all the essential nutrients, the organism grows. In batch culture, the cells multiply in a closed system (there is no addition to, or subtraction from, the fermenter once it has been inoculated) until either some nutrient becomes exhausted or until some product accumulates to inhibit further growth, or the number of cells reaches such a level that there is no further space available for new ones to occupy. During the growth of the cells, the various components of the cell alter in relative amounts (Fig. 2.23) and the cell even changes in size as it elongates before cell division. During the initial exponential growth period when the cells are growing at maximum rate, the RNA content of the cell increases rapidly due to the cells synthesizing more proteins on the ribosome (see Fig. 2.19). The content of DNA may fall, as indicated in Fig. 2.23 although the exact extent of this fall will depend upon the rate at which the cell is able to replicate its DNA (see Section 2.8.3).

The rate at which an organism can grow is expressed either as the doubling time (t_d), i.e. the time taken for one cell to become two, or as the specific growth rate (μ) which is the rate of synthesis of new cell material expressed per unit weight of existing cell material. These two values are related by the equations

$$\mu = \frac{1}{x} \cdot \frac{dx}{dt} = \frac{d(\ln x)}{dt} = \frac{\ln 2}{t_d}$$

(where x = cell concentration, t = time), from which $\mu = 0.69/t_d$.

In batch culture the value of μ varies throughout, due to the continually decreasing concentration of nutrients; in many practical situations with aerobic organisms the rate of supply of oxygen eventually governs the rate

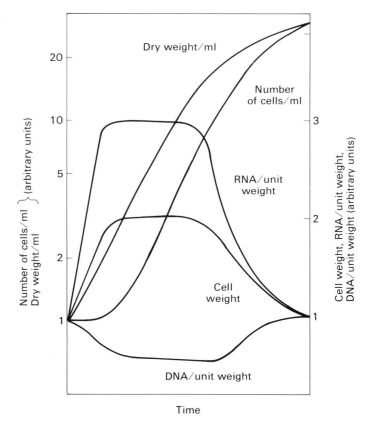

Fig. 2.23   Idealized representation of the changes in cell size (i.e. cell weight), cell numbers, total dry weight and chemical composition during growth of bacteria in batch culture. Note that the scale on the left is logarithmic.

of their growth. Only in continuous culture can the specific growth rate be held constant through the continuous introduction of fresh nutrients (and an ensured supply of oxygen). What governs the maximum growth rate that an organism can attain probably varies from organism to organism, and is quite unknown for most. It may be the ultimate rate at which DNA can be synthesized, or the rate at which a particular nutrient can be taken up into the cell, or the rate at which some part of the cell, such as the walls, can be assembled. Doubling times can range from about 10 minutes with *Benekea natriegens* through to several hours with yeasts and fungi. Most bacteria have doubling times of 30 minutes or more, and some have doubling times which extend to several days. Examples are listed in Table 2.3.

Table 2.3   Growth rates of various microorganisms under optimal conditions

Microorganism	Maximum specific growth rate, μ_m (h^{-1})	Doubling time, t_d (h) ($= 0.693/\mu_m$)
Escherichia coli	2.1	0.33
Saccharomyces cerevisiae	0.45	1.5
Candida utilis	0.40	1.7
Schizosaccharomyces pombe	0.17	4.0
Penicillium chrysogenum	0.28	2.5
Geotrichum lactis	0.35	2.0
Fusarium graminearum	0.28	2.5
Chlorella pyrenoidosa	0.08	8.5

2.8.3   The cell cycle and DNA replication

The processes of cell division are different in prokaryotes and eukaryotes, although both use essentially similar mechanisms for controlling the expression of genes and of regulating the activities of the gene products (enzyme proteins). In a rapidly growing bacterium DNA synthesis takes place more or less continuously, but in a eukaryotic cell it forms only part of the cell cycle. The bacterial genome is but two molecules of DNA in the double helix conformation, linked head to tail to form a circular chromosome. The eukaryotic cell contains several physically separate chromosomes.

In the eukaryote, the cell cycle is divided into stages, each of variable duration depending on the growth conditions. The cycle culminates with the replication of all the chromosomes which then divide between daughter and mother cell by a process known as *mitosis*. The process becomes more complex the higher one advances through the microbial kingdom when sexual reproduction, rather than the simple binary fission of bacteria or the budding of yeasts, may occur. The process of chromosomal division then occurs by *meiosis* in which chromosomes are distributed, after replication of the DNA, into germ cells.

In all microorganisms the DNA probably replicates by similar mechanisms. The double-stranded DNA unwinds with each strand forming a new strand complementary to itself. This is shown in Fig. 2.24. Each replication fork operates bidirectionally, that is unwinding and synthesis of the complementary DNA strand occurs at both ends of the separation.

In bacteria, there can be more than one replication fork functioning at any one time so that the genome can be reproduced with great rapidity. This happens when the time for complete DNA replication is longer than the time for cell division, and under these conditions each cell will contain more than one copy of the replicated parts of the chromosome. Cell division

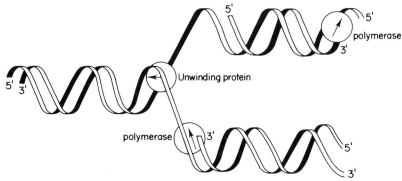

Fig. 2.24   DNA replication. In the two complementary strands of DNA the phospho-3'-deoxyribotyl-5' 'backbone' runs in opposite directions. In replication, the complementary strands are first separated ('unwinding') so that additional deoxyribonucleotides can complement the bases in each strand. On one strand DNA polymerase then acts directly to assemble the new DNA, but on the other it must act in short 'reverse' sections. The whole set of operations constitutes a 'replication fork'. In replication of the circular chromosome of bacteria, pairs of such forks are initiated and move in opposite directions around the chromosome.

is however synchronized to the complete replication of the chromosomes, so that each daughter cell receives its DNA before the dividing septum is formed. Apparently a 'termination protein' is synthesized when replication of DNA has been completed, and it is this protein which triggers septum formation and cell division.

Cell duplication in eukaryotic cells is complicated by the requirement for organelles also to divide during the course of the cell cycle. Mitochondria and chloroplasts (the photosynthetic apparatus of photosynthetic microorganisms and plants) have their own DNA and divide independently of the cell nucleus. In this way the number of mitochondria and chloroplasts can change according to the environmental conditions. For example, a yeast growing under anaerobic conditions does not rely on its mitochondria for provision of ATP (see Sections 2.5 and 2.6) and the number of mitochondria is minimal.

Although the genetic information carried by the DNA is accurately reproduced so that each daughter cell carries the same chromosomal programme as its parent, mistakes can be made. Such mistakes lead to the formation of a *mutant*. Mutants can arise both spontaneously, usually at a very low frequency (about 1 in every 10^8–10^{10} cell divisions), or can be induced by exposure of the organism to some DNA-damaging agent (mutagen). Mutants having altered capacities of enzyme activity can be of considerable benefit in increasing the productivity of a fermentation process. However, finding the one desired mutant amongst the many thousands of unwanted mutants can be a very long process, which is discussed for example in Chapter 8. In the vast majority of cases the mutants have an impaired capacity to carry out some activity; in many cases this will be lethal and the cell will be unable to grow. In general very few mutants will survive, but because large numbers of microorganisms are handled in an experiment a 0.001% survival still may represent over 10 000 organisms.

2.8.4   Efficiency of microbial growth

The overall efficiency of microbial growth is discussed in strict thermo-dynamic terms in Chapter 3 (p. 57). Empirically, it is usually expressed in terms of the yield of cells formed from unit weight of carbon substrate consumed. The *molar growth yield* Y_s is the cell yield (dry weight) per mole of substrate, while the *carbon conversion coefficient*, which allows more meaningful comparisons between substrates of different molecular sizes, is the cell yield per gram of substrate carbon. Some typical values for both expressions are collected in Table 2.4; all are maximum values, because under certain conditions (in particular, at low growth rates: see Chapter 3, p. 67) the use of substrates for cell growth becomes less than fully efficient.

A particular feature in Table 2.4 which can be readily understood by reference to previous discussions is the lower growth yields attained when facultative organisms are transferred from aerobic to anaerobic conditions, a phenomenon which is obviously connected with the reduced energy flow, and lower yields of ATP, in anaerobic processes.

Empirically, the actual growth yield will depend on many factors:

(1) the nature of the carbon source.
(2) the pathways of substrate catabolism.
(3) any provision of complex substrates (obviating some anabolic path-ways).
(4) energy requirements for assimilating other nutrients, especially nitrogen (less if amino acids are supplied than if NH_3 is used, and considerably more if nitrate ion is the nitrogen source).
(5) varying efficiencies of ATP-generating reactions.
(6) inhibiting substances, adverse ionic balance, or other medium com-ponents imposing extra demands on transport systems.
(7) the physiological state of the organism; nearly all microorganisms modify their development according to the external environment, often very considerably (e.g. spore formation), and the different development processes will entail different mass and energy balances.

Table 2.4   Growth yields of microorganisms growing on different substrates

Substrate	Organism	Molar growth yield (g organism dry wt per g-mol substrate)	Carbon conversion coefficient (g organism dry wt per g substrate carbon)
Methane	*Methylomonas methanooxidans*	17.5	1.46
Methanol	*Methylomonas methanolica*	16.6	1.38
Ethanol	*Candida utilis*	31.2	1.30
Glycerol	*Klebsiella pneumoniae (Aerobacter aerogenes)*	50.4	1.40
Glucose	*Escherichia coli:*		
	aerobic	95.0	1.32
	anaerobic	25.8	0.36
	Saccharomyces cerevisiae:		
	aerobic	90	1.25
	anaerobic	21	0.29
	Penicillium chrysogenum	81	1.13
Sucrose	*Klebsiella pneumoniae (A. aerogenes)*	173	1.20
Xylose	*Klebsiella pneumoniae (A. aerogenes)*	52.2	0.87
Acetic acid	*Pseudomonas* sp.	23.5	0.98
	Candida utilis	21.6	0.90
Hexadecane	*Saccharomycopsis (Candida) lipolytica*	203	1.06
	Acinetobacter sp.	251	1.31

In continuous culture systems, in which the growth rate and nutritional status of the cells are controlled (see Chapter 4, p. 105), further factors can be identified:

(8) the nature of the limiting substrate; carbon-limited growth is often more 'efficient' than, for example, nitrogen-limited growth, in which catabolism of excess carbon substrate may follow routes which are energetically 'wasteful' (however useful to the biotechnologist!).

(9) the permitted growth rate; the significance of the 'maintenance' effect at low growth rates is discussed in Chapter 4, p. 94).

As a final factor governing all aspects of microbial performances, one might usefully add:

(10) "the inclinations of the microbe, and the competence of the micro-biologist".

Reading list

Anthony, C. (1982). 'The Biochemistry of Methylotrophs'. Academic Press: London, and New York.

Dawes, I. W. and Sutherland, I. W. (1976). 'Microbial Physiology', Vol. 4, Basic Microbiology. Blackwell Scientific Publications: Oxford.

Edwards, E. (1981). 'The Microbial Cell Cycle', Aspects of Microbiology. Nelson/Van Nostrand Reinhold (UK) Co. Ltd: Wokingham.

Evans, C. T. and Ratledge, C. (1984). Induction of xylulose-5-phosphoketolase in a variety of yeasts grown on D-xylose: the key to efficient xylose metabolism. *Arch. Microbiol.* **139,** 48–52.

Gottschalk, G. (1979). 'Bacterial Metabolism'. Springer-Verlag: New York, Heidelberg and Berlin.

Linton, J. D. and Stephenson, R. J. (1978). A preliminary study on growth yields in relation to the carbon and energy content of various organic growth substrates. *FEMS Microbiol. Lett.* **3,** 95–98.

Mandelstam, J. and McQuillen, K. (Eds) (1982). 'Biochemistry of Bacterial Growth', 3rd Edn. Blackwell Scientific Publications: Oxford.

Nagai, S. (1979). Mass and energy balances for microbial growth kinetics. *Adv. Biochem. Engng* **11,** 49–83.

Neijssel, O. M. and Tempest, D. W. (1979). The physiology of metabolite over-production. *Symp. Soc. Gen. Microbiol.* **29,** 53–82.

Quayle, J. R. (Ed.) (1979). 'Microbial Biochemistry', Vol. 21, International Review of Biochemistry. University Park Press: Baltimore.

Stouthamer, A. H. and Van Verseveld, H. W. (1985). Stoichiometry of microbial growth. In 'Comprehensive Biotechnology' (C. L. Cooney and E. A. Humphrey, eds.) pp. 215–238. Pergamon Press: Oxford, New York.

Strathern, J. N., Jones, E. W. and Broach, J. R. (Eds) (1982). 'The Molecular Biology of the Yeast Saccharomyces: Metabolism and Gene Expression'. Cold Spring Harbor Laboratory Monograph Series 11B.

3 | Thermodynamics of Growth

J. A. ROELS

List of symbols

A_r	affinity of a chemical reaction	$(kJ\ mol^{-1})$
D	dissipation per unit biomass produced	$(kJ\ C\text{-}mol^{-1})$
g_i	partial molar free enthalpy of compound i	$(kJ\ mol^{-1})$
$\Delta \bar{g}^o_{ci}$	partial molar free enthalpy of combustion of compound i at standard conditions	$(kJ\ mol^{-1})$
Δg^0_{ci}	partial free enthalpy of combustion per C-mole of compound i at standard conditions	$(kJ\ C\text{-}mol^{-1})$
$\Delta \bar{g}^{0'}_{ci}$	partial molar free enthalpy of combustion of compound i at pH = 7	$(kJ\ mol^{-1})$
$\Delta g^{0'}_{ci}$	partial free enthalpy of combustion per C-mole of compound i at a pH = 7	$(kJ\ C\text{-}mol^{-1})$
$\Delta g^{0'}_{av/e}$	free enthalpy gained on transfer of one mole of electrons to the electron acceptor at a pH of 7	$(kJ\ mol^{-1})$
h_i	partial molar enthalpy of compound i	$(kJ\ mol^{-1})$
$\Delta \bar{h}^0_{ci}$	partial molar heat of combustion of compound i at standard conditions	$(kJ\ mol^{-1})$
Δh^0_{ci}	partial heat of combustion per C-mole of compound i at standard conditions	$(kJ\ C\text{-}mol^{-1})$
$N_{av/e}$	electrons available for transfer to oxygen on combustion of one mole of a compound	$(-)$
S	system's entropy	$(kJ\ K^{-1})$
s_i	partial molar entropy of compound i	$(kJ\ mol^{-1}\ K^{-1})$
Δs^0_{ci}	partial molar entropy of combustion per C-mole of compound i at standard conditions	$(kJ\ C\text{-}mol^{-1}\ K^{-1})$

57

T	absolute temperature	(K)
U	system's internal energy	(kJ)
$Y_{av/e}$	yield on available electrons	(g dry matter mol^{-1})
Y_{ATP}	cell yield per mole ATP	(g mol^{-1})
α_i	stoichiometric coefficient	(—)
γ_i	degree of reduction	(—)
η_{th}	thermodynamic efficiency	(—)
μ_i	thermodynamic potential of compound i	(kJ mol^{-1})
Π_S	rate of entropy production	(kJ K^{-1} s^{-1})
Π_U	rate of energy production	(kJ s^{-1})
Φ_U	rate of flow of energy to the system	(kJ s^{-1})
Φ_Q	rate of flow of heat to the system	(kJ s^{-1})
Φ_i	rate of flow of compound i to or from the system	(mol s^{-1})
Φ_S	rate of flow of entropy to the system	(kJ K^{-1} s^{-1})

3.1 Introduction

The absolute description of any real process, even if such a process is of elementary simplicity at the macroscopic level, is a hopelessly complex task. The behaviour of a mole of iron, i.e. 56 grams, is governed by the behaviour of at least 6×10^{23} atoms; the description of the detailed behaviour of such a vast number of particles is well beyond the potential of even the largest computers.

The macroscopic approach towards systems, which ignores the detailed corpuscular structure of reality, treats systems in terms of average quantities, attributed to a region containing a sufficiently large number of particles. Quantities like energy, entropy, concentrations of chemical substances and temperature are typical variables in the macroscopic approach.

Macroscopic methods, based on a continuum rather than a corpuscular approach, are a powerful tool towards the description of real systems and have become widely accepted in the engineering disciplines. In the present chapter some aspects of the macroscopic treatment will be adopted in treating the processes occurring in living systems. The approach developed is of a mainly thermodynamic nature and discusses the energy transformations which fuel life. The extent and detail in which the thermodynamic theory is treated will be kept to the minimum and the results of the thermodynamic theory will have to be introduced without proper proof or derivation. The reader is referred to the specialized literature for more detail (Glansdorff & Prigogine, 1971; Haase, 1969; Prigogine, 1967).

3.2 The methodology of thermodynamics

3.2.1 Energy and entropy, the first and second laws of thermodynamics

Thermodynamics deals with energy transformations and its methodology rests on the possibility of defining the quantity and the quality of the energy contained in a system by two so-called functions of state, i.e. internal energy, U, and entropy, S. The internal energy relates to the *quantity* of energy, the entropy to its *quality* with respect to it being suited to drive a given process in the system.

The application of thermodynamics to processes occurring in a living system can be based on the fact that living organisms are energy transducers, which convert a given source of energy, e.g. radiation in photosynthetic processes or chemical energy in chemotrophs, to another source of energy. In this process the energy flows from a form of higher quality, i.e. of a lower entropy, to a form of a lower quality, i.e. of higher entropy; throughout, in accord with the first law of thermodynamics, the total *amount* of energy is unaffected by the processes taking place in the system.

In Fig. 3.1 a system is depicted which could, for the purpose of thermo-

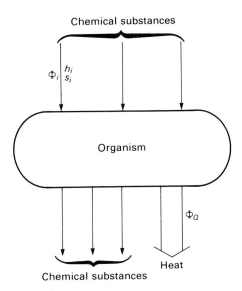

Fig. 3.1 A schematic representation of a microbial system for the purpose of a macroscopic analysis.

dynamic analysis, represent an organism engaged in metabolism, exchanging energy with the environment in the form of chemical substances and heat (in the present treatment organisms feeding on radiational energy or performing mechanical or electrical work are excluded, though the basic theory can be quite easily extended to include these cases). The exchange flows of chemical substances are Φ_i (moles h^{-1}), their energy content and quality are represented by the partial molar enthalpy, h_i, and the partial molar entropy, s_i, respectively. Furthermore, a flow of heat, Φ_Q, is exchanged with the environment. For consistency, all flows are defined positive if they are directed towards the system.

In the macroscopic description of a system the description of the time evolution of the amount of so-called extensive quantities (i.e. quantities like mass and energy, which are additive with respect to parts of a system) is undertaken by the formulation of balance equations, which relate the rate of change of the amount of a property present in the system to the amount produced or consumed in conversion processes taking place *in* the system and the amount gained or lost by net transport *to or from* the system. In mathematical terms, taking internal energy U as an example, this results in the equation

$$\dot{U} = \Pi_U + \Phi_U \tag{3.1}$$

In this equation $\dot{U}$ is the overall rate of change in U, Π_U is the rate of energy production in the processes taking place in the system, Φ_U is the rate of transport of energy towards the system. For a system in a steady state the overall term $\dot{U}$ vanishes. Furthermore, the first law of thermodynamics states that energy is a so-called conserved quantity, i.e. the total amount in a closed system cannot change due to transformation processes, so $\Pi_U = 0$, and Equation 3.1 reduces, for a system in steady state, to

$$\Phi_U = 0 \tag{3.2}$$

Equation 3.2 shows that the net flow of energy towards a system in steady state is necessarily zero, so that the flows of energy to the system must be compensated by flows of equal magnitude from the system.

A fundamental result of the thermodynamic method is that different contributions can be distinguished in Φ_U, the total flow of energy that a system exchanges with the environment. In the present analysis where only the heat and the energy contained in chemical substances are considered, the energy flow takes the form

$$\Phi_U = \Phi_Q + \sum_i h_i \Phi_i \tag{3.3}$$

If Equations 3.2 and 3.3 are combined the fundamental equation for the calculation of the heat exchange between a steady state system and the environment results

$$\Phi_Q = -\sum_i h_i \Phi_i \tag{3.4}$$

Heat Φ_Q is the rate of heat flow to the system; h_i and Φ_i are the partial molar enthalpy and the rate of mass flow, respectively, of substance i into or out of the system.

By analogy to Equation 3.1 a balance equation for the rate of change of entropy, $\dot{S}$, can be similarly formulated as

$$\dot{S} = \Pi_S + \Phi_S \tag{3.5}$$

in which Π_S is the rate of entropy production in the system and Φ_S is the rate of exchange of entropy between the system and the environment. For the system in steady state $\dot{S} = 0$ and Equation 3.5 becomes

$$\Pi_S = -\Phi_S \tag{3.6}$$

Now as long as any processes actually take place, Π_S *must* exceed zero (second law of thermodynamics) and Φ_S, and entropy flow towards the system, must be negative. A system can only keep its entropy at a value lower than that corresponding to thermodynamic equilibrium by 'feeding on negentropy', a phase coined by Schrödinger (1944).

Two different contributions to Φ_S can be distinguished in the entropy flow a system exchanges with the environment

$$\Phi_S = (\Phi_Q + T\sum_i \Phi_i s_i)/T < 0 \tag{3.7}$$

The first term appearing on the right-hand side of Equation 3.7 accounts for the entropy exchange associated with heat, the second term for exchange of entropy due to transport of chemical substances, the s_i being the partial molar entropies of the compounds. T is (absolute) temperature.

Equation 3.7 clearly shows that the second law of thermodynamics ($\Phi_S < 0$) does not necessarily imply that heat transport to the environment (i.e. a negative Φ_Q) must result from a system in which processes take place (e.g. an organism engaged in a chemical reaction), processes in which heat is taken up from the environment (endothermic processes, Φ_Q positive), are possible as long as the second contribution to the entropy flow makes the total flow negative.

If Equation 3.7 is combined with Equation 3.4 it follows that

$$\Phi_S = -\sum_i \Phi_i g_i / T \tag{3.8}$$

where g_i is the partial molar *free* enthalpy, defined by

$$g_i = h_i - Ts_i \tag{3.9}$$

If Equation 3.8 is combined with Equation 3.6 and the restriction posed by the second law it follows that

$$\sum_i \Phi_i g_i > 0 \tag{3.10}$$

Equation 3.10 summarizes the restrictions the first and second laws pose to the exchange of chemical substances between a steady state system and the environment. As such, it also applies to a metabolizing organism. The nature of Equation 3.10 can be illustrated by the following example. Consider a system in which a single chemical reaction takes place according to the following stoichiometry

$$\alpha_a A + \alpha_b B \rightleftharpoons \alpha_c C + \alpha_d D \tag{3.11}$$

It is a well-known result of thermodynamics that a reaction according to Equation 3.11 will proceed if and only if its affinity, A_r, is positive, the affinity A_r being given by

$$A_r = \alpha_a \mu_a + \alpha_b \mu_b - \alpha_c \mu_c - \alpha_d \mu_d \tag{3.12}$$

where the μ_i are the thermodynamic potentials of the compounds and are actually equal to the partial molar free enthalpies, g_i.

Assume the reaction according to Equation 3.11 takes in a continuous stirred tank reactor in a steady state. The net rates of consumption of each of the compounds must equal the net rate of transport to the system and Equation 3.12 can also be written as

$$A_r = (\Phi_a g_a + \Phi_b g_b + \Phi_c g_c + \Phi_d g_d)/r_1 \tag{3.13}$$

in which the Φ_i are net rates of flow to the system and r_1 is the rate of reaction of the reaction from A and B to C and D. As A_r has to be positive, a positive rate of reaction can only result if

$$\sum_i \Phi_i g_i > 0 \tag{3.14}$$

The summation in Equation 3.14 extends over compounds A through D.

It is clear that the derivation given directly above, which results in Equation 3.14, leads to the same result as the reasoning underlying Equation 3.10. At this stage the reader may wonder why the development underlying Equation 3.10 was necessary, whilst the same result can apparently be obtained in a much more simple way. The answer rests in the fact that Equation 3.14 only follows in a straightforward manner if just one reaction is assumed to take place, while the derivation underlying Equation 3.10 is perfectly general, and the reaction pattern inside the system may be of arbitrary complexity—as in microorganisms.

Another conclusion implicit in Equation 3.10 is of great importance. Compounds exhibiting no net exchange with the environment, i.e. compounds for which Φ_i equals zero, do not influence the evaluation of Equation 3.10 and hence even their existence need not be considered in formulating it. This is very relevant to the description of the processes taking place in microbial systems, in which there are a multitude of compounds which in normal situations are not subject to net exchange with the environment (notable and relevant examples are the energy carriers ATP and NADH). In a detailed thermodynamic analysis of the various metabolic pathways in an organism these compounds do, of course, appear, but a macroscopic treatment according to Equation 3.10 does not involve them.

3.2.2   The thermodynamic efficiency

One of the important applications of Equation 3.10 is that it allows definition of the thermodynamic efficiency of the processes taking place in a system. In principle the thermodynamic efficiency can be straightforwardly defined. The free enthalpy contained in the flows of matter leaving the system can never surpass the free enthalpy of the combined flows of matter entering the system, and it becomes logical to define the thermodynamic efficiency as the ratio of these free enthalpy contents. However, a problem is encountered if this is attempted. Energy, and hence also a derived state function such as free enthalpy, cannot be attributed a unique value, but can only be determined as a difference with respect to a given base level. Consequently, useful numerical values of the thermodynamic efficiency are only obtained if such a base level is chosen in a realistic way. For organisms a realistic base level is the energy contained in a mixture of CO_2, H_2O, N_2 and O_2 as (in absence of energy forms other than chemical energy) organisms cannot derive useful metabolic energy from such a mixture. The free enthalpy or enthalpy content of a chemical compound with respect to that base level is then equal to its free enthalpy or enthalpy ($\equiv$ heat) of combustion to CO_2, H_2O and N_2.

Once the reference state is defined, the thermodynamic efficiency can be straightforwardly calculated, and this is illustrated in Fig. 3.2. The thermodynamic efficiency is defined as the ratio of the free enthalpies gained

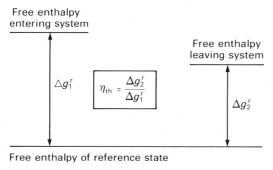

Fig. 3.2   The thermodynamic efficiency.

if the compounds leaving and entering the system respectively are transformed to the reference state.

3.2.3   The degree of reduction; electrons available for transfer to oxygen

In treating energy transformations in a generalized way the concept of the degree of reduction (Erickson *et al.*, 1978; Roels, 1981) becomes quite useful. The degree of reduction definition is based on the concept of the number of moles of electrons available for transfer to oxygen on combustion of a compound to CO_2, H_2O and N_2 (Kharasch, 1929). From simple stoichiometry considerations it is quite easily shown that a compound of elemental composition $C_{d_i}H_{a_i}O_{b_i}N_{c_i}$ contains a number of moles of electrons for transfer to oxygen, $N_{av/e}$ which is equal to

$$N_{av/e} = 4d_i + a_i - 2b_i \qquad (3.15)$$

As each mole of oxygen can be assumed to accept four moles of electrons in a combustion process, the oxygen consumption on combustion equal $N_{av/e}/4$.

The degree of reduction, γ_i, of a compound is now defined as the number of moles of electrons available for transfer to oxygen on a per C-mole base, i.e. expressed for the amount of substance containing 12 grams of carbon. Thus, γ_i becomes equal to

$$\gamma_i = (4d_i + a_i - 2b_i)/d_i \qquad (3.16)$$

3.2.4   The enthalpies and free enthalpies of combustion of common substrates

A numerical evaluation of the energetics of microbial processes is only possible if data on the enthalpy and free enthalpy content of the compounds engaged in metabolism are known. As was argued in Section 3.2.1 heats and free enthalpies of combustion are well suited to this purpose.

In principle, the treatment has to be based on partial molar quantities, which cannot be considered a property of a given substance, but rather are related to the properties of the integral mixture and hence to the concentrations of each and every component present at the site of the chemical reactions. Thus the detailed composition of a microbial system would enter the calculations, posing unsurmountable problems, so an approximate treatment must be used. In many treatments of energetics the analysis is performed in terms of thermodynamic quantities at standard conditions (i.e. at a standardized temperature and pressure and at unit molar concentration of the reactants involved). However, such quantities are by no means ideally suited to describe microbial energetics; in particular the concentration of the H^+-ion and the concentration of water are often markedly different from a molar concentration equal to unity. Therefore the free enthalpy and enthalpy of water are often taken to be equal to that of free liquid water, whilst the concentration of the H^+-ion is taken equal to a value corresponding to a pH of 7. These considerations in fact only affect free enthalpy changes, as the enthalpy changes are much less dependent on the detailed values of the concentrations of the reactants involved. In the present treatment, therefore, heats of combustion, $\Delta \bar{h}_c^0$, will be expressed at standard conditions. For free enthalpies of combustion, two values are used, one referring to standard conditions in the strict sense, $\Delta \bar{g}_c^0$, and one referring to standard conditions at which the concentrations of the

Table 3.1   Thermodynamic data of some representative compounds

Compound	Formula	$\Delta \bar{g}_c^0$	$\Delta \bar{g}_c^{0'}$ (kJ mol^{-1})	$\Delta \bar{h}_c^0$
Formic acid	CH_2O_2	281	236	255
Acetic acid	$C_2H_4O_2$	894	844	876
Palmitic acid	$C_{16}H_{32}O_2$	9800	9683	9989
Lactic acid	$C_3H_6O_3$	1377	1322	1369
Gluconic·acid	$C_6H_{12}O_7$	2661	2593	
Oxalic acid	$C_2H_2O_4$	327	238	246
Succinic acid	$C_4H_6O_4$	1599	1500	1493
Malic acid	$C_4H_6O_5$	1444	1345	1329
Citric acid	$C_6H_8O_7$	2147	1998	1963
Glucose	$C_6H_{12}O_6$	2872	2843	2807
Methane	CH_4	818	813	892
Ethane	C_2H_6	1467	1458	1562
Pentane	C_5H_{12}	3385	3367	3533
Methanol	CH_4O	693	682	728
Ethanol	C_2H_6O	1319	1308	1369
Glycerol	$C_3H_8O_3$	1643	1629	1663
Ammonium	NH_4^+	356	316	383
Nitrous acid	HNO_2	81.4	41.5	
Nitric acid	HNO_3	7.3	-32.6	-30
Hydrogen sulphide	H_2S	323	243	247
Sulphurous acid	H_2SO_3	-249	-329	-329
Sulphuric acid	H_2SO_4	-507	-587	-602
Biomass	$CH_{1.8}O_{0.5}N_{0.2}$	541	536	560

H^+-ion and water are modified in the sense discussed above, $\Delta \bar{g}_c^{0'}$. Table 3.1 provides an overview of these thermodynamic data for some relevant compounds.

There are some global regularities which allow general expressions to be fomulated, from which the thermodynamic data can be estimated with reasonable accuracy. The following correlations have been developed to this purpose

$$\Delta h_{ci}^0 = 115\,\gamma_i \tag{3.17}$$

$$\Delta g_{ci}^0 = 94.4\,\gamma_i + 86.6 \tag{3.18}$$

Equations 3.17 and 3.18 relate the heats and free enthalpies of combustion of a C-mole of a given substance, Δh_{ci}^0 and Δg_{ci}^0 respectively, to the degree of reduction of that compound, and are graphically illustrated in Fig. 3.3. Equation 3.17 implies that the heats of combustion of compounds (per

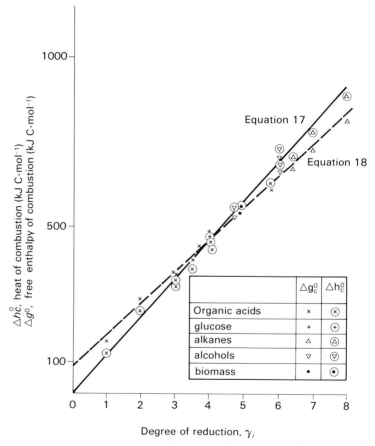

Fig. 3.3 Free enthalpy and heat of combustion per C-mole plotted against degree of reduction. Correlations and data for organic compounds.

C-mole) are as a general rule directly proportional to their degrees of reduction. This has far-reaching consequences; as the degree of reduction is also proportional to the amount of oxygen involved in complete combustion of a substance, it follows that, in any process involving, apart from oxygen, only compounds to which Equation 3.17 applies, the heat production is directly proportional to the oxygen consumption. In general 460 kJ are generated per mole of oxygen consumed.

Equation 3.18 shows that direct proportionality between free enthalpy of combustion and degree of reduction does not apply. Comparison of Equations 3.17 and 3.18 shows that the free enthalpies of combustion of lowly reduced compounds, e.g. oxalic acid, exceed the heats of combustion, whilst the opposite is true for highly reduced compounds, e.g. methane. This is due to the entropy contribution to the free enthalpy of combustion, $T\Delta s_{ci}^0$, which is equal to $\Delta h_{ci}^0 - \Delta g_{ci}^0$. The relevance of this observation is discussed later.

3.3 The efficiency of microbial growth

In the foregoing, a view of the thermodynamic theory and the energy content of compounds involved in metabolism has been developed to a level which allows some understanding of the energetics of growth. In general, a microorganism can be considered as a chemical energy transducer in a steady state, and most cases of growth are well depicted by the diagram in Fig. 3.4. The organism obtains a carbon and energy source and a nitrogen

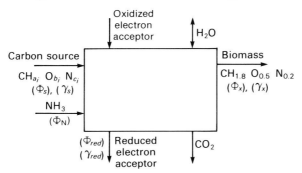

Fig. 3.4 Schematic representation of a metabolizing organism for the general case.

source from the environment (for convenience, the nitrogen source is assumed to be NH_3, but the basic formalism is easily extended to other nitrogen sources). The flows are Φ_s C-moles h^{-1} and Φ_N N-moles h^{-1} respectively. The organism also obtains an oxidized electron acceptor from the environment, and its flow is Φ_{ox} moles h^{-1}. The electron acceptor accepts electrons from the substrate or the nitrogen source, a process in which energy is generated that can be used to drive the processes of life, and is transported to the environments in a reduced form. As a result of the processes taking place in the organism a flow of Φ_x moles h^{-1} of new biomass is generated.

Figure 3.4 also illustrates the well-known fact that biomass from a variety of sources is generally well represented by the composition formula $CH_{1.8}O_{0.5}N_{0.2}$. Finally CO_2 and H_2O are also (generally) exchanged with the environment.

In order to clarify the notation, two representative and specific examples of the general case depicted in Fig. 3.4 will be discussed; first, the case of aerobic growth. In aerobic growth the electron acceptor is oxygen, and its reduced form is water. Hence, for this case, only six flows of chemical substances can be distinguished, the reduced form of the electron acceptor being water. As a second case anaerobic growth without the apparent presence of an electron acceptor will be discussed. In such processes the electron acceptor is (at least formally) carbon dioxide derived from the substrate itself, which is transformed into a product of a higher degree of reduction than CO_2. A typical example of such a process would be the anaerobic formation of ethanol from glucose by yeast. The reduced form of the electron acceptor in this case is the product ethanol, whilst the oxidized form is not supplied separately but is derived from the substrate.

To understand microbial energetics it is important to stress some other important principles. Firstly, electrons available for transfer to oxygen are conserved in the metabolism of organisms. With reference to Fig. 3.4, it follows that

$$\Phi_s \gamma_s = \Phi_{ox}(\gamma_{red} - \gamma_{ox}) + 4.2\,\Phi_x \qquad (3.19)$$

In formulating Equation 3.19 we take account of the fact that the ammonia flow to the system has to equal $0.2\,\Phi_x$ in order for the system to be in steady state.

We shall look at two examples of the application of Equation 3.19. In the aerobic growth of yeast the reduced form of the electron acceptor (water) has a degree of reduction zero, whilst the oxidized form has a degree of reduction of -4. Equation 3.19 can be reworked to form the following expression allowing the calculation of the oxygen flow, Φ_O, to the system

$$\Phi_O = 1/4(\Phi_s \gamma_s - 4.2\,\Phi_x) \qquad (3.20)$$

In the anaerobic growth of yeast with production of ethanol, the reduced form is ethanol, with a degree of reduction of 6, whilst the oxidized form is nominally CO_2 with a degree of reduction of zero. In this way the following equation for the estimation of the ethanol flow, Φ_p(C-moles h^{-1}), is obtained

$$\Phi_p = 1/6(\Phi_s\gamma_s - 4.2\,\Phi_x) \tag{3.21}$$

The similarity between the two cases now becomes obvious.

Secondly, it is important to note that in the case depicted in Fig. 3.4, a proportion of the electrons available for transfer to oxygen in the substrate, and all of the electrons available in ammonia, can be assumed to be conserved in the biomass; in general, the free enthalpy changes involved turn out to be comparatively small! Table 3.2 gives two representative

Table 3.2 The free enthalpy change on transfer of available electrons from substrate and nitrogen source to biomass

$$2.1\,CH_2O_2 + 0.2\,NH_3 \longrightarrow CH_{1.8}O_{0.5}N_{0.2} + 1.1\,CO_2 + 1.5\,H_2O$$
$$\Delta g_r^{0'} = -22.8\,kJ\,C\text{-mol}^{-1}$$

$$0.175\,C_6H_{12}O_6 + 0.2\,NH_3 \longrightarrow CH_{1.8}O_{0.5}N_{0.2} + 0.05\,CO_2 + 0.45\,H_2O$$
$$\Delta g_r^{0'} = -24.7\,kJ\,C\text{-mol}^{-1}$$

examples of this general tendency; for almost all practical purposes the free enthalpy change can be assumed to be zero, a procedure which will be adopted here.

The electrons which are not conserved in the biomass must be transferred to the electron acceptor so the number of electrons transferred to that compound is

$$\Phi_{av/e} = \Phi_s\gamma_s - 4.2\,\Phi_x \tag{3.22}$$

The electrons transferred from the substrate to the electron acceptor in the process according to Equation 3.22 are the source of the energy which turns up as dissipation in the process of growth. If a free enthalpy change $\Delta g_{av/e}^{0'}$ is the result of the transfer of one mole of electrons from the substrate to the electron acceptor, we can define the energy dissipation in the process, $T\Pi_S$, as

$$T\Pi_S = \Delta g_{av/e}^{0'}(\Phi_s\gamma_s - 4.2\,\Phi_x) \tag{3.23}$$

Equation 3.23 can be reworked into a more simple form if the concept of the yield on available electrons, $Y_{av/e}$, as introduced by Payne (1970) is adopted; this expresses the observed biomass yield as grams of dry matter per mole of electrons in the substrate available for transfer to oxygen. Recognizing that 1 C-mole of 'average' biomass equals 25.8 grams, it follows that

$$D = \Delta g^{0'}_{av/e}\left(\frac{25.8}{Y_{av/e}} - 4.2\right) \tag{3.24}$$

where D is the dissipation per C-mole of biomass produced.

The concept of the thermodynamic efficiency developed in Section 3.2.2 can now be introduced. The thermodynamic efficiency η_{th} is defined with respect to the production of biomass, and hence only the energy contained in the biomass leaving the system is taken into account in the evaluation of the energy content of the flows of matter leaving the system. The free enthalpy contained in the flows entering the system is, by virtue of the very definition of the dissipation $T\Pi_s$, equal to the sum of $T\Pi_s$ and the free enthalpy of the biomass leaving the system, from which it follows that

$$\eta_{th} = \frac{\Delta g^{0'}_{cx}}{\Delta g^{0'}_{cx} + D} \tag{3.25}$$

in which $\Delta g^{0'}_{cx}$ is the free enthalpy of combustion of a C-mole of biomass.

Equations 3.24 and 3.25 can now be used to compare the dissipation and the thermodynamic efficiency in various growth processes. If the thermodynamic efficiency of growth η_{th} were constant at a given value, a definite correlation would be expected between the yield on the substrate's available electron, $Y_{av/e}$, and the free enthalpy obtained per mole of electrons transferred from the substrate to the electron acceptor. A rearrangement of Equations 3.24 and 3.25 results in

$$Y_{av/e} = \frac{25.8\Delta g^{0'}_{av/e}}{4.2\Delta g^{0'}_{av/e} + \Delta g^{0'}_{cx} \times \dfrac{1 - \eta_{th}}{\eta_{th}}} \tag{3.26}$$

To check experimental data for the existence of such a relationship some earlier data are compiled in Table 3.3, referring to growth on a variety of substrate–acceptor combinations. The $\Delta g^{0'}_{av/e}$ values are seen to vary by a factor of 30. In Fig. 3.5, the data are shown as a plot of the observed $Y_{av/e}$ against $\Delta g^{0'}_{av/e}$ and the expected correlation according to Equation 3.26 for η_{th} values of 0.5, 0.6 and 0.7 is also shown. As can be seen, the observed yields are not very different from those predicted for such values of η_{th}; in other words there is a noticeable tendency for the thermodynamic efficiency to be independent of the nature of the energy-supplying process.

As an example of the correlation given by Equation 3.26 we may take growth of the yeast *Saccharomyces cerevisiae*. This can grow both aerobically and anaerobically on glucose as the carbon source, and in the latter case ethanol is produced. In aerobic growth on glucose a yield of typically

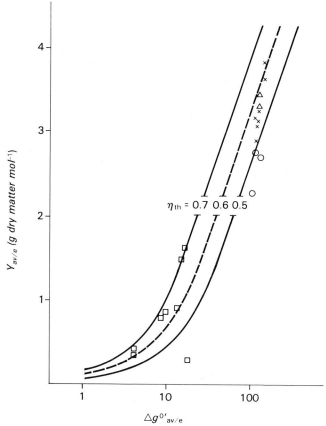

Fig. 3.5 The constant efficiency hypothesis. Theoretical curves of the yield on available electrons, $Y_{av/e}$, plotted against the free enthalpy gained in the transfer of one mole of electrons to the electron acceptor, $\Delta g^{0'}_{av/e}$, for three values of the thermodynamic efficiency. Data for aerobic growth, $\times$; growth supported by succinate with O_2, NO_2^- and NO_3^- as electron acceptor, O; growth supported by gluconate with O_2 and NO_3^- as electron acceptor, $\triangle$; various cases of growth without an externally supplied electron acceptor, $\square$.

0.55 grams biomass per gram of substrate is obtained. This corresponds to a $Y_{av/e}$ of 4.12; that is, to a thermodynamic efficiency of 0.69 and a dissipation of about 250 kJ per C-mole of biomass produced. If the same thermodynamic efficiency were obtained on anaerobic growth the $Y_{av/e}$ would be expected equal to 0.87 g biomass per mole available electrons or 0.12 grams of biomass per gram substrate, and this is, in fact, a quite realistic value for anaerobic growth. With its limitations, taking a rather crude rule-of-thumb thermodynamic efficiency around 0.60, and a dissipation of about 350 kJ per C-mole of biomass produced, provides a useful first estimate of the expected growth yield for a given substrate–electron acceptor combination.

Another feature illustrated in Table 3.3 relates to the contributions of

Table 3.3   Yields on available electrons and heat and free enthalpy dissipation per mole of electron transferred for various substrate/electron acceptor combinations

Substrate	Electron acceptor	$\Delta h_{av/e}^0$	$\Delta g_{av/e}^{0'}$	$Y_{av/e}$
Methane	O_2	111.5	101.6	1.56
Methanol	O_2	121.3	113.7	2.27
Ethanol	O_2	114.1	109.0	2.17
Glycerol	O_2	118.8	116.4	3.65
Lactic acid	O_2	114.1	110.1	3.26
Glucose	O_2	116.7	118.5	3.84
Succinic acid	O_2	106.6	107.1	2.90
Citric acid	O_2	109.1	111.0	3.09
Malic acid	O_2	110.8	112.1	3.18
Formic acid	O_2	127.5	118.0	3.17
Succinic acid	$NO_3^- \rightarrow N_2$	100.6	100.6	2.28
Succinic acid	$NO_2^- \rightarrow N_2$		121.0	2.71
Succinic acid	O_2	106.6	107.1	2.76
Gluconate	O_2		117.8	3.26
Gluconate	$NO_3^- \rightarrow N_2$		111.3	3.44
Glucose	CO_2 (ethanol)[a]	2.9	9.5	0.88
Glucose	CO_2 (lactic acid)	2.9	8.3	0.81
Glucose	CO_2 (propionic/acetic acids)	7.6	12.9	1.48
Glucose	CO_2 (methane/acetic acid)	6.7	14.4	1.63
Acetic acid	CO_2 (methane)	−2.0	3.9	0.36
Methanol	CO_2 (methane)	9.8	12.0	0.91
Formic acid	CO_2 (methane)	16.0	16.4	0.27
Propionic acid	CO_2 (methane)	−2.3	4.0	0.41

[a] Product of anaerobic metabolism given in parentheses.

enthalpy and free enthalpy to the total dissipation observed in the growth process. For aerobic growth (and growth supported by other efficient electron acceptors) the free enthalpy change on transfer of a mole of electrons to the electron acceptor is almost equal to the enthalpy change. This implies that the contribution of heat to the total entropy exchange flow, as given by Equation 3.7, is of overwhelming importance. The heat production almost equals the free enthalpy dissipation, and a treatment in terms of heat of combustion becomes sufficiently accurate.

For anaerobic growth the situation is seen to be completely different. The $\Delta g_{av/e}^{0'}$ values are much larger than the $\Delta h_{av/e}^{0'}$ values and a treatment in terms of $\Delta h^{0'}_{av/e}$ values would lead to unrealistically low values of the dissipation. There would then be a large apparent discrepancy between the dissipation in anaerobic and in aerobic processes. In fact, in anaerobic growth the contribution to the total entropy flow of the second terms appearing at the right-hand side of Equation 3.7, the exchange of 'chemical entropy', becomes of major importance. *Whilst aerobic processes are driven by production of heat, anaerobic processes are mainly driven by difference in entropy of the chemical substances exchanged between the system and the environment.* Since the total dissipation in anaerobic processes is certainly no greater than that in aerobic processes, the heat production per unit biomass produced will be lower in anaerobic processes. In fact, typical aerobic processes involve a heat production of about 350 kJ per C-mole of biomass produced, whilst anaerobic growth of *S. cerevisiae* with production of ethanol from glucose leads to a heat production of about 100 kJ per C-mole biomass produced. It is also clear from Table 3.3 that aerobic growth on highly reduced substrates, such as methane, methanol and ethanol, seems to involve significantly higher heat production and to proceed at a much lower η_{th}. This is ascribed to factors other than available energy, in particular the limited amount of carbon present in these substrates, which may limit the maximum attainable efficiency in growth on these substrates. This is more extensively discussed by Roels (1981) and Payne (1970).

3.4 Biochemical approaches towards the energetics of growth

The efficiency of the growth process can also be approached in a more biochemical way. Such a discussion is often based on a so-called Y_{ATP}-value, the yield (in grams dry matter) per mole of ATP obtained in catabolism of the substrate (Stouthamer, 1979). Such an approach should in principle be much more reliable than the crude macroscopic analysis presented above. Unfortunately, the amount of ATP obtained in catabolism is not always known with certainty and is even subject to much debate in aerobic growth (Stouthamer, 1979). Furthermore, discrepancies of at least a factor 2 exist between the Y_{ATP} values to be expected in the light of the biochemical theory and those actually observed, typically between 10 and 15. In our view, an approach based on an assumption of a constant Y_{ATP} is not necessarily superior. Only in cases where the efficiency of ATP generation in catabolism is very different between different pathways, for example as between aerobic growth on methane and on glucose, or as between anaerobic ethanol production in the yeast *S. cerevisiae* and the bacterium *Z. mobilis*, is the Y_{ATP}-based approach more informative.

In all other cases the two approaches are equivalent and provide at present a means of predicting growth yields (and, by extension, other stoichiometrically linked quantities), with an accuracy of generally ±25%. The macroscopic approach treated here has the advantage that it does not involve knowledge—or dispute—of the actual ATP generation is catabolism.

Reading list

Erickson, L. E., Minkevich, I. G. and Eroshin, V. K. (1978). *Biotechnol. Bioengng.* **20,** 1595.

Glansdorff, P. and Prigogine, I. (1971). 'Thermodynamic Theory of Structure Stability and Fluctuations'. Wiley: New York.

Haase, R. (1969). 'Thermodynamics of Irreversible Processes'. Addison-Wesley: Reading, MA.

Kharasch, M. S. (1929). *J. Res. Nat. Bur. Stand.* **2,** 359.

Payne, W. J. (1970). *Ann. Rev. Microbiol.* **24,** 17.

Prigogine, I. (1967). 'Introduction to the Thermodynamics of Irreversible Processes', 3rd edn. Wiley: New York.

Roels, J. A. (1981). *Ann. New York Acad. Sci.* **369,** 113.

Schrödinger, E. (1944). 'What is Life?' Cambridge University Press: New York.

Stouthamer, A. H. (1979). *Microbial Biochem.* **21,** 1.

4 | Microbial Process Kinetics

C. G. SINCLAIR

4.1 Introduction

4.1.1 Setting in context

To quantify processes by which microorganisms grow, consume substrates and form products, there are two basic questions which must be answered:

(1) What are the ratios of the various amounts of material and energy involved in the process; for a given quantity of any one, what quantities of other materials or energy must be supplied or removed?
(2) At what rates do the growth of cells, the consumption of substrates and the production of products proceed; hence what volume of equipment is required for a given level of production?

Question 1 deals with stoichiometry and is covered in other chapters, especially Chapter 2. Question 2 leads to the concepts of process kinetics considered here.

4.1.2 The design process

To see how kinetics fits into the overall design and operation of a process it is useful to show the relationship in diagrammatic form as in Fig. 4.1. This covers all aspects from the birth of an idea through bench experiments and full-scale process design to routine operation. The mathematical model is an abstract realization of the idea of a process, and it requires representations of:

(1) the physical reality of the equipment, the 'abstracted physical model'; and

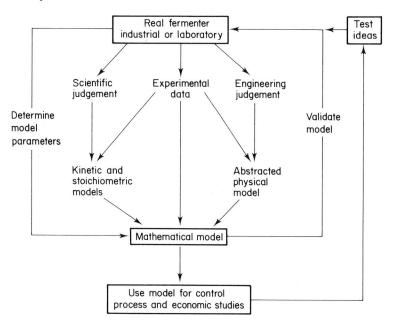

Fig. 4.1   The design process.

(2) the chemical reactions of cell growth, substrate utilization and product formation.

As Fig. 4.1 shows, the formalization of the stoichiometric and kinetic models requires a knowledge of the underlying science, in this case the relevant biochemistry; this makes it possible using experimental data for the particular process being studied to set up reasonably adequate expressions for the stoichiometric and rate equations. Similarly, the whole body of engineering knowledge of mixing, aeration, mass and heat transfer, etc., is used in compiling the abstracted physical model.

When the two are combined, we have a mathematical model of the process, which serves a number of purposes:

(a) *Arising from construction of the model.*
     Aid to thought
     Organization of knowledge
(b) *Arising from use of the model.*
     Organization of data
To carry out 'hypothetical experiments'
   Quantification of concepts
   Sensitivity studies
   Parameter determination

4.1.3 Mathematical models

Students without a mathematical background may be discouraged from considering or using mathematical models by the apparent complexity of the mathematical equations which are their formal expression and by the facility—often more apparent than real—with which many authors and lecturers appear to manipulate them; much of the published literature is confusing (to say the least) and by no means all of it is correct. In fact the work required to grasp the essential techniques of modelling is not excessive and the main steps in writing models are all set out in Section 4.2.

Once the three groups of equations discussed in Section 4.2 can be identified and understood, half the problem is overcome. The remainder of the problem is sorted out by following through examples, by filling in the logical steps which authors tend to omit, and by tackling some problem of interest to the student with the assistance of a skilled practitioner.

The other essential requirements for modelling are more practical than mathematical—clear pictorial diagrams of the microbial process and an unambiguous and easily understood nomenclature; it is often helpful to substitute a familiar nomenclature for any unusual symbols used by an author. A final requirement is that the dimensions of every term in an equation should be checked for consistency (this often exposes errors in formulation) and that the physical meaning of every term can be understood.

4.2 Formulating mathematical models

4.2.1 Introduction

A model is a *set of relationships* between the *variables of interest* in the *system being studied*.

A *set of relationships* may be in the form of equations, graphs, tables, or even, as for many skilled plant operators, an unexpressed set of cause/effect relationships which are his picture of the process and which determine his actions in controlling it. All such representations constitute the basic structure of every model.

The variables of interest depend upon the use to which the model is to be put. For example, when confronted with a fermenter, a biotechnologist will be concerned with the feed rate, state of agitation, temperature, viability of the microorganism, etc.; an electrical engineer would be interested in the motor currents, voltages, etc.; a mechanical engineer with the stresses and strains in the structure; an accountant with the costs of the daily inputs and outputs of material and energy. Each specialist sees the same equipment but looks at it in quite a different manner.

The abstracted physical model selects from the real physical object those physical and gometrical properties which are of importance for the particular way in which the modeller is treating the system.

The system being studied must be defined in some detail; in biotechnology this is usually a reactor containing microorganisms, or the processes downstream of such a reactor where the product is separated into its constituents. For our purposes, an abstracted physical model is defined as a region in space throughout which all the variables of interest (e.g. temperature, concentration, pH, dissolved oxygen), are uniform. This is called the 'control region' or 'control volume'. The control region may be of constant volume as is usually assumed for a chemostat or simple batch fermenter, or it may vary in size as for example in a fed-batch fermenter. It may be finite as in the normal representation of a well-mixed fermenter, or infinitesimal as in a tower fermenter in which the concentrations of substrate and product vary continuously throughout the liquid space, so that only over an infinitesimally thin slice can they be considered uniform.

The boundaries of the control region may be:

(1) boundaries across which no exchange of material takes place, e.g. the containing vessel walls;
(2) phase boundaries across which exchange of mass or energy takes place, e.g. bubble–liquid interface;
(3) geometrically defined boundaries within one phase across which exchanges take place either by bulk flow or by molecular diffusion, e.g. nutrient inlet and outlet pipes.

An example of an abstracted physical model for a fermenter is shown in Fig. 4.2.

To construct a conventional mathematical model, we write a set of equations for each control region. This set consists of:

(a) *Balance equations* for each extensive property of the system, i.e. mass, energy or individual elements or species; where the extensive property is also a conserved property which can neither be created nor destroyed (such as mass, energy or chemical elements) then the balance equations become conservation equations.
(b) *Rate equations.* These are of two types:
   (i) Rates of transfer of mass, energy, individual components or species *across* the boundaries of the region;
   (ii) Rates of generation or consumption of individual species *within* the control region.

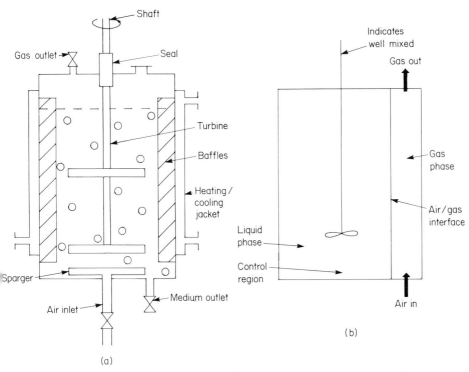

Fig. 4.2 (a) Simplified drawing of a batch fermenter. (b) Abstracted physical model of a batch fermenter.

(c) *Thermodynamic equations* relate thermodynamic properties (pressure, temperature, density, concentration) either within the control region (e.g. gas laws) or on either side of a phase boundary (e.g. Henry's law).

Models are constructed to be used. The simpler they are, the easier they are to use, and so the golden rule of the model maker must be:

Always use the simplest adequate model firmly rooted in known fundamental physical, chemical and biochemical ideas.

4.2.2   Balance equations

Balance equations are written for every extensive property of interest in each control region. An extensive property is one for which the total amount of that property in a given system is the sum of the amounts in the separate parts of the system. Thus mass and energy are extensive properties, but pressure, temperature and concentration are not. The only point to watch is that each balance equation is linearly independent of the others, and cannot, for example, be produced simply by summing any group of the other balance equations.

The balance equation is written:

$$
\begin{array}{ccc}
\text{Rate of accumulation} & \text{Rate of input} & \text{Rate of output} \\
\text{in the control} & = \text{into the control} + & \text{from the control} \\
\text{region} & \text{region} & \text{region}
\end{array}
$$

The input and output terms are further subdivided into:

Input terms:
bulk flow across geometrical boundaries
diffusion across geometrical boundaries (only important for infinitesimal regions)
transfer across phase boundaries
generation *within* the control region

Output terms:
bulk flow across geometrical boundaries
diffusion across geometrical boundaries
transfer across phase boundaries
consumption *within* the control region.

Rate of accumulation
Writing this term seems to give beginners more trouble than any other term in the balance equation. The first thing to remember is that it is not a kinetic, mass transfer of heat transfer rate term; it is simply the rate at which the amount of the extensive property within the control region changes with respect to time.

For example, if x_v is the concentration of cells in a well-mixed fermenter in $kg\,m^{-3}$ and the volume of the control region is $V\,m^3$, then the amount of the extensive property 'mass of cells' is $Vx_v\,kg$ and the rate of accumulation of cells in the control region is $d\,(Vx_v)/dt$, $kg\,h^{-1}$. This formulation makes no assumptions about the constancy or otherwise of the control region volume. To express the rates as amount per unit volume then we must divide the term by V giving:

$$
\text{Rate of accumulation} = \frac{1}{V}\frac{d}{dt}(Vx_v) \quad kg\,m^{-3}\,h^{-1}
$$

Input and output terms
(a) Bulk flow. This is an expression for material or energy carried by or with the bulk flow of fluid into or out of the control region and is thus

equal to the flow rate in $m^3 h^{-1}$ multiplied by the concentration of the extensive property in $kg\,m^{-3}$ or $J\,m^{-3}$. If all the terms are to be based on unit volume then this quantity is divided by the control region volume as before. Thus for example taking cell mass as the extensive property, if F is the medium input flow rate in $m^3 h^{-1}$ then the bulk flow input term is $Fx_v\,kg\,h^{-1}$ based on total volume or $Fx_v\,V^{-1}\,kg\,m^{-3}\,h^{-1}$ based on unit volume of the control region.

(b) Transfer across phase boundaries. This is usually expressed on a unit volume basis and is given by the product of three terms, a transfer coefficient, a phase boundary area term and a driving force term. The driving force term is the potential which causes the transferring species to move across the phase boundary; strictly this should be chemical potential for chemical species. In practice a concentration or partial pressure driving force is used.

The equation is, therefore, of the form:

$$
\begin{array}{ccccc}
\text{Rate of} & = & \text{Transfer} & \times & \text{Area per} & \times & \text{Driving force} \\
\text{transfer} & & \text{coefficient} & & \text{unit volume} & & \text{(intensity gradient)} \\
(kg\,h^{-1}\,m^{-3}) & & (kg\,h^{-1}\,m^{-2}\,DF\,unit) & & (m^2\,m^{-3}) & & (DF\,unit)
\end{array}
$$

The most common example is the equation for the rate of mass transfer of oxygen in a fermenter which is written:

$$N_\alpha = k_L\,a\,(C_g^* - C_L) \qquad (4.1)$$

(c) Generation or consumption terms. These rate expressions are the main subject of this chapter. They are always written as r with an appropriate subscript; for example, r_x is the rate of growth of cells, r_p is the rate of production of some product, r_s is the rate of consumption of substrate and so on; in each case the units will be $kg\,m^{-3}\,h^{-1}$. To get the rate on a total control region volume basis each term is multiplied by the volume.

The actual form of these expressions are dealt with very briefly in the next section and in detail in Section 4.4.

4.2.3 Rate equations

The basic form of the kinetic rate expressions as used in the balance equation are either a specific rate on a cell mass basis multiplied by the concentration of cells in the control region, or a stoichiometric constant multiplied by a rate expression, or a linear combination of these two. An example of the first is the well-known expression for cell growth $r_x = \mu x_v$ where μ is the specific growth rate of cells in kg cells per kg cells per hour.

An example of the second is the common expression for growth-related product formation $r_p = \alpha r_x$. In this α is the stoichiometric coefficient (kg product produced per kg cells produced). An example of the third combined type is the Luedeking and Piret expression for product formation $r_p = \alpha r_x + \beta x_v$.

4.2.4   Thermodynamic relationships

Thermodynamic relationships are used to relate variables which cannot be measured to those which can. The best example of this in fermentation technology is in the relationship for the rate of oxygen transfer across a liquid–gas interface. In deriving this relationship it is assumed that the driving force is the difference between the oxygen concentration in the liquid at the interface, C_g^*, and that in the bulk of the liquid, C_L. Now it is possible to measure oxygen concentration in the bulk directly, but there is no probe sufficiently small or non-invasive to measure it at the interface. In this case, Henry's law (the thermodynamic relationship relating the concentration of gas in a liquid to the partial pressure of that gas above the liquid surface under equilibrium conditions) is invoked; the partial pressure of the oxygen in the gas phase can of course be measured.

4.2.5   Constraints and initial conditions

Each time we write a balance equation according to Section 4.2.2, for any extensive property, we are specifying the manner in which the extensive property varies through time. This is done by solving the differential equation which results from the substitution of the rate equations (Section 4.2.3) and thermodynamic relationships (Section 4.2.4) into the balance equation. Each of the extensive properties is called a state variable and all of their values at any point in time determine the state of the system. The way they vary through time depends upon starting conditions and the equations, i.e. the set of differential equations which govern their variation in time. For each state variable, it is, therefore, necessary to specify the initial conditions, one for each balance equation. Usually they are the values at the time of inoculation of all the extensive properties in the balance equations.

When solving a set of equations there is nothing inherent in the mathematics which forbids such things as negative concentrations, flows, etc., which do not make any physical sense. It is, therefore, necessary to include appropriate constraints on the values of the variables (especially when relying on computer solutions, since the computer may generate negative intermediate solutions which are not printed out, but can lead to spurious final solutions). The constraints are written as a set of inequalities on the variables, the most common being that all concentrations are greater than zero.

4.2.6   Steady state and unsteady state models

The method of model writing we have presented makes no conditions as to the constancy of the variables through time. This is called an unsteady state model since the state of the system can vary. With most physical systems encountered in biotechnology, if the inputs to a system are fixed, the system will eventually come to a steady state where all the variables are constant (it may, of course, be that all the cells are dead, but nevertheless the cell concentrations and the substrate concentration will be uniform!). One way of determining what this steady state is would be to solve the equations and allow time to proceed until every variable had reached its steady state value. Indeed this is a recognized method of solving for steady state values where the steady state equations are particularly difficult to solve. However, since all the variables are constant at the steady state, the rates of accumulation (i.e. the rate of change of the extensive variables) must all be zero. We can, therefore, reduce the unsteady state model to a steady state model by setting all the derivatives with respect to time to zero. The balance equations now become:

$$\text{Input} - \text{Output} = 0 \tag{4.2}$$

The balance equations are now algebraic instead of differential and in simple cases, some of which will be discussed later, they can be solved analytically. In more general cases both forms may be equally difficult to solve.

4.2.7   Checking the model

It is not unusual for the first attempt at writing a mathematical model to result in a set of equations which do not have a solution which makes physical sense, or which are not solvable even by computer methods. Normally this is because the problem has been wrongly specified in the first place. Although this may be remedied by rechecking the physical and conceptual background of the model, a brief check of the mathematical consistency of the set of equations will often reveal some of the errors.

The model consists of a set of balance equations, one for each extensive property of the system considered to be among the variables of interest, or state variables. All the other variables in the equations must either be independent variables, i.e. those which are fixed externally to the system by an operator or some external control mechanism (e.g. dilution rate, temperature, inlet substrate concentration, etc.) or else can be expressed as functions of the state variables or the independent variables. As an example of the latter type, consider a kinetic rate expression such as

$$r_x = \frac{\mu_m S}{k_s + S} x_v$$

Here the variable r_x is a function of two state variables S and x_v.

The consistency check is to apply the equality

$$I = V - E \tag{4.3}$$

where I is the number of independent variables which must be externally specified, V is the total number of variables, and E is the number of equations we have written.

Having written the model and checked it for mathematical consistency, the next step is to check it for physical sense. To do this the model is solved for a wide variety of conditions, analytically or by computer, and the physical sense of the solutions is verified. If the model is at all complex, considerable time will often be saved by checking that simplifying assumptions reduce it to models which are already well known and proven in the literature. A number of obvious checks are listed below.

(1) Reduce to the *steady state form* by setting all differentials to zero.
(2) Reduce to the *pure batch form* by setting all inflows and outflows to zero.
(3) Write all kinetic expressions as first-order expressions and check the solutions.
(4) Check that the limit of the steady state solution as the dilution rate is reduced to zero is the limit of the pure batch solution as time approaches infinity.
(5) Check that the model behaves reasonably at extremes of the independent variables.

Finally we should reiterate the golden rule: start simply and work up to the more complex, and *always use the simplest adequate model firmly rooted in known fundamental physical, chemical and biochemical ideas.*

4.3   Basic kinetic mechanisms

4.3.1   Unstructured mechanisms

Kinetic expressions have to be used or developed for the various steps in the overall biological process, and to do this it is necessary to agree on a basic biological mechanism. This may be at various levels of detail. For example, it would be possible to set out every one of the reaction steps that occur within the cell and then write the overall set of equations to describe all of them. Such a model would be far too detailed for technological purposes, although, of course, it might be of interest to biochemists. A less detailed mechanism might identify one or two key chemical species within the cell, perhaps those which were to be extracted on a commercial scale, and lump all other components of the cell under one or two general classes such as proteins, lipids, etc. Such mechanisms are the basis of *structured* models, i.e. those which take into account some basic aspects of cell structure. However, an *unstructured* mechanism of cell operation is sufficient for many technological purposes, and is sufficiently simple to give rise to sets of equations which can be understood in a physical sense and are likely to be generally helpful.

In an unstructured mechanism of cellular operation, the microorganism is regarded as a single reacting species, possibly with a fixed chemical composition; the basic biological reaction scheme is shown in Fig. 4.3. Here

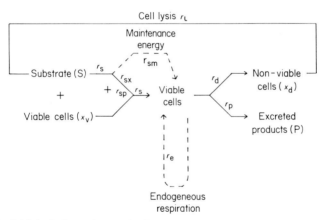

Fig. 4.3 Basic biological reaction mechanism.

we identify substrate, viable cells, non-viable cells and excreted products as the four major components. Substrate reacts with viable cells to form more viable cells, which may be harvested, which may excrete products into the medium, or may be converted into non-viable forms of the cell. The non-viable cells may form part of the harvested cell mass or may themselves lyse to release constituents into the medium.

As discussed in Chapters 2 and 3, the energy necessary for growth and product formation is obtained either by diversion of some of the substrate for complete oxidation (leading to some forms of product formation) or (as in the absence of an oxidizable substrate) by breakdown of cell material itself. In addition to the energy required for synthetic reactions, energy is required to maintain the cell in a viable state. As considered later, this requirement is termed maintenance energy if it is assumed to come from the substrate, or endogeneous respiration if from breakdown of cell constituents.

In Fig. 4.3, each path is labelled with a subscripted letter r to designate the rates at which the labelled reaction proceeds (in units of 'amount of reactant consumed or product formed per hour per unit volume'). The subscript is mnemonically related to the particular reaction step.

To handle the complex series of reactions responsible for the growth of cells implied in Fig. 4.3, the idea of the limiting substrate or substrates is invoked. Of all the substrate constituents necessary for growth, it is assumed that all except one or at most two are in excess and that it is the concentration of these one or two 'limiting' substrates which control the overall rate of reaction. When all the substrates are in excess as at the beginning of a batch culture, it is assumed that the cells grow at some maximum rate determined by their intrinsic nature and by environmental conditions other than soluble substrate concentrations (see also Section 4.4.2).

4.3.2   Structured mechanisms

The unstructured type of model discussed in the previous section has played the major part in the development of our understanding of fermentation processes; it is still the basis of much research and of the technologist's view of biological processes. Its limitations are that it does not make proper use of our considerable knowledge of the processes which occur within the cell, or of our ability to analyse cells for particular constituents. Consequently, although models based on an unstructured mechanism are adequate for interpolation between experimental results on a particular fermenter, they may not allow for extrapolation either to larger scales of operation or to radically different environmental conditions.

It is, therefore, necessary to look at the cellular processes in some detail. A normal procedure is to consider some individual constituents of the cell and to group others into a small number of general classes. One of the earliest and simplest of these mechanisms for cell growth is that due to Williams. Figure 4.4 shows the mechanism in diagrammatic form. In this the circled letter 'S' represents the limiting substrate, 'D' the *synthetic* apparatus of the cell (i.e. enzymes), and 'G' the *genetic* part. Thus the cell is considered to consist of two constituents only, called D-mass and G-mass. The solid lines in this figure represent the reaction steps, and the broken

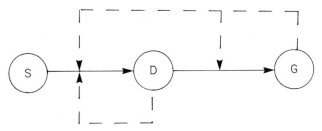

Fig. 4.4 Williams' model for microbial growth.

lines the control over the reaction rates exerted by the fractional concentrations of the D-mass and G-mass. In the model which Williams derived using this structured mechanism, he assumed that the fraction of G-mass could only lie between two limits:

- a lower limit below which the cell is dead;
- an upper limit at which it divides.

His model was found to be useful in predicting the lag phase in batch culture. By assigning part of the D-mass to product formation and part to cell growth, the mechanism can also be adapted to describe product formation.

Another useful structured mechanism considers the ATP in the cell as a separate component. We can then write a conservation equation with the principal mechanisms for consumption and generation of ATP as the main terms. This approach organizes a large amount of information which is available on energy turnover in the cell and has proved very useful in such diverse problems as modelling alcohol production, citric acid formation, polysaccharide formation, etc.

Some of the models resulting from these structured mechanisms are discussed in more detail in Section 4.5.

4.4 Kinetic rate equations

4.4.1 General principles

Microorganisms require substrates for three main functions:

(1) to synthesize new cell material,
(2) to synthesize extracellular products,
(3) to provide the energy necessary:
    (a) to drive the synthetic reactions;
    (b) to maintain concentrations of materials within the cells which are different from those in the environment;

(c) to drive the recycling (turnover) reactions within the cell.

Thus, growth, substrate utilization, maintenance and product formation are all intimately related, and as will be shown later, the various rate expressions are also mathematically related.

The energy required to drive the cell processes is the chemical energy of ATP or similar substances provided either by the oxidation of substrate by molecular oxygen to CO_2 and water (oxidative phosphorylation) or by the degradation of substrate to simpler products such as ethanol, lactic acid, citric acid (substrate level phosphorylation), CO_2 and water, etc., which are excreted by the cell. The products of substrate level phosphorylation are called type 1 products.

Extracellular products are such compounds as

- exoenzymes (for breaking down substrates which cannot pass through the cell wall),
- polysaccharides (for cell aggregation),
- special metabolites (whose function may be to prevent competing microorganisms but in the general case is unknown, e.g. antibiotics).

These products are called type 2.

What may be a third kind of product, type 3, is sometimes produced in situations where carbon substrate is in excess and other substrates such as nitrogen or magnesium are limiting. These are possible energy storage compounds such as glycogen or fat, etc., which are stored within the cell, or similar polysaccharides excreted by the cell. These type 3 products are thought by some to act as an energy sink; excess ATP is produced so as to use the limiting substrate more efficiently, and the formation of type 3 products dissipates the chemical energy of the excess.

The total energy required for functions 3b and 3c above is usually referred to as the maintenance energy requirement, used only to maintain the cell in a viable state and not to produce cell material or products of types 2 or 3.

The general relationship between the flows of cells, substrates, energy and products is represented in Fig. 4.5, showing the processes which occur in the cell diagrammatically. The solid lines represent material flows and the broken lines energy flows in the form of ATP. Carbon substrate (represented as CHO) forms endogeneous products type 3 and with nitrogen substrate (N), combines to form cell mass and type 2 products. All of these processes require energy. Carbon substrate is broken down either anaerobically to type 1 products or aerobically to CO_2 and water to provide energy for the previously mentioned syntheses and for maintenance. Each stream is labelled with its rate.

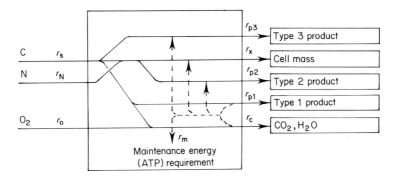

Fig. 4.5 Cell processes.

Since we have specified three entering material streams (CHO, N, O_2), we can write three independent material balances, and we can also write the internal ATP balance; thus we have four independent equations. Having specified eight material flows and one energy flow, that is nine rates in all, we can independently specify the kinetic expressions for any five of the rates. The other four kinetic expressions must then be related to these specified five.

It is normal to specify kinetic expressions for r_m, the maintenance energy requirement, and for r_x, the growth rate, independently. Other kinetic expressions which might be specified are those for the rate of oxygen uptake r_O and for production of type 2 and type 3 products, r_{p2} and r_{p3}.

In practice it is not possible to distinguish the CO_2 and water resulting from oxidative phosphorylation from the CO_2 which is a type 1 product and, therefore, both this flow and the corresponding oxygen balance are omitted from the analysis. The two remaining material balances and the one ATP balance are therefore:

$$r_s = r_{sx} + r_{sp1} + r_{sp2} + r_{sp3} + r_{sO}$$

$$= \frac{r_x}{Y_{x/s}} + \frac{r_{p1}}{Y_{p1/s}} + \frac{r_{p2}}{Y_{p2/s}} + \frac{r_{p3}}{Y_{p3/s}} + \frac{r_O}{Y_{O/s}} \tag{4.4}$$

$$r_N = \frac{r_x}{Y_{x/N}} + \frac{r_{p2}}{Y_{p2/N}} \tag{4.5}$$

and

$$a_x r_x + a_{p2} r_{p2} + a_{p3} r_{p3} + r_{mATP} = a_{p1} r_{p1} + a_O r_O \tag{4.6}$$

This last equation is the ATP balance and it assumes that no ATP is accumulated in the cell and that no ATP is lost by slippage reactions leading to the release of energy as heat.

We thus have three equations and seven rates so that the kinetic expressions for four of the rates are independently specified and the other three rate expressions follow automatically.

The independently specified kinetic rate expression relates the rate of consumption of a substrate or the rate of formation of a product (including cell mass) to the concentrations of the various components in the environment of the cell and the other environmental conditions such as pH, temperature, etc. Such an equation will have the form:

$$r = f(x_v: S_1, S_2, S_3, \ldots; C_1, C_2, C_3, \ldots) \tag{4.7}$$

where r is the rate (usually in amount per unit time per unit volume of medium); x_v is the concentration of microorganism in amount per unit volume; $S_1, S_2, \ldots$ are the concentrations of the various substrates in the medium in amount per unit volume; and $C_1, C_2, \ldots$ are the other environmental variables.

The function has to have some explicit form, and it is one of the objects of research to discover these forms. Several possible expressions for the growth rate are given in the following section.

4.4.2   Cell growth and inhibition

Many of the kinetic models for cell growth and inhibition are based on those used in enzyme kinetics. The rationale behind this is the accepted picture of a cell as a miniature chemical reactor in which a complex network of enzyme-catalysed reactions converts substrates into living cell matter and externally excreted biochemicals. Part of such a network is represented in Fig. 4.6. As in Section 4.3.1, it is assumed that all but one of the substrates are in excess of the cells capacity for absorption, and this one substrate is limiting. Again, of all the various routes by which this substrate is incorporated into the cell mass, it is assumed that one route is the slowest and thus will limit the overall reaction rate. Lastly within the limiting path, there is one reaction step which governs the overall rate, so that this is the limiting reaction step. Hence in this simplified model the overall rate of growth of the cell depends on this one enzymatic reaction step and on the effect of the concentration of the limiting substrate on the rate of that step. The simplest of the enzyme reaction rate expressions is the Michaelis–Menten expression:

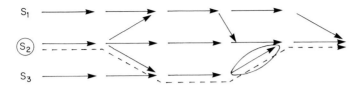

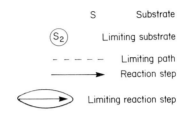

S	Substrate
S_2	Limiting substrate
$---$	Limiting path
$\longrightarrow$	Reaction step
$\Longleftrightarrow$	Limiting reaction step

Fig. 4.6   Schematic diagram of reaction network inside a cell.

$$v = \frac{kE_0 S}{k_m + S} \tag{4.8}$$

in which v is the reaction velocity, k is the rate constant, E_0 is the total amount of enzyme, k_m is the Michaelis constant, and S is the substrate concentration.

The product kE_0 is the maximum rate at which the reaction can proceed and is often written as v_m.

If we identify an enzyme reaction of this type with the rate-controlling step and assume that while the concentration of the rate-controlling enzyme is proportional to the cell concentration, the concentration of the substrate for the rate-controlling step is proportional to the limiting substrate concentration, then we can write an analogous expression, the classical Monod equation for cell growth, as

$$r_x = \frac{\mu_m S}{k_s + S} x_v \tag{4.9}$$

Here r_x is the rate of cell growth (kg cells $m^{-3} h^{-1}$); μ_m is the maximum specific growth rate (h^{-1}); S is the limiting substrate concentration (kg substrate m^{-3}); k_s is the saturation constant (kg substrate m^{-3}); and x_v is the viable cell concentration (kg cells m^{-3}).

This expression is often written:

$$r_x = \mu(S)x_v$$

where

$$\mu(S) = \mu_m S/(k_s + S) \tag{4.9a}$$

The specific growth rate $\mu(S)$ is often abbreviated to μ and has units h^{-1}. The various properties of this expression are shown in Fig. 4.7. Note

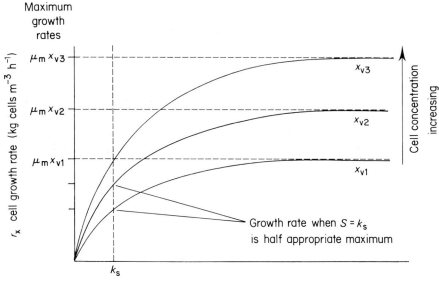

Fig. 4.7   A plot of cell growth rate against limiting substrate concentration for Monod kinetics.

that when $S \gg k_s$, $\mu(S)$ approaches μ_m and r_x becomes independent of S and simply proportional to x_v. Variants of Equation 4.9a have been devised which are commonly found useful in specific situations and some of these are listed below.

Double substrate limitation
Where two substrates are present in such low concentrations that the cell growth rate is limited by both (i.e. a small increase in either concentration will increase the growth rate), we can write:

$$\mu = \mu_m \frac{S_1}{k_{s1} + S_1} \times \frac{S_2}{k_{s2} + S_2} \qquad (4.9b)$$

(this expression reduces to Equation 4.9 if either of the limiting substrates becomes large compared to its associated saturation constant).

Substrate inhibition
The above expressions imply that the growth rate always continues to increase with substrate concentration up to any value. In practice the growth

rate usually begins to decline above some particular value of the substrate concentration. This is modelled by the expression:

$$\mu = \frac{\mu_m S}{k_s + S + (k_i S)^2} \qquad (4.9c)$$

or by any of the inhibition expressions listed below.

Growth inhibition
There is a wide range of substances which inhibit growth and the effects of increasing concentrations of these inhibitors mirror the effects of enzyme reaction inhibitors. Therefore, all of the expressions that have been developed for enzyme inhibition can be applied to model cell growth inhibition. Amongst these are:

$$\mu = \mu_m \frac{S}{k_s + S}(1 - k_i I) \qquad (4.9d)$$

$$\mu = \frac{\mu_m S}{k_s + S}\frac{k_i}{k_i + I} \qquad (4.9e)$$

$$\mu = \frac{\mu_m S}{k_s + S}\exp(-k_i I) \qquad (4.9f)$$

where I is the inhibitor concentration.

All of the expressions given above are first order with respect to cell concentration; note that the coefficients k_i do not have the same meanings in the different expressions. The choice between them is partly a matter of convenience (one equation may be easier to manipulate than another) and partly requires reference to actual observations, made with sufficient accuracy over a sufficiently wide range of values of the inhibitor concentration. When the observations are not very accurate, or when only a limited range of data is available, there is little to choose between these (and other) expressions—as the reader may quickly verify by expanding Equations 4.9e and 4.9f in series form. In general it is important to remember basic modelling principles—use the expression which fits closest to the observed facts and is also intelligible in terms of physical and biological thinking.

4.4.3   Maintenance and endogeneous respiration

Maintenance energy is that part of the energy requirements of the cell that is used to maintain the cell in a viable state, for example, for resynthesis of cell constituents which are continuously being degraded (turnover), and for maintaining concentration gradients between the interior and exterior of the cell (osmotic work). The rate of consumption of substrate to provide the energy for maintenance is written r_m (kg substrate $m^{-3} h^{-1}$), and the only generally accepted kinetic expression for the maintenance energy treatment is:

$$r_m = m_s x_v \qquad (4.10)$$

We can alternatively write the maintenance equation in terms of ATP as

$$r_{mATP} = m_{ATP} x_v \qquad (4.11)$$

where r_{mATP} is the ATP requirement (moles ATP $m^{-3} h^{-1}$) and m_{ATP} is the rate constant (moles ATP kg-cells$^{-1} h^{-1}$). The value of the rate coefficient m_s can range from as little as 0.02 to as high as 4 kg substrate kg-cells$^{-1} h^{-1}$, and values of m_{ATP} range from 0.5 to over 200. Clearly maintenance energy can be a substantial fraction of the total energy consumption.

The value of the maintenance energy coefficient, m_s, will depend upon the environmental conditions surrounding the cell and on its rate of growth. A large part of the maintenance energy is required for osmotic work; thus increasing the external salt concentration increases m_s substantially, and pH has a pronounced effect. The most rapid turnover of cell protein occurs when the cell is adapting its enzyme spectrum to changing environmental conditions. Under constant growth conditions the amount of maintenance energy for turnover is low but when growth stops or when the cell is adapting to changes in substrate or to cessation of growth then this becomes more significant. So far there has been little quantification of these effects, and there are no kinetic expressions for m_s as a function of medium composition comparable to the kinetic expressions for cell growth.

Endogeneous respiration is an alternative way of looking at the provision of energy for maintenance of cell viability. It is assumed that the energy for maintenance is provided by the oxidation or degradation of some of the cell mass itself. This is reasonable where there is no other source of energy, as when the external substrate supply has run out, but is less so when there is still an excess of external energy-providing substrate. The rate expression is written:

$$r_e = k_e x_v \qquad (4.12)$$

where r_e is the rate of endogeneous respiration (kg-cells $m^{-3} h^{-1}$) and k_e is the rate constant (kg-cell matter kg cells$^{-1} h^{-1}$).

The two consumption terms r_m and r_e appear in different balance equations. If we use the maintenance concept then r_m appears in the substrate balance, whereas if we use the endogeneous respiration concept then r_e

appears in the cell mass balance. There is no practical difference between the resulting sets of equations (we can convert from one to the other by making the substitution $k_e = m_s Y_{x/s}$) except in the case where the external substrate supply is less than the maintenance requirement. In this case the maintenance concept breaks down since it predicts a cessation of growth whereas the endogeneous concept predicts what is also observed experimentally, that is a decline in cell mass with time. More complex models involving a switch in the energy-supplying mechanisms according to external conditions should eventually reconcile the two approaches, but meanwhile the maintenance energy approach seems more generally useful.

4.4.4 Cell death

There seems to be a natural rate at which cells become non-viable, i.e. incapable of growth and reproduction. Sometimes the cells are truly dead and their only fate is lysis, while in other cases the cells are in a state of suspended animation but do not revert to the viable state in times of interest in commercial biotechnological processes. The rate of conversion to the non-viable form is assumed to be directly proportional to the mass of viable cells, and is written:

$$r_d = k_d x_v \qquad (4.13)$$

where r_d is the rate of conversion to non-viable form (kg cells $m^{-3} h^{-1}$), x_v is the concentration of viable cells (kg cells m^{-3}), and k_d is the rate constant (kg cells kg-cells$^{-1} h^{-1}$). Little is known of the influence of environmental conditions on the rate constant k_d.

4.4.5 Production formation

Classification
We have already partially classified products in Section 4.4.1 on the basis of their relationship to the cell processes and their appearance as flows of material out of the cell under normal conditions. This approach can now be extended.

Product classification
Type 1 Products of energy metabolism, i.e. by-products of the basic energy production processes in the cell.
Type 2 Extracellular products released by the cell.
Type 3 Energy storage compounds
Type 4 Cell constituents.

Many other classification schemes have been proposed in the literature, one of the most useful is due to Gaden who distinguished Type I products, which are the result of primary energy metabolism, similar to type 1 above; Type II products, which arise indirectly from energy metabolism such as some intermediate metabolites, e.g. citric acid, and Type III complex molecules not resulting directly from energy metabolism, similar to type 2 above.

A useful kinetic classification is based on simple kinetic models which have been found to be of practical use:

Growth-associated products	$r_p = \alpha r_x$	(4.14)
Non-growth associated products	$r_p = \beta x_v$	(4.15)
Mixed kinetics	$r_p = \alpha r_x + \beta x_v$	(4.16)

Kinetic models of product formation
The simplest kinetic model for product formation is that which is the basis of the kinetic classification of products and was suggested by Luedeking and Piret for a lactic acid fermentation:

$$r_p = \alpha r_x + \beta x_v$$

which, using Equation 4.9 for cell growth, expands to:

$$r_p = \frac{\alpha \mu_m S}{k_s + S} \times x_v + \beta x_v \qquad (4.16a)$$

Originally suggested as an empirical model (i.e. one which fits the experimental data but is not based on any theoretical principles) this expression can be derived from our basic picture of the cell process (Section 4.4.1) as follows:

The ATP balance for a type 1 product ($r_{p2} = r_{p1} = 0$) formed under anaerobic conditions ($r_O = 0$) is

$$a_x r_x + r_{mATP} = a_{p1} r_{p1} \qquad (4.17)$$

but since $r_{mATP} = m_{ATP} x_v$ from Equation 4.11 above

$$r_{p1} = \frac{a_x r_x}{a_{p1}} + \frac{m_{ATP} x_v}{a_{p1}} \qquad (4.18)$$

This theoretically based version of Equation 4.16a suggests that for high rates of type 1 product formation under these conditions we need an organism with a high maintenance requirement (m_{ATP}) and a low production of ATP per unit of type 1 product produced (a_{p1}). This equation explains why *Z. mobilis* is a more efficient producer of ethanol than *S. cerevisiae* since reported values of m_{ATP} for *Z. mobilis* are between 14 and 25 times the corresponding value for *S. cerevisiae* and a_{p1} for *Z. mobilis* is 0.5 kmol ATP per kg ethanol produced as compared to a value of 1 for *S. cerevisiae*.

The production of other products can be approached in a similar manner. For example for type 2 product formation we obtain:

$$r_{p2} = \frac{a_O r_O}{a_{p2}} - \frac{m_{ATP} x_v}{a_{p2}} - \frac{a_x r_x}{a_{p2}}$$

(4.19)

which suggests that for maximum product formation rate, cell growth should be suppressed, oxygen uptake rate maximized and sufficient substrate provided only for maintenance and product formation according to:

$$r_s = \left\{ \frac{1}{Y_{p2/s}} + \frac{a_{p2}}{a_O Y_{O/s}} \right\} r_{p2} + \frac{m_{ATP} x_v}{a_O Y_{O/s}}$$

(4.20)

Kinetic models for product formation can also be adapted from the structured mechanisms for cell growth discussed in Section 4.3.2. Some work has been done in this area, and it is likely to become of increasing importance in the next few years.

Product inhibition and degradation
Products which reach a sufficiently high concentration in the medium may inhibit and eventually stop their own production. This phenomenon is well known in ethanol production and limits the maximum strength of alcoholic beverages that can be produced by fermentation. The kinetic expressions for such inhibition will mirror the expression for cell growth inhibition given in Section 4.4.2. Indeed, many products are also inhibitors of cell growth and it is only necessary to replace the inhibitor concentration I in these expressions by the product concentration P. A very common expression for modelling the inhibition of ethanol production is derived from Equation 4.16 in the form:

$$r_p = (1 - P/P_m) \cdot (\alpha r_x + \beta x_v)$$

(4.21)

where P_m is the maximum attainable ethanol concentration.

Products may be removed from the medium either by degradation, usually enzymatic, or by their utilization as substrate by the cell when the normal substrate becomes depleted. The consumption of a product as a substrate follows the normal kinetics for substrate uptake, while, in the absence of detailed knowledge, degradation of product is usually assumed to be a first-order process.

4.4.6 Substrate utilization

Substrate is used to form cell material and metabolic products as discussed in Section 4.4.1, and the rate of substrate utilization is related stoichiometrically to the rates of formation of these materials. In some cases, it is possible to write the chemical equations; for example, the equation for ethanol production from glucose is:

$$C_6H_{12}O_6 = 2C_2H_5OH + 2CO_2$$

from which it is easy to calculate that 0.51 kg of ethanol are formed from 1 kg of glucose. If we write r_p as the rate of product (ethanol) formation and r_{sp} as the rate of substrate uptake for product formation then:

$$r_{sp} = r_p/0.51 \tag{4.21}$$

$$r_{sp} = r_p/Y_{p/s} \tag{4.21a}$$

where $Y_{p/s}$ is the yield coefficient for product on substrate and has the units kg product formed per kg substrate utilized *for product formation*.

Another equation for anaerobic cell growth might be:

$$CH_2O + 0.2\,NH_4^+ + e^- = CH_{1.8}O_{0.5}N_{0.2} + 0.5\,H_2O \tag{4.22}$$

where CH_2O represents the carbohydrate substrate, NH_4^+ is the nitrogen substrate, and $CH_{1.8}O_{0.5}N_{0.2}$ is the approximate formula for many cells. (We have neglected phosphorous and other elements which occur in much smaller proportions.)

Again we can calculate that we form 0.81 kg of cells per kg of carbohydrate substrate utilized for cell formation and 8.8 kg of cells per kg of nitrogen. Thus we would write as above:

$$r_{sx} = r_x/0.81 \text{ or } = r_x/Y_{x/s} \tag{4.23}$$

and

$$r_{Nx} = r_x/8.8 \text{ or } = r_x/Y_{x/N} \tag{4.24}$$

where r_{sx} is the rate of carbohydrate utilization for cell growth, r_{Nx} is the rate of nitrogen utilization for cell production, r_x is the rate of cell growth, $Y_{x/s}$ is the yield coefficient for cells on carbohydrate, and $Y_{x/N}$ is the yield coefficient for cells on nitrogen.

It is important that we do not confuse yield coefficients with yields as normally reported in the literature. A yield coefficient is a stoichiometric constant which depends on the chemical equation relating the reactants and the products, whereas a yield is a ratio of one product to one reactant which may be entering into reactions which form a variety of products. Thus in the anaerobic growth of yeast, the energy required for synthesis is provided by the reaction forming ethanol from carbohydrate. Using the nomenclature of Section 4.4.1, we have

$$a_x = \text{kmoles of ATP required per kg of cells formed}$$

and

$$a_{p1} = \text{kmoles of ATP formed per kg of ethanol produced}$$

hence in forming 1 kg of cells we will have produced a_x/a_{p1} kg of ethanol. Therefore, for the 1 kg of cells we utilize $1/0.81$ kg of glucose to form the cells and $a_x/(0.51a_{p1})$ kg of glucose to form the ethanol. The yield of cells on glucose is therefore

$$\frac{1}{(1/0.81) + (a_x/0.51a_{p1})}$$

which for typical values of $a_x = 0.095$ and $a_{p1} = 1/46$ is 0.102. Similarly for the yield of ethanol on carbohydrate, using these same values, we get 0.446. This is often stated as being 87% (0.446/0.51) of the theoretical yield. (We have not included any term for the maintenance energy requirement which would further reduce the effective yields of both cells and ethanol.)

The overall substrate utilization rates are expressed as the sum of the rates of substrate utilization for the various products, including new cells as shown in Section 4.4.1. Where the energy for cell and product synthesis as well as for maintenance is provided by oxidative phosphorylation, then CO_2 and water are amongst the products. Unfortunately these may also be among the products of the synthesis reactions and it is not possible to discover how much arises from the different sources. However, if the oxygen uptake rate is known then the substrate utilization for oxidative phosphorylation can be determined by the stoichiometry of the combustion reaction. Thus for oxidation of glucose to CO_2 and water with molecular oxygen, we use 32 kg of oxygen for each 30 kg of glucose consumed. Therefore, the yield coefficient for substrate required for oxygen, $Y_{O/s}$, is 1.067 and the rate of substrate uptake for this purpose is given by

$$r_{sO} = r_O/Y_{O/s} \tag{4.25}$$

Where the oxygen uptake rate for oxidative phosphorylation is not known, then the substrate needed to generate the energy required for synthesis is included with the substrate requirement for the synthesis itself according to the chemical equation. The two substrate requirements are lumped together and to distinguish the quantity $Y_{a/s}$ from a true stoichiometric coefficient we refer to it as a yield factor since it includes both the elemental requirements and the energy requirements for the synthesis of component a. The context usually determines which of these two descriptions applies and the symbol $Y_{a/s}$ is used indiscriminately for either. If it is required to distinguish between the two then the yield factor has the tick superscript, i.e. $Y'_{a/s}$. The substrate requirement to provide energy for maintenance is assumed, in those cases where we do not explicitly use the oxygen uptake rate, to be first-order with respect to cell concentration, i.e.

$$r_{sm} = m_s x_v \tag{4.26}$$

Thus the more common expression for substrate utilization (Equation 4.4) is

$$r_s = r_{sx} + r_{sp1} + r_{sp2} + r_{sp3} + r_{sm}$$

$$= \frac{r_x}{Y'_{x/s}} + \frac{r_{p1}}{Y'_{p1/s}} + \frac{r_{p2}}{Y'_{p2/s}} + \frac{r_{p3}}{Y'_{p3/s}} + m_s x_v$$

Any energy losses due to the slippage reaction ATP→ADP→AMP with consequent additional heat release affects the values of the parameters $Y'_{a/s}$ and m_s.

4.4.7   Environmental effects

Temperature effects
The individual reactions which occur within the cell are influenced by temperature in the usual way, i.e. the rate constant for the reaction is given by the Arrhenius equation in the form:

$$k = A \exp\left(-E/RT\right) \tag{4.27}$$

where E is an activation energy and T the absolute temperature.

Since a very large number of cooperating reactions influence the growth and product formation within the cell, temperature will have a complex effect overall. An additional complicating factor arises from the effects of temperature on the conformation of the enzyme proteins, in which very rapid changes in activity can take place over a small range of temperatures.

It is normal to express the parameters in the rate equations as Arrhenius functions of temperature; for example, the Monod equation for cell growth (Equation 4.9):

$$r_x = \frac{\mu_m S}{k_s + S} x_v$$

is written

$$r_x = \frac{\mu_m(T) S}{k_s(T) + S} x_v \tag{4.28}$$

By curve fitting on the experimental data relating μ_m and k_s to temperature, it is found that:

$$\mu_m(T) = A_1 \exp(-E_1/RT) - A_2 \exp(-E_2/RT) \qquad (4.29)$$

and

$$k_s(T) = A_3 \exp(-E_3/RT) \qquad (4.27a)$$

In the expression for maximum specific growth rate the first term on the right accounts for the general increase in reaction rate with temperature, whilst the second term, which has a much higher activation energy than the first, has been associated with a rapid reduction in the activity of the enzymes as temperature increases beyond the optimum. A typical plot of $\mu_m(T)$ against T is shown in Fig. 4.8. If an Arrhenius relation applies,

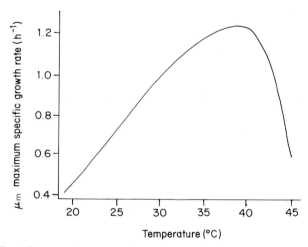

Fig. 4.8   Effect of temperature on growth rate.

then a plot of ln (kinetic parameter) against $1/T$ (K^{-1}) should be either a straight line or be made up from the sum or differences of two or more straight lines. Figure 4.9 shows these relationships.

Since the activation energies of the various reaction steps and conformational changes can differ widely it is to be expected that optimum temperatures for growth and product formation may be quite different and indeed, in a well-balanced medium, the limiting substrate may also vary with temperature. Control to within $\pm 0.5°C$ is usually required for efficient optimisation.

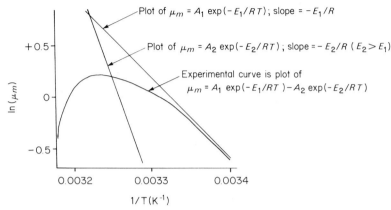

Plot of $\mu_m = A_1 \exp(-E_1/RT)$; slope $= -E_1/R$

Plot of $\mu_m = A_2 \exp(-E_2/RT)$; slope $= -E_2/R$ $(E_2 > E_1)$

Experimental curve is plot of
$\mu_m = A_1 \exp(-E_1/RT) - A_2 \exp(-E_2/RT)$

Fig. 4.9   Arrhenius plots of growth rate constant.

pH effects

pH has less of an effect on biological activities than does temperature, because the cell is reasonably well able to regulate its internal hydrogen ion concentration in the face of adverse external concentrations, though the maintenance energy required to do this is obviously affected. In addition, the pH of the external medium probably has an important effect on the structure and permeability of the cell membrane. A typical plot of specific growth rate of a cell against the external pH is shown in Fig. 4.10. There is a fairly wide range of pH over which the growth rate varies very little. This is often centred around pH 7 for bacteria and pH 4–5 for

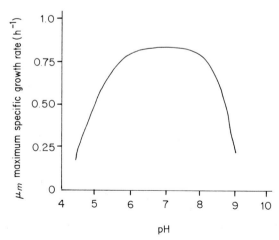

Fig. 4.10   Effect of pH on growth rate.

fungi. Such curves can be modelled by expressions used in enzyme kinetics, of which one example might be:

$$\mu_m(pH) = \frac{\mu_m}{1 + (k_1/[H^+]) + k_2[H^+]} \tag{4.30}$$

where $[H^+]$ is the hydrogen ion concentration $(= \text{antilog}(-pH))$.

4.5 Fermentation process models

4.5.1 The general model for a single vessel

This section is limited to a discussion of a general unstructured model based on the reaction mechanism outlined in Section 4.3.1, but ignoring cell lysis. The following extensive properties (concentrations) are considered to be important:

Viable cells	x_v
Non-viable cells	x_d
Limiting substrate	S
Product	P

The subscript 'i' refers to the input stream and 'o' to the output stream. A sketch of the process with nomenclature is shown in Fig. 4.11. Writing

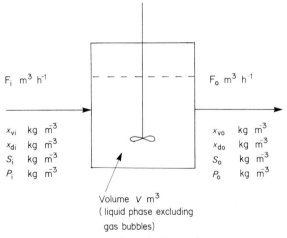

Fig. 4.11   General fermentation process for a single vessel.

balance equations for each extensive property as described in Section 4.2.2 above we obtain:

Viable cells

$$\frac{d(Vx_{vo})}{dt} = Vr_x - Vr_d + F_i x_{vi} - F_o x_{vo} \qquad (4.31)$$

Non-viable cells

$$\frac{d(Vx_{do})}{dt} = Vr_d + F_i x_{di} - F_o x_{do} \qquad (4.32)$$

Substrate

$$\frac{d(VS_o)}{dt} = -V(r_{sx} + r_{sm} + r_{sp}) + F_i S_i - F_o S_o \qquad (4.33)$$

Product

$$\frac{d(VP_o)}{dt} = Vr_p + F_i P_i - F_o P_o \qquad (4.34)$$

For total volume

$$\frac{dV}{dt} = F_i - F_o \qquad (4.35)$$

Notice that the first four balance equations can be written on a unit volume basis in the general form:

$$\frac{1}{V} \frac{d(Vy_o)}{dt} = \Sigma r_{gen} - \Sigma r_{cons} + Dy_i - \gamma Dy_o \qquad (4.36)$$

where y stands for the general extensive property, in this case x_v, x_d, S or P; Σr_{gen} means the sum of all the rates of generation; Σr_{cons} means the sum of all the rates of consumption; $D = F_i/V$; and $\gamma = F_o/F_i$.

Remembering the equations in this form makes it very easy to write them down by rote.

The simplest set of kinetic rate expressions, assuming only one limiting nutrient, is as follows:

Cell growth and death (Equations 4.9 and 4.13)

$$r_x = \mu x_{vo} = \frac{\mu_m S_o}{k_s + S_o} x_{vo}; \qquad r_d = k_d x_{vo}$$

Substrate uptake for cell growth and product formation (from Equation 4.4)

$$r_{sx} = r_x/Y_{x/s}; \qquad r_{sp} = r_p/Y_{p/s}$$

Substrate uptake for maintenance (Equation 4.10)

$$r_{sm} = m_s x_{vo}$$

Product formation (Equation 4.16)

$$r_p = \alpha r_x + \beta x_{vo}$$

The limiting nutrient is assumed to be supplied in the liquid feed, so there are no terms for mass transfer of any of the extensive quantities across a phase boundary and no such transfer can be a limiting factor.

These equations, together with initial conditions for the extensive properties, the volume of medium in the vessel, and the constraints on the variables, constitute the general mathematical model for this simple system. The next two sections consider particular aspects of this model.

4.5.2   Chemostat and simple batch processes at constant volume

If the volume is constant, in the general form of the balance equation (Equation 4.36) we have:

$$\frac{1}{V}\frac{d(Vy_o)}{dt} = \frac{dy_o}{dt} \qquad (4.37)$$

and since $dV/dt = 0$, we also have $F_i = F_o = F$, and $\gamma = 1$. The general form of the balance Equation 4.36 therefore becomes:

$$\frac{dY_o}{dt} = r_{gen} - r_{cons} + D(Y_i - Y_o) \qquad (4.36a)$$

Chemostat

The classical chemostat is a well-stirred fermenter which operates with a constant inflow and outflow, and is assumed to be so well mixed that all concentrations are uniform throughout the whole of the medium volume. Evaporation of water or other low boiling materials from the medium is assumed to be negligible so that the volume of medium within the fermenter is constant at the initial conditions. In the general equations V represents the volume of the liquid phase excluding any gas bubbles. The final requirement is that conditions have been uniform for sufficient time so that any transient changes due to fluctuations in the flow rates, medium compositions or environmental conditions have died out and the fermenter is operating at steady state.

Under such conditions the rate of change of any extensive property is zero so that all the terms on the left-hand side of the balance Equations 4.31–4.35 can be set equal to zero to give:

Viable cells $0 = r_x - r_d$ $+ D(x_{vi} - x_{vo})$ (4.31a)

Non-viable cells $0 = r_d$ $+ D(x_{di} - x_{do})$ (4.32a)

Substrate $0 = -(r_{sx} + r_{sm} + r_{sp}) + D(S_i - S_o)$ (4.33a)

Product $0 = r_p$ $+ D(P_i - P_o)$ (4.34a)

Here $D = F/V$ is the dilution rate (h^{-1}); its reciprocal is the mean residence time of all materials flowing through the fermenter.

Note again the symmetry of the equations, and that the equation for total volume disappears completely since the volume does not change.

The above set of equations together with the kinetic rate equations, and the constraints represent the mathematical model of a chemostat.

At this stage it is useful to list and classify the variables and parameters. The variables are:

(a) *State variables.* These define the state of the process and there is one for each extensive property:

x_{vo} Outlet and fermenter viable cell concentration;
x_{do} Outlet and fermenter non-viable cell concentration;
S_o Outlet and fermenter limiting substrate concentration;
P_o Outlet and fermenter product concentration.

(b) *Operating variables.* These are variables whose values can be set by the operator of the process:

   D Dilution rate;
   S_i, x_{vi}, x_{di}, P_i (Inlet values of the four state variables).
Usually x_{vi}, x_{di}, and P_i are equal to zero.

(c) *Intermediate variables.* These are all the rates r_x, r_d, r_{sx}, r_{sm} and r_p which can be expressed in terms of the state variables listed above.

(d) *Kinetic parameters.* $\mu_m, k_d, k_s, m_s, \alpha, \beta$.

(e) *Stoichiometric parameters.* $Y_{x/s}, Y_{p/s}$.

It is possible to solve this particular set of equations analytically with x_{vi}, x_{di}, and P_i zero, to give the following relationships between the state variables and the operating variables. The reader should verify for him or herself that the set of equations is mathematically consistent as defined in Section 4.2.7.

$$S_o = \frac{k_s(D + k_d)}{\mu_m - (D + k_d)} \qquad (4.38)$$

$$x_{vo} = \frac{D(S_i - S_o)}{(D + k_d)/Y_{x/s} + m_s + [\alpha(D + k_d) + \beta]/Y_{p/s}} \quad (4.39)$$

$$x_{do} = \frac{k_d x_{vo}}{D} \quad (4.40)$$

$$P_o = \frac{[\alpha(k_d + D) + \beta]x_{vo}}{D} \quad (4.41)$$

Constraints:

$$0 \leqslant S_o \leqslant S_i; \qquad 0 \leqslant x_{vo}; \qquad 0 \leqslant x_{do} \leqslant x_{vo}; \qquad 0 \leqslant P_o$$

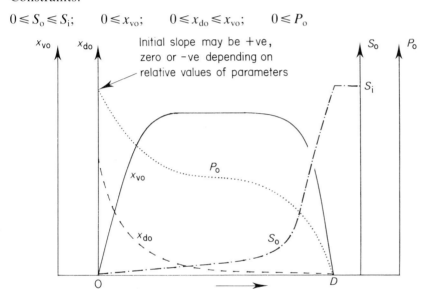

Fig. 4.12 State variables plotted against dilution rate with constant S_i.

A sample of these solutions are plotted in Figs 4.12 and 4.13 for variations in the dilution rate and the inlet limiting substrate concentration. A study of the solutions and the plots gives valuable insight into the operation of a chemostat and the reader should consider some of the implications.

An analytical solution is not always obtained so readily, and then we have to resort to computer solutions and be content with graphical plots. It is because computer solutions can generate utter rubbish if not handled carefully, that constraints on the variables are a necessary part of the model.

One aim in using models of this kind is to allow prediction of the behaviour of the chemostat when conditions are altered, whether by changing the operating variables or by the influence of changed environmental conditions on the values of the parameters. This, of course, requires that the effect of changes in environmental conditions on the parameters is already

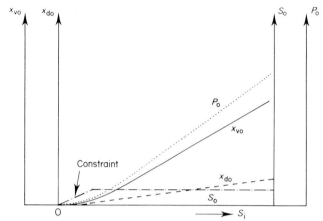

Fig. 4.13　State variables plotted against inlet substrate concentration at constant D. (Caution: these curves only apply if no other substrate, including oxygen, is limiting.)

known, but conversely, the chemostat is ideally suited for determining these effects. The next major section illustrates how models may be used for this purpose.

Batch fermenter
The batch fermenter is the simplest example of an unsteady state process. Here there is no input or output so that the volume is constant and we can put $D = 0$ in Equations 4.31–4.39 which then become:

Viable cells
$$\frac{dx_{vo}}{dt} = r_x - r_d \tag{4.31b}$$

Non-viable cells
$$\frac{dx_{do}}{dt} = r_d \tag{4.32b}$$

Substrate
$$\frac{dS_o}{dt} = -(r_{sx} + r_{sm} + r_{sp}) \tag{4.33b}$$

Product
$$\frac{dP_o}{dt} = r_p \tag{4.34b}$$

These, together with the kinetic rate equations and a set of initial conditions (the values of all the state variables [concentrations] at the time of inoculation) and constraints, are the process model for the batch fermenter.

An analytical solution of this set of equations is possible but is too complex to be useful. Computer solution by numerical integration is more usual and a typical solution is presented in Fig. 4.14. Two features that this plot

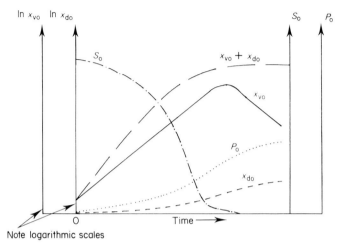

Fig. 4.14   Batch growth and production curves.

does not show but which are present in most fermentations are the lag phase and the phase of decline. The lag phase arises from the need for cells in the inoculum to adapt to the prevailing conditions in the fermenter, and the model as constructed does not incorporate any mechanism for this adaptation. (Structured models are required to do this, see Section 4.3.2.) The phase of decline would be shown if we had included cell lysis in the model; in Section 4.5.1. we assumed that the non-viable cells remain unchanged in the culture so that the total mass of viable and non-viable cells remains constant after the limiting substrate has been exhausted.

4.5.3   Chemostat with recycle

One consequence of the equations describing the simple chemostat is that the maximum dilution rate is limited to a value slightly less than the maximum specific growth rate μ_m. This is called the critical dilution rate for washout, D_c. Washout occurs when cells are being removed from the fermenter at a rate $(D_c x_{vo})$ which is just equal to the maximum rate $(\mu x_{vo} - d_c x_{vo})$ at which they can grow in the fermenter, so that the only stable value for the cell concentration is zero. This maximum rate of growth in the fermentation occurs when $\mu = \mu_m S_i/(k_s + S_i)$ since S_i is the maximum substrate concentration that can occur in the fermenter. As the value of k_s is small compared to S_i it follows that:

$$D_c = \mu_m \frac{S_i}{k_s + S_i} - k_d \simeq \mu_m - k_d \qquad (4.42)$$

Of course at washout the steady state is one in which no cells exist or grow and the substrate passes unchanged through the fermenter.

This limitation on the dilution rate limits the productivity of the fermenter (Dx_{vo} or DP_o kg m^{-3} h^{-1}); the use of cell recycle overcomes this limitation to some extent.

In a chemostat with recycle, all or part of the outflowing medium is removed from the system either completely free of cells or with a cell concentration less than that which exists in the chemostat. This may be achieved either within the chemostat by filtering or settling part of the outflowing stream or externally using a separating device such as a filter, settler or continuous centrifuge; different ways of doing this are shown in Fig. 4.15. However, for the purposes of mathematical analysis, they are equivalent to having an outflowing stream with a cell concentration equal to the internal cell concentration multiplied by a constant which we will call the separation constant, δ. An abstracted physical model with nomenclature is shown in Fig. 4.16. If the separation constant, δ, is zero than all the cells recycle in the fermenter and without cell lysis or endogeneous respiration the biomass in the chemostat would increase without limit. (In practice gas transfer limitation or product inhibition would terminate this increase.) If δ equals unity then we have the simple chemostat without recycle.

The steady state equations are, modifying those given in Section 4.5.2 (4.31a–4.34a), for zero values of x_{vi}, x_{di}, and P_i:

Viable cells	$0 = r_{xv} - r_d$	$- \delta Dx_{vo}$	(4.31c)
Non-viable cells	$0 = r_d$	$- \delta Dx_{do}$	(4.32c)
Substrate	$0 = -(r_{sx} + r_{sm} + r_{sp}) + D(S_i - S_o)$		(4.33c)
Product	$0 = r_p$	$- DP_o$	(4.34c)

Following the procedure set out in Section 4.5.2. for the simple chemostat, we obtain the analytical solutions:

$$S_o = \frac{k_s(\delta D + k_d)}{\mu_m - (\delta D + k_d)} \qquad (4.38a)$$

$$x_{vo} = \frac{D(S_i - S_o)}{(\delta D + k_d)/Y_{x/s} + m_s + [\alpha(\delta D + k_d) + \beta]/Y_{p/s}} \qquad (4.39a)$$

$$x_{do} = \frac{k_d x_{vo}}{\delta D} \qquad (4.40a)$$

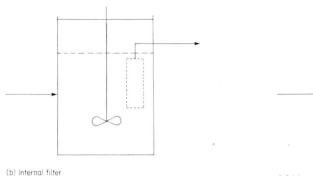

(b) Internal filter

(a) Internal settling

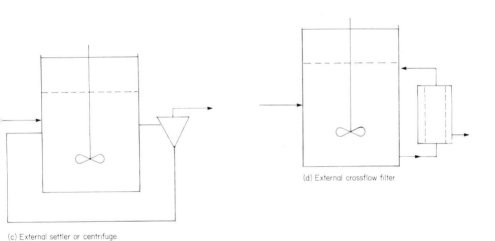

(d) External crossflow filter

(c) External settler or centrifuge

Fig. 4.15   Methods of recycling cells.

$$P_o = \frac{[\alpha(k_d + \delta D) + \beta]}{D} x_{vo}$$

(4.41a)

Constraints:

$$0 \leq S_o \leq S_i; \qquad 0 \leq x_{vo}; \qquad 0 \leq x_{do} \leq x_{vo}; \qquad 0 \leq P_o.$$

Note that these equations reduce to those for the simple chemostat when $\delta = 1$. The washout dilution rate is now given by:

$$\delta D_c x_{vo} = \text{the maximum value of } (\mu x_{vo} - k_d x_{vo})$$

i.e.

$$D_c = \frac{\mu_m S_i}{\delta(k_s + S_i)} - \frac{k_d}{\delta} \simeq \frac{\mu_m - k_d}{\delta} \tag{4.42a}$$

Thus the smaller δ, the higher is the maximum possible dilution rate. Again at any given dilution rate, cell concentration x_{vo}, and hence the product concentration, are greater the smaller the value of δ.

These quantities are plotted against dilution rate for various values of the separation constant in Fig. 4.17 using representative values of the parameters. The corresponding productivities are plotted in Fig. 4.18.

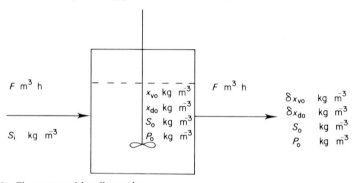

Fig. 4.16   Chemostat with cell recycle.

4.5.4   Fed-batch processes (variable volume)

In a variable volume process, the volume of medium in the fermenter varies because the inflow is not equal to the outflow; in practical cases V increases, and often the outflow is zero.

Since V is not constant, we can write using elementary calculus,

$$\frac{d(Vy_o)}{dt} = \frac{V dy_o}{dt} + y_o \frac{dV}{dt} \tag{4.43}$$

and since from Equation 4.35

$$\frac{dV}{dt} = F_i - F_o$$

then

$$\frac{d(Vy_o)}{dt} = \frac{V dy_o}{dt} + F_i y_o - F_o y_o \tag{4.44}$$

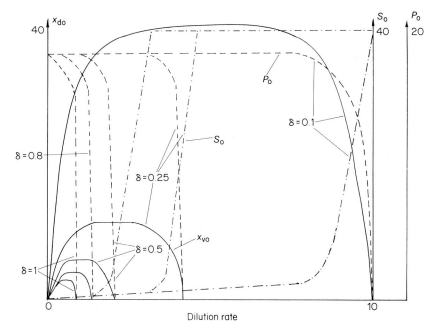

Fig. 4.17 Effects of separation factor on chemostat performance. $k_s = 0.5$, $\mu_m = 1.05$, $k_d = 0.01$, $Y_{x/s} = 0.5$, $Y_{p/s} = 0.51$, $\alpha = 4.4$, $\beta = 0.03$, $m_s = 0$, $S_i = 40$.

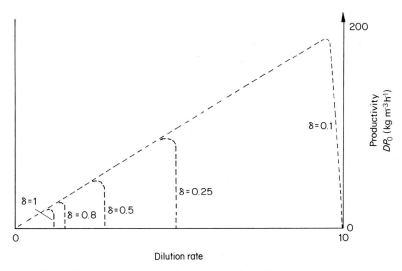

Fig. 4.18 Effect of separation factor on chemostat productivity.

and

$$\frac{1}{V}\frac{d(Vy_o)}{dt} = \frac{dy_o}{dt} + D(1-\gamma)y_o \tag{4.44a}$$

where, as before,

$$D = F_i/V \text{ and } \gamma = F_o/F_i$$

Combining Equations 4.36 and 4.44a, we can write the general balance equation as

$$\frac{dy_o}{dt} + D(1-\gamma)y_o = \Sigma r_{gen} - \Sigma r_{cons} + Dy_i - \gamma Dy_o \tag{4.45}$$

which gives the surprising final result

$$\frac{dy_o}{dt} = \Sigma r_{gen} - \Sigma r_{cons} + D(y_i - y_o) \tag{4.36a}$$

which is identical in appearance to the general form of the balance equation for the constant volume process given in Section 4.5.2. However, although the term $D(y_i - y_o)$ is identical in both of the equations, it does not have the same physical significance. In both the constant volume and the variable volume process, the term Dy_i represents the amount of the extensive quantity brought in with the outflow. The term Dy_o in the constant volume process is the amount of the extensive quantity leaving with the outflow, but in the variable volume process it is a combination of the term representing the amount leaving with the outflow $Dy_o = (F_o/V)Y_o$, and the term arising from the volume dilution effect of the difference between the input and output, $D(1-\gamma)y_o = (F_i - F_o/V)Y_o$, so that the outflow effect cancels out part of the dilution effect. The full set of balance equations for the variable volume process is:

Viable cells $\qquad \dfrac{dx_{vo}}{dt} = r_{xv} - r_d \qquad\qquad + D(x_{vi} - x_{vo}) \tag{4.31d}$

Non-variable cells $\dfrac{dx_{do}}{dt} = r_d \qquad\qquad + D(x_{di} - x_{do}) \tag{4.32d}$

Substrate $\qquad \dfrac{dS_o}{dt} = -(r_{sx} + r_{sm} + r_{sp}) + D(S_i - S_o) \tag{4.33d}$

Product $$\frac{dP_o}{dt} = r_p + D(P_i - P_o) \qquad (4.34d)$$

Volume $$\frac{dV}{dt} = F_i - F_o \qquad (4.35)$$

Note that $D = F_i/V$

These equations, together with the set of kinetic rate equations, the initial conditions and the constraints form the mathematical model of the process.

If we assume that $x_{vi} = x_{di} = P_i = F_o = 0$, we have the classical fed-batch process with continuous input of the limiting substrate, usually a carbon energy source, typical of type 2 and 3 product formation processes.

As a possible set of kinetic rate equations we could take

$$r_{xv} = \mu x_{vo} = \frac{\mu_m S_o}{k_s + S_o} x_{vo} \qquad \text{(Equation 4.9)}$$

$$r_d = k_d x_{vo} \qquad \text{(Equation 4.13)}$$

$$r_{sx} = r_x/Y_{x/s} \qquad \text{(Equation 4.4)}$$

$$r_{sm} = m_s x_{vo} \qquad \text{(Equation 4.26)}$$

$$r_{sp} = r_p/Y_{p/s} \qquad \text{(Equation 4.21a)}$$

$$r_p = \beta x_{vo} \qquad \text{(Equation 4.15)}$$

Note that the operator can normally manipulate S_i and F_i, and as will be seen later, from a purely mathematical point of view it is better to have high values of S_i and low values of F_i. However, there are considerable practical difficulties in achieving a uniform (and low) sugar concentration in the fermenter (due to mixing limitations) and so there is a limit on the maximum value of S_i.

This particular model is an example of the use of mathematical models for 'what if' types of computer experiments as there are a variety of objectives in running a fed-batch project which can be very easily explored with simple extensions of the model to take into account some of the equipment factors. One possible objective might be to have zero growth and high product formation rates. The high product formation rate depends on having a high viable cell concentration as can be seen from the product balance equation (Equations 4.15 and 4.34d). The first stage of the process would, therefore, be to grow up a high cell concentration, followed by a phase where growth is suppressed and only sufficient of the substrate is supplied for maintenance and product formation. For this second stage we have:

zero growth, ($\mu = 0$), hence $S_o = 0$, $dS_o/dt = 0$ and $r_{sx} = 0$. Combining the kinetic rate equations with the balance equations gives:

$$\frac{dx_{vo}}{dt} = -k_d x_{vo} - D x_{vo} \tag{4.31e}$$

$$\frac{dx_{do}}{dt} = k_d x_{vo} - D x_{do} \tag{4.32e}$$

$$0 = k S_i x_{vo} + D S_i \tag{4.33e}$$

$$\frac{dP_o}{dt} = \beta x_{vo} - D P_o \tag{4.33e}$$

$$\frac{dV}{dt} = F_i \tag{4.46}$$

where

$$k = \left\{ m_s + \frac{\beta}{Y_{p/s}} \right\} \frac{1}{S_i} \tag{4.47}$$

The solution of these equations is simplest if we fix S_i and allow F_i to vary. From the substrate balance, we obtain:

$$D = F_i/V = k x_{vo} \tag{4.48}$$

Since V is given directly by integrating the volume Equation 4.46 and k is fixed provided we fix S_i, then F_i can be easily computed providing x_{vo} is known. The solution for x_{vo} is given directly by substitution for $D = k x_{vo}$ from Equation 4.48 into Equation 4.31e and direct integration of the balance equation for viable cells to give:

$$x_{vo} = \frac{k_1 k_d \exp(-k_d t)}{1 - k_1 k \exp(-k_d t)} \tag{4.49}$$

where

$$k_1 = \frac{x_{vo}(0)}{k_d + k x_{vo}(0)} \tag{4.50}$$

and where $x_{vo}(0)$ is the value of the viable cell concentration when the batch feeding phase begins and t is the time from the beginning of the feeding phase.

The expressions for x_{do} and P_o are more complicated, but if we write

$$f(t) = \int_0^t x_{vo} dt \tag{4.51}$$

then the equation for P_o is

$$P_o = \frac{\beta}{k}\{1 - \exp[-kf(t)]\}$$

(4.52)

Typical plots of these expressions are given in Fig. 4.19.

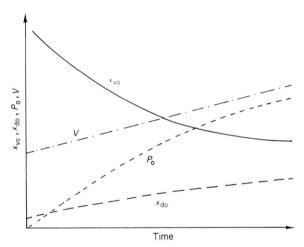

Fig. 4.19   Fed-batch curves.

In the case where there is no death of cells, i.e. $k_d = 0$ and the second equation in the model is eliminated, the analytical solutions are much simpler and are given below:

$$x_{vo} = x_{vo}(0)/[1 + kx_{vo}(0)t]$$

(4.53)

$$F_i = kx_{vo}(0)V(0)$$

(4.54)

$$P_o = \beta x_{vo}(0)t/[1 + kx_{vo}(0)t]$$

(4.55)

where, as before, $x_{vo}(0)$ is the cell concentration at the beginning of the fed batch phase and $V(0)$ is the corresponding volume.

4.5.5   Mass transfer limitations (oxygen transfer)

We have already encountered one problem of mass transfer limitation in dealing with fed batch processes in the previous section (the dispersion of a concentrated substrate solution in a mixed fermenter). Another, important in waste treatment processes, is the limitation on the rates of diffusion of substrates into and products out of mycelial pellets, fungal mats, bacterial slimes, etc. In aerobic fermentation processes, the greatest problem is that of limitations on the transfer of oxygen across the boundary between the gas phase and the liquid phase (see Chapter 5). Here we will consider the modelling aspects of this problem; for simplicity we will ignore death of cells and product formation.

Because of its low solubility in liquids, oxygen is not supplied in the feed, so in the substrate balance for oxygen, there is no bulk flow input term. The only term for input of oxygen into the control region (the liquid phase) is the mass transfer rate of Equation 4.1:

$$N_a = k_L a(C_g^* - C_L)$$

where N_a is the oxygen transfer rate (kg oxygen $m^{-3} h^{-1}$), $k_L a$ is the mass transfer coefficient (h^{-1}), C_g^* is the concentration of oxygen which would exist in the bulk liquid phase if it were in equilibrium with the gas phase (kg oxygen m^{-3}), and C_L is the actual concentration of oxygen in the bulk liquid phase (kg oxygen m^{-3}).

If we assume that both oxygen and one of the substrates supplied in solution in the liquid medium are limiting, then we get as a possible set of equations:

(a) Balance equations.
For cells

$$\frac{d(Vx_{vo})}{dt} = Vr_{xv} + F(x_{vi} - x_{vo}) \tag{4.31f}$$

For soluble substrate

$$\frac{d(VS_o)}{dt} = -V(r_{sx} + r_{sm}) + F(S_i - S_o) \tag{4.33f}$$

For oxygen

$$\frac{d(VC_{Lo})}{dt} = -V(r_{ox} + r_{om}) + VN_a - FC_{Lo} \tag{4.56}$$

(b) Rate equations. Cell growth (double limiting substrate kinetics, Equation 4.9b):

$$r_{xv} = \mu x_{vo} = \mu_m \frac{S_o}{(k_s + S_o)} \frac{C_{Lo}}{(k_o + C_{Lo})} x_{vo}$$

Substrate uptake for cell growth and maintenance

$$r_{sx} = r_x/Y'_{x/s}; \qquad r_{sm} = m_s x_{vo}$$

Oxygen transfer across the gas–liquid phase boundary

$$N_a = k_L a(C_g^* - C_{Lo})$$

Oxygen consumption for cell growth and maintenance

$$r_{ox} = r_{xv}/Y'_{x/O}; \qquad r_{om} = r_{sm}/Y_{s/O}$$

(c) Thermodynamic equation.
Henry's law

$$C_g^* = \frac{p_o}{H} \tag{4.58}$$

where p_o is the partial pressure of oxygen in the gas phase (bar) and H is Henry's law constant (bar m^3 kg-oxygen^{-1}).

(d) Constraints.

$$0 \leqslant x_{vo}; \qquad 0 \leqslant S_o \leqslant S_i; \qquad 0 \leqslant C_{Lo} \leqslant C_g^*$$

It should be noted that in the rate equations, the yield factor for substrate and oxygen uptake are not true stoichiometric coefficients since they include their respective requirements for generation of energy for cell synthetic reactions. The yield factor relating oxygen uptake for maintenance to substrate uptake for maintenance is a true yield coefficient since it can be calculated from the stoichiometry of the oxidation reaction for the substrate (see Section 4.4.6).

These equations, together with the initial conditions of the state variables x_{vo} S_o and C_{Lo} constitute a model of the process.

Chemostat
At steady state all the derivatives are zero as indicated in Section 4.5.2 (Equations 4.31a–4.34a). For a sterile feed x_{vi} is also zero and we assume constant volume. With these assumptions we can derive the algebraic relationships in the same way as previously. From the cell balance we have:

$$r_{xv} = Dx_{vo} = \mu x_{vo} = \mu_m \frac{S_o}{k_s + S_o} \times \frac{C_{Lo}}{k_o + C_{Lo}} x_{vo}$$

which has the non-trivial solution

$$D = \mu = \mu_m \frac{S_o}{k_s + S_o} \times \frac{C_{Lo}}{k_o + C_{Lo}} \tag{4.59}$$

Unfortunately this cannot be solved directly for S_o or C_{Lo} as in the previous case with only one limiting substrate.

From the soluble substrate balance we obtain:

$$x_{vo} = \frac{DY'_{x/s}(S_i - S_o)}{D + m_s Y'_{x/s}} \tag{4.60}$$

and from the oxygen balance

$$x_{vo} = \frac{Y'_{x/o}[k_L a(C_g^* - C_{Lo}) - DC_{Lo}]}{D + m_s(Y'_{x/o}/Y_{s/o})} \tag{4.61}$$

This last equation may be simplified to:

$$x_{vo} = \frac{Y'_{x/o} k_L a(C_g^* - C_{Lo})}{D + m_s(Y'_{x/o}/Y_{s/o})} \tag{4.62}$$

since $D \ll k_L a$ in all normal circumstances.

Note that there are three independent equations and it is therefore possible to solve for the three variables x_{vo}, S_o and C_{Lo} in terms of D and S_i. However, there is not a simple easy solution as in the single limiting substrate case and a more instructive approach is to consider separately the two extreme conditions of soluble substrate limitation and oxygen limitation.

When only oxygen is limiting, we assume that S_o is so much greater than k_s that the ratio $S_o/(k_s + S_o)$ is unity. Similarly for soluble substrate limiting conditions, C_{Lo} is sufficiently greater than k_o so that the ratio $C_{Lo}/(k_o + C_{Lo})$ is also unity. Under these conditions the algebraic solution of the equations is simple and is as follows:

(a) Oxygen limiting, soluble substrate in excess.

$$D = \mu = \mu_m C_{Lo}/(k_o + C_{Lo}) \tag{4.63}$$

giving the solutions

$$C_{Lo} = \frac{k_o D}{\mu_m - D} \tag{4.64}$$

$$x_{vo} = \frac{Y'_{x/o} k_L a(C_g^* - C_{Lo})}{D + m_s(Y'_{x/o}/Y_{s/o})} \tag{4.62}$$

and

$$S_o = S_i - x_{vo}\frac{D + m_s Y'_{x/s}}{DY'_{x/s}} \tag{4.65}$$

The important feature of these equations compared to the case where the limiting substrate is dissolved in the feed is the dependence of cell concentration on the dilution rate. As can be seen, this declines in a hyperbolic fashion instead of rising to a broad maximum before dropping to zero at washout. Washout under conditions of oxygen limitation occurs when $C_{Lo} = C_g^*$ and is given by:

$$D_c = \mu_m \frac{C_g^*}{k_o + C_g^*} \cong \mu_m \qquad (4.66)$$

A plot of the three curves for C_{Lo}, x_{vo} and S_o against the dilution rate D is shown in Fig. 4.20. In this plot we have shown several curves for

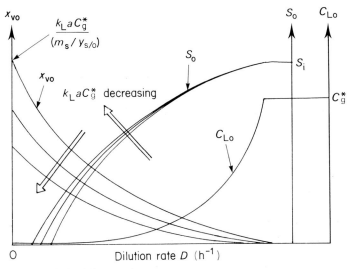

Fig. 4.20   Oxygen limited chemostat.

different values of $k_1 a$ to give an idea of the effect of this most important operating parameter on the behaviour of the chemostat. The hyperbolic reduction in x_o as D increases is most noticeable. At low dilution rates the soluble substrate concentration S_o drops to zero, i.e. the lower constraint. The value of D at which this occurs is given roughly by

$$D = Y'_{x/o} k_L a C_g^* / Y'_{x/s} S_i \qquad (4.67)$$

and decreases as either $k_L a$ or C_g^* decrease or S_i increases.

Obviously at low dilution rates the original assumption that only the oxygen substrate is limiting breaks down, so to investigate this region it is necessary to look at the alternative case (soluble substrate limiting).

(b) Soluble substrate limiting, oxygen in excess.

$$D = \mu = \mu_m S_o / (k_s + S_o) \qquad (4.68)$$

giving the solutions

$$S_o = \frac{k_s D}{\mu_m - D} \qquad (4.69)$$

$$x_{vo} = \frac{DY'_{x/s}(S_i - S_o)}{D + m_s Y'_{x/s}} \tag{4.60}$$

$$C_{Lo} = C_g^* - x_{vo}\frac{D + m_s(Y'_{x/O}/Y_{s/O})}{Y'_{x/O}k_L a} \tag{4.70}$$

The plot of the three state variables against dilution rate is shown in Fig. 4.21, again with different values of $k_L a$. As $k_L a$ decreases the minimum

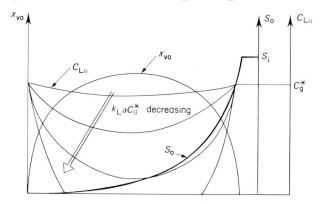

Fig. 4.21 Soluble substrate limited chemostat.

value of C_{Lo} decreases until at some stage it will hit the constraint and be zero according to this model. This occurs when

$$C_g^* = x_{vo}\frac{D + m_s(Y'_{x/O}/Y_{s/O})}{Y'_{x/O}k_L a} \simeq \frac{x_{vo}D}{Y'_{x/O}k_L a} \tag{4.71}$$

Thus the oxygen will become limiting in this central region of dilution rate and again the assumption that oxygen is not limiting will no longer apply.

(c) Double substrate limitation. The situation where both substrates are limiting at the same time will only occur over a narrow range of dilution rate. At dilution rates other than in this narrow range we can determine which is limiting by calculating the two values of x_{vo} as given by the assumptions of oxygen limiting and soluble substrate limiting. Whichever gives the lower value is the equation which applies.

Formally we write:

x_{vo} is equal to the lower of:

$$\frac{DY'_{x/s}}{D + m_s Y'_{x/s}}\left\{S_i - \frac{k_s D}{\mu_m - D}\right\} \quad or \quad \frac{Y'_{x/O}k_L a}{D + m_s(Y'_{x/O}/Y_{s/O})}\left\{C_g^* - \frac{k_o D}{\mu_m - D}\right\}$$

The corresponding values of the limiting substrate are:

$$S_o = \frac{k_s D}{\mu_m - D} \quad or \quad C_{Lo} = \frac{k_o D}{\mu_m - D}$$

and of the non-limiting substrate

$$C_{Lo} = C_g^* - x_{vo}\frac{D + m_s(Y'_{x/O}/Y_{s/O})}{Y'_{x/O}k_L a} \quad or \quad S_o = S_i - X_{vo}\frac{D + m_s Y'_{x/s}}{DY'_{x/s}}$$

Figure 4.22 shows how this approach appears graphically. The true curve switches from one set to the other when the limitation changes. In the correct solution of the original set of equations using the double substrate kinetics the changeover from insoluble substrate to soluble substrate limiting occurs gradually but the errors in adopting the simpler approach are very small, as is shown in Fig. 4.22a, where both solutions are shown superimposed.

Where only one of the two substrates is limiting it is obvious as can be seen from either Figs 4.20 or 4.21 that a considerable amount of the other substrate leaves with the exit stream; its recovery will seldom be either practicable or economic, and even in providing oxygen we have incurred significant costs. Therefore it does not pay to work with any significant excess of any substrate. The influence of the inlet soluble substrate concentration and the $k_L a$ or C_g^* value, both of which are related to the cost of providing oxygen, are shown in Fig. 4.23. It is obvious that for a particular value of $k_L a$ or C_g^* there is, for any value of dilution rate, a maximum value of S_i above which it would be wasteful to operate, and similarly for a given value of S_i then depending upon the desired dilution rate there is some maximum value of $k_L a$ or C_g^* at which the chemostat should be run. In a small-scale experimental situation excess of oxygen carries no financial penalty, but in the industrial situation, of course, the chemostat would be run with a slight excess of oxygen and the designer would also have to take into account problems associated with the dispersion and mixing of the substrates in the bulk of the medium.

If it is required that both substrates should be limiting, then we can determine the conditions for this by equating the values of x_{vo} from Equations 4.60 and 4.62 as follows:

$$x_{vo} = \frac{Y'_{x/O}k_L a(C_g^* - C_{Lo})}{D + m_s(Y'_{x/O}/Y_{s/O})} = \frac{DY'_{x/s}(S_i - S_o)}{D + m_s Y'_{x/s}}$$

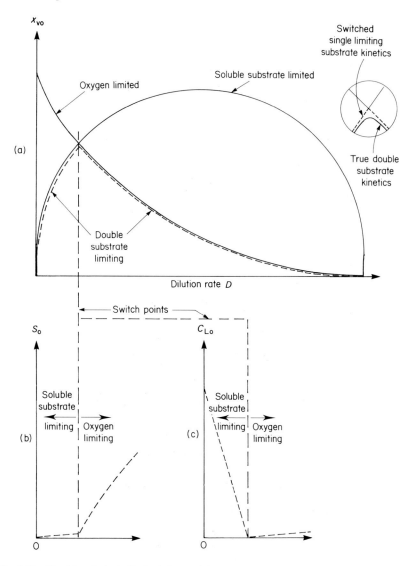

Fig. 4.22   Double substrate limited chemostat.

If we make the reasonable assumption that when both substrates are limiting.

$$C_{Lo} \ll C_g^* \qquad \text{and} \qquad S_o \ll S_i$$

and also that

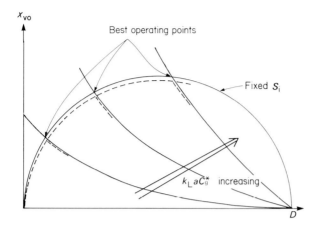

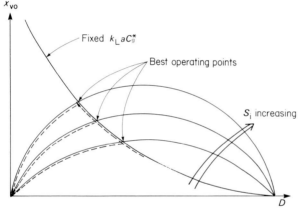

Fig. 4.23   Effect of S_i and $k_L a C_g^*$ on behaviour of chemostat.

$$D \gg m_s(Y'_{x/O}/Y_{s/O}) \qquad \text{and} \qquad D \gg m_s Y'_{x/s}$$

then we obtain the very simple relationship

$$\frac{Y'_{x/O} k_L a C_g^*}{D} = Y'_{x/s} S_i \tag{4.72}$$

or
$$k_L a C_g^* = (Y'_{x/s}/Y'_{x/O})D S_i \tag{4.73}$$

which allows for a quick estimation of any one of $k_L a$, C_g^*, S_i or D, given that the others are known.

Batch fermenter
The analysis for the batch fermenter proceeds along the same lines but is simple to deal with since it is reasonable to assume that the soluble

substrate is non-limiting for the greater part of the fermentation time. We will, therefore, ignore the soluble substrate and concentrate only on the cell and the dissolved oxygen amounts.

The batch equations are, therefore:

(a) Balance equations.
For cells

$$\frac{d(Vx_{vo})}{dt} = Vr_{xv} \tag{4.31g}$$

For oxygen

$$\frac{d(VC_{Lo})}{dt} = -V(r_{sx} + r_{sm}) + VN_a \tag{4.56a}$$

(b) Rate equations.

$$r_{xv} = \mu x_{vo} = \mu_m \frac{C_{Lo}}{k_o + C_{Lo}} x_{vo}$$

$$r_{sx} = r_{xv}/Y'_{x/s}; \qquad r_{sm} = m_s x_{vo} \qquad \text{and} \qquad N_a = k_L a(C_g^* - C_{Lo})$$

Other equations are as before.

Making the various substitutions we have to solve:

$$\frac{dx_{vo}}{dt} = r_{xv} = \mu_m \frac{C_{Lo}}{k_o + C_{Lo}} x_{vo} \tag{4.74}$$

$$\frac{dC_{Lo}}{dt} = -\left\{\frac{r_x}{Y'_{x/s}} + m_s x_{vo}\right\} + k_L a(C_g^* - C_{Lo}) \tag{4.75}$$

given the appropriate initial conditions and values of $k_L a$ and C_g^*. As before it is most convenient to solve this set of equations using a computer and the solution is shown in Fig. 4.24 for a fermentation which starts with an oxygen saturated medium.

Notice that the growth of cells is exponential only in the initial stages when the oxygen in solution is being used up and soon becomes almost linear under mass transfer control. (The non-oxygen limited growth curve is shown for comparison.) A simplified explanation of this may be seen by considering the oxygen balance equation. When the oxygen concentration in solution, C_{Lo}, drops to a very low value, the derivative dC_{Lo}/dt will be almost zero. It is, therefore, easily shown that:

$$r_{xv} = Y'_{x/s}(k_L a C_g^* - m_s x_{vo}) \tag{4.76}$$

Since $k_L a C_g^*$ will remain constant during the fermentation, or possibly decrease for the reasons given in Chapter 5, and the term $m_s x_{vo}$ increases as the fermentation proceeds, the growth rate slowly declines. In industrial aerobic fermentations, the supply of oxygen is usually the limiting factor.

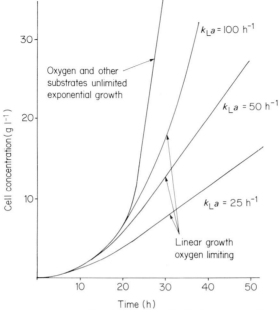

Fig. 4.24   The effect of oxygen gas/liquid mass transfer limitation on cell growth.

4.6   Measuring and quantifying kinetic parameters

Having constructed a model, one of the tasks as indicated in Fig. 4.1 is to determine the model parameters. These parameters (Section 4.5.2: Chemostat) can be subdivided into stoichiometric (those concerned with the material and energy balance relationships between the various flows) and kinetic (those concerned with the rates of consumption, generation or rates of transfer of species). To determine the values of the parameters, experimental data are required. It goes without saying that these should be as accurate, consistent, and as reliable as possible. They should provide sets of values of the variables of interest, either determined when the system has settled down at a steady state or determined simultaneously at some particular instant in time (unsteady state, dynamic experiments). The method of quantifying parameters is to fit the experimental data to the appropriate mathematical equations.

   There are two principal variants of the method:

   (a) Straight line fitting. This usually involves transformation of the variables but can be done with the minimum of equipment (calculator and graph paper, etc.).

(b) Curve fitting. This is usually carried out using the sets of data in their normal form but requires the use of a computer and software which may need to be tailored for the particular class of model being fitted.

The most familiar example of straight line fitting is the Lineweaver–Burke plot used to determine the maximum rate and saturation constant of a Michaelis–Menten or Monod type rate equation. Equations 4.8 and 4.9 (Section 4.4.2) are both of this form. Taking Equation 4.9, we have:

$$r_x = \frac{\mu_m S}{k_s + S} x_v \tag{4.76}$$

where r_x, S and x_v are the variables and μ_m and k_s are the parameters. As it stands the relationship between the variables is non-linear (involves multiplication or division). The first step is to transform the variables, i.e. manipulate them in some way to give a linear relationship between the new transformed variables. This procedure is not formalized and depends on the ingenuity of the modeller. In this case division of r_x by x_v simplifies the expression by removing a multiplication. Thus by defining the new variable $\mu = r_x/x_v$ we have transformed the equation to:

$$\mu = \frac{\mu_m S}{k_s + S}$$

where now we have just two variables, μ and S. The next step is to take the reciprocal of either side of the equation, i.e.

$$\frac{1}{\mu} = \frac{k_s + S}{\mu_m S} = \frac{k_s}{\mu_m} \frac{1}{S} + \frac{1}{\mu_m} \tag{4.77}$$

or in terms of new transformed variables:

$$p = \frac{k_s}{\mu_m} q + \frac{1}{\mu_m} \tag{4.78}$$

where $p = x_v/r_x$ and $q = 1/S$.

This equation is now linear in the transformed variables, and a plot of p against q should give a straight line on arithmetic graph paper with slope k_s/μ_m and intercept $1/\mu_m$. A typical plot is shown in Fig. 4.25. Geometrical considerations show that the intercept on the $1/S$ axis is equal to $-1/k_s$.

This type of plot is very useful and because the experimenters carry out the transformation and plotting themselves they can take account of the relative accuracy of the data and the distortions introduced by the

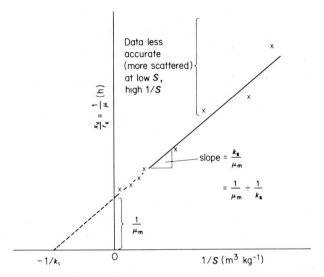

Fig. 4.25 Lineweaver–Burke type plot.

transformations. Thus, for example, low values of S are less accurate in a relative sense than are high values. Consequently the high values of $1/S$ will be relatively inaccurate. This affects the estimate of k_s more than that of μ_m and indicates the experimental necessity of obtaining as high a relative accuracy as possible by replicate measurements, etc., at low substrate concentrations. Two further examples of other transformations of equations from this chapter are given below. Equation 4.16 from Section 4.4.5 is transformed to:

$$\frac{r_p}{x_v} = \alpha \frac{r_x}{x_v} + \beta \qquad (4.79)$$

to obtain values of α and β by plotting r_p/x_v against r_x/x_v. Similarly Equation 4.41 from Section 4.5.2.1 is transformed to

$$\frac{P_o}{x_{vo}} = (\alpha k_d + \beta)\frac{1}{D} + \alpha \qquad (4.80)$$

to obtain any two of α, β or k_d by plotting P_o/x_{vo} against $1/D$. Note how this last example allows for the estimation of two parameters only if the third is known. Thus we can estimate α from the intercept and either β or k_d from the slope given that the other is known. Equation 4.38 in Section 4.5.2 is another example where there are three parameters and thus by analogy with the above case we would expect that we could only estimate two of the parameters given that the third was known. However, this is an example of an equation where one of the parameters (k_d in this case) occurs only in a linear association with one of the variables (D). We can, therefore, use the observation that only for one value of this parameter will we be able to obtain a straight line plot. The transformed equation is:

$$\frac{1}{S_o} = \frac{\mu_m}{k_s} \frac{1}{(D + k_d)} - \frac{1}{k_s} \tag{4.81}$$

i.e. a Lineweaver–Burke type of plot. The procedure is to plot $1/S_o$ against $1/(D + k_d)$ for various values of the parameter k_d. For only one of these k_d values will we get a straight line; all others give a curve. Hence this gives us the value of k_d which best fits the data, and subsequently μ_m and k_s. Figure 4.26 illustrates the technique.

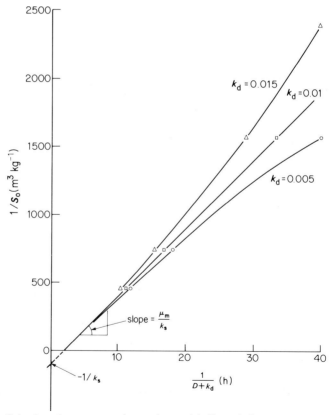

Fig. 4.26   Selection of parameter value to give straight line relation.

Where the number of parameters to be fitted using one equation or set of equations is too large to adopt the straight line fitting technique, then some form of curve fitting must be used. The techniques are beyond the scope of this chapter and appropriate mathematical or computer texts should be consulted. Whatever program is used it is essential that the experimenter should understand the basic principles of operation before using it and should make sure that the values of the parameters generated make sense. As many of the parameters as possible should have been determined singly or in pairs by experiments designed solely to elucidate these values, either by the experimenter himself or by others as published in the literature.

Reading list

This reading list mentions only five books in order to make it of reasonable size and to make it of some use. The fact that a book or periodical is not mentioned does not mean that it lacks merit; on the contrary there is much good work spread throughout the whole of the biotechnological literature. The books mentioned, however, expand in detail most of the material presented in this chapter and form a useful first source of reference.

A classic text which repays study is:

Dean, A. C. R. and Hinshelwood, C. (1966). 'Growth, Function and Regulation in Bacterial Cells.' Oxford University Press: Oxford.

Two basic texts which cover much of the ground in this and other chapters are:

Bailey, J. F. and Ollis, D. F. (1977). 'Biochemical Engineering Fundamentals'. (McGraw-Hill): New York.
Pirt, S. J. (1975). 'Principles of Microbe and Cell Cultivation'. (Blackwells): Oxford.

The formulation of mathematical models for biological systems is treated rigorously in:

Roels, J. A. (1983). 'Energetics and Kinetics in Biotechnology'. (Elsevier): Amsterdam.

This is a good starting point for a detailed review of the relevant literature.

Finally, for those who wish to delve into the mathematical basis of modelling at all levels of complexity, the following comprehensive text is recommended:

Himmelblau, D. M. and Bischoff, K. B. (1968). 'Process Analysis and Simulation'. (John Wiley): New York.

5 | Transport Phenomena and Bioreactor Design

M. MOO-YOUNG and H. W. BLANCH

List of symbols

Roman letters

a	interfacial area per unit liquid volume
C_L	concentration of solute in bulk liquid (generally)
C_L	concentration of solute in bulk media (for intraparticle diffusion case)
C_0	initial concentration of solute
c_P	heat capacity
$C_g{}^*$	saturation (equilibrium) concentration of solute
D	impeller diameter
D_L	liquid phase diffusivity
d	diameter of particle as an equivolume sphere; characteristic length
d_B	bubble diameter
d_o	orifice diameter
E_f	effectiveness factor
G	molar gas flow rate
g	acceleration due to gravity
H	Henry's constants
H_L	liquid height in reactor
h	heat transfer coefficient
K	consistency coefficient of power-law fluids
k	thermal conductivity
k_L	local liquid phase mass transfer coefficient
$k_L a$	volumetric liquid phase mass transfer coefficient
L	scale of primary eddies
l	scale of terminal eddies
m	an exponent

N speed of agitator
N_a oxygen transfer rate
n fluid behaviour index of power-law fluids; an exponent
P agitator power input in ungassed systems
P_g power input in gassed systems
P_i pressure in liquid at sparger
P_o pressure at surface of liquid
p partial pressure
Q volumetric gas flow rate
q specific respiration rate in bulk media
R universal gas constant
T temperature; tank diameter
t time
u_B bubble velocity
u bulk velocity
V volume of reactor contents
V_s superficial gas velocity
VVM   volume of air per unit volume of medium per minute
x concentration of biomass

Greek letters
β coefficient of thermal expansion (volumetric)
$\dot{\gamma}$ shear rate
μ viscosity (dynamic) of continuous phase
μ_a apparent viscosity (dynamic)
μ_D viscosity (dynamic) of dispersed phase
ν kinematic viscosity of continuous phase
ρ density of continuous phase
ρ_D density of dispersed phase
$\Delta\rho$ density difference between the two phases
σ interfacial tension
τ_0 yield stress
τ shear stress
φ hold-up of dispersed phase
Δ difference

Abbreviations for dimensionless groups
Fr   Froude number
Fr_o orifice Froude number
Gr   Grashof number for mass transfer (based on particle–environment density difference)
Gr_H Grashof number for heat transfer

N_A	aeration number
Nu	Nusselt number
Pe	Peclet number (single particles)
Po	power number
Pr	Prandtl number
Re	Reynolds number for particles
Re_I	Reynolds number for impellers
Re_e	Reynolds number for isotropic turbulence
Re_o	orifice Reynolds number (based on liquid properties)
Sh	Sherwood number
Sc	Schmidt number
We	Weber number

5.1 Introduction

5.1.1 Reactor types and transfer implications

A bioreactor is a device in which materials are treated to promote biochemical transformation by the action of living cells or by *in vitro* cellular components such as enzymes. Bioreactors are widely employed in the food and fermentation industries, in waste treatment, and in many biomedical facilities. In industrial fermentations, they are invariably at the heart of the process (see Fig. 5.1). Broadly speaking, there are two types of bio-

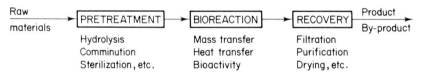

Fig. 5.1 Outline of generalized biotechnological process.

reactors, for fermentation (live-cell) and enzyme (cell-free) transformations. Depending on the process requirements (aerobic, anaerobic, solid state, immobilized), numerous subdivisions of this classification are possible.

In fermentation reactors or fermenters, cell growth is promoted or maintained to allow formation of products, either a metabolite (e.g. antibiotic, alcohol, citric acid), or biomass (e.g. Baker's yeast, SCP), or transformed substrate (e.g. steroids) or purified solvent (e.g. in water reclamation). Systems based on cultures of mammalian or plants cells (see Chapters 19 and 20) are usually referred to as 'tissue cultures' while those based on dispersed non-tissue-forming cultures of microorganisms (bacteria, yeast, fungi) are loosely referred to as 'microbial' reactors (bioreactors, fermenters).

In enzyme reactors, substrate transformation is promoted without the life-support system whole cells provide. Frequently, these reactors employ 'immobilized enzymes' where solid or semi-solid supports are used to entrap (internal) or attach (external) the biocatalyst so that it is not lost, and may be reused.

Virtually all bioreactors of technical importance deal with heterogeneous systems involving at least two, or usually more, phases. To provide optimal conditions for the required biochemical changes, interphase transfers of mass, heat and momentum must occur.

In industrial practice, theoretical explanation frequently lags behind technological realization, and many biochemical process developments are good examples of this apparent paradox. In this chapter, the basic mass transfer concepts, which govern bioreactor performance, are generalized so that the rationale for traditional empiricisms as well as for recent developments and potential innovations can be identified in terms of unifying fundamental principles.

5.1.2   Systems and operating constraints

The application of chemical engineering principles is useful in an analysis of the design and operation of bioreactors. However, classical approaches to the analysis are limited by the following special constraints:

(1) The reactant mixture is relatively complex. Microbial biomass often increases in parallel with the biochemical transformations it carries out—in effect, the catalyst is synthesized as the reactions proceed.

(2) The bulk densities of suspended microbial cells and substrate particles are generally similar to those of their liquid (usually aqueous) environment, and the sizes of the individual microbial cells are relatively small (if compared to chemical catalyst particles). As a result, relative flow between the dispersed and continuous phases is normally low and it is generally difficult to promote high particle velocities and attain turbulent-flow transfer conditions.

(3) Polymeric substrates or metabolites (see Chapter 7) and the mycelial growth of many microorganisms can produce very viscous reaction mixtures which are generally non-Newtonian; again, these conditions tend to limit desirably high flow dynamics in bioreactors.

(4) The growth mode of some microorganisms, especially of mycelial fungi, can lead to the formation of relatively large multicellular aggregates (clumps or pellets) in which intraparticle diffusional resistances

are often serious, leading to partial anaerobiosis, nutrient starvation, or product intoxication.

(5) Bioreactors operate under very mild conditions compared to those frequently employed in chemical processes, but within those conditions they frequently require critically close control of solute concentrations, pH, temperature, and local pressures in order to avoid damage or destruction of live or labile components essential to the process.

(6) Concentrations of reactants and/or products in aqueous media are normally low so that the concentration driving forces for mass transfer are correspondingly limited.

(7) For similar reasons, and because microbial growth rates are substantially lower than chemical reaction rates, relatively large volumes and long residence times are often required.

(8) The requirement that a desired microorganism may flourish in the reactor while all undesired microorganisms are excluded and/or eliminated imposes many special design constraints (see also Chapter 7).

To illustrate some of the problems imposed by the above constraints, consider the supply of oxygen to cells growing in a fermenter. Because of its low solubility, gaseous oxygen, usually in the form of air freed from unwanted microbes, must be supplied continuously to the fermentation medium and dispersed in such a way that the oxygen uptake rate by the system will at least match the oxygen consumption rate by the cells. Even a temporary depletion of dissolved oxygen might mean irreversible cell damage. Experiments generally show that the oxygen uptake rate is independent of the dissolved oxygen concentration over a large concentration range down to a critical value. Below this critical dissolved oxygen concentration the decrease in uptake rate follows a hyperbolic pattern compatible with Michaelis–Menten kinetics, see Fig. 5.2. The rate-controlling step in a fermentation may shift from the oxygen supply rate into the bulk liquid to the demand inside the cell itself, even if quite small cell aggregates are formed. Usually, this is seen as an increase in the *apparent* value of the critical dissolved oxygen concentration, and an increase in the range over which the oxygen supply to the cells is a function of the concentration driving force at the cell surface. Some investigators have found that to obtain a desired fermentation product, maintaining a given dissolved oxygen concentration in the fermenter gave optimal results. However, the total picture is usually far more complex; for example, another effect of the air sparging and mechanical mixing in a fermenter is to remove carbon dioxide or other gaseous metabolic products from the broth; this may or

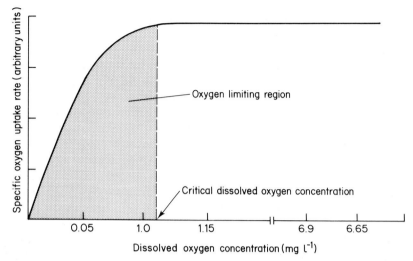

Fig. 5.2   Relationship between oxygen uptake rate and dissolved oxygen concentration.

may not be a desirable effect to promote. Moreover, since the metabolism of cells often changes considerably as a batch fermentation progresses, the optimum oxygen supply criteria will also change.

5.2   Physical pathways in bioreactors

5.2.1   Rate-controlling steps

A generalized physical representation of a bioreactor subsystem involving two or more phases can be set out, as in Fig. 5.3. For example, this representation can be applied to an aerobic fermentation in which a microbe utilizes oxygen (supplied by air bubbles) and other dissolved nutrients (sugars,

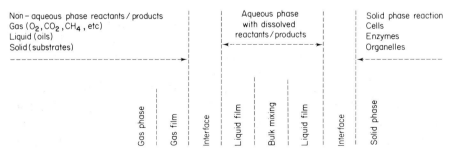

Fig. 5.3   Generalization of biochemical reactor conditions illustrating the heat and mass transfer steps.

etc.), to grow and also produce soluble extracellular metabolites. Eight resistances in the mass transfer pathways, for the nutrient supply and utilization and for metabolite excretion and removal, are possible, and appear, as indicated in the illustration, at the following locations: (1) in the gas film; (2) at the gas–liquid interface; (3) in the liquid film at the gas–liquid interface; (4) in the bulk liquid; (5) in the liquid film surrounding the solid; (6) at the liquid–solid interface; (7) in the solid phase containing the cells; (8) at the sites of the biochemical reactions. It should be noted that all these obstacles are purely physical, except for (8) which may well introduce its own biological complexities. The scheme of Fig. 5.3 can also be used to represent a wide range of other practical situations. The continuous phase may be liquid or gas [the latter in such special cases as 'solid-state' fermentations (e.g. composting, trickle-bed reactors, and 'Koji' processes)] while the dispersed phase may be one or more of the following: solid (e.g. microbial cells, immobilized enzyme particles, solid substrates); liquid (e.g. insoluble or slightly soluble substrates); gas (e.g. air, carbon dioxide; methane).

In most bioreactors an aqueous liquid phase resistance almost invariably controls the overall interparticle mass transfer rate. For example, in Fig. 5.3 one of the following four liquid phase resistances is usually rate-controlling:

(a) Interparticle resistances
(1) A combined liquid phase resistance near and at a gas–liquid interface: this resistance is often rate-controlling in oxygen requiring fermentations because of the relatively low solubility of oxygen in aqueous solutions and through retardation effects of adsorbed materials at the interface.
(2) A liquid phase resistance in the bulk aqueous medium separating the dispersed phases (the significance of this resistance is often minimized by good mixing).

(b) Intraparticle resistances
(3) A liquid phase resistance near the solid–liquid interface: this resistance can be significant because of a low density difference between the continuous aqueous medium and the dispersed phase (e.g. microbes, gel-entrapped enzymes, liquid drops).
(4) A liquid phase resistance inside the dispersed solid phase: this resistance can be significant in cell flocs or pellets, or in immobilized enzyme or insoluble substrate particles.

The complex interaction of these mass transfer steps and the biological reactions is illustrated in Fig. 5.4. This figure also indicates a relationship between aspects of 'genetic engineering' and biochemical engineering in

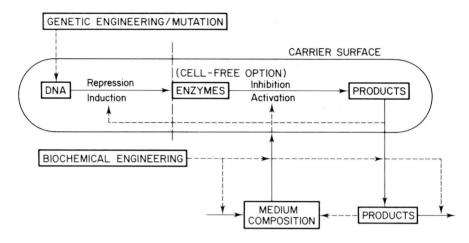

Fig. 5.4 Factors affecting bioreactor performance.

bioreactor design for whole-cell or cell-free systems. The overall and specific aims (biological/biochemical and engineering) of bioreactor design are summarized in Table 5.1, and Table 5.2 lists a range of possible systems and operational variables.

Table 5.1 Aims of bioreactor design

Overall:
To provide the most economical means (capital and operating costs) to optimise bioreactor productivity within the context of an integrated process

Biological/biochemical specifics:
(1) Design or select most appropriate biocatalyst (living cells, non-viable enzyme systems, etc.) for bioreaction(s)
(2) Design or modify medium composition for (1)

Biochemical engineering specifics:
(1) Design transport processes (mass, heat, momentum) to allow intrinsic bioreaction kinetics or to maximize global bioreaction kinetics
(2) Design hardware to carry out (1)

Constraints:
(1) Technical feasibility (e.g. in construction and installation)
(2) Reliability
(3) Biological (sterile operation, materials compatability, etc.)

Table 5.2 Examples of bioreactor systems

Biological/biochemical specifications
(1) Viable vs non-living enzymic systems
(2) Aerobic vs anaerobic systems
(3) Aseptic vs non-aseptic conditions
(4) Mono vs mixed cultures
(5) Single cells vs multicellular aggregates
(6) Single vs multi-substrate media

Physical operational modes
(1) Suspended vs immobilized cells
(2) Dissolved vs immobilized enzymes
(3) Soluble vs insoluble substrates
(4) Monomeric vs polymeric substrates
(5) Batch vs continuous operations
(6) Stirred vs plug flow processing
(7) Single vs multi-stage systems
(8) Single stream vs recycle operations

5.2.2 Definition of mass transfer coefficient

A mass transfer coefficient relates transfer rates to concentration terms and can be defined by a mass balance for a given reactant or product species in the bioreactor. For example, considering the oxygen solute of the air bubbles passing through a fermenter subsystem, the oxygen transfer rate N_a will be given by:

$$N_a = k_L a \, (C_g{}^* - C_L) \qquad (5.1)^*$$

where C_L is the local dissolved oxygen concentration in the bulk liquid at any time t, $C_g{}^*$ is the oxygen concentration in the liquid at the gas–liquid interface at infinite time (equivalent to the saturation concentration) and a is the interfacial area; k_L is the local liquid phase mass transfer coefficient.

Where necessary, methods of evaluating heat transfer rates between the dispersed and continuous phases in bioreactors can be calculated in the same way, as seen later.

In Equation 5.1, the mass transfer rate is dependent on the transfer coefficient, the interfacial area and the concentration driving force. Interfacial area is controlled by factors discussed later, while the concentration driving force is affected by partial pressures (Henry's law). Let us now consider the effects of the operating conditions on the magnitude of the

---

*In strict thermodynamic terms this equation should probably be replaced by $N_a = k_L a R T \ln (C_g{}^*/C_L)$ (Sinclair, 1984), but the linear equation 5.1 has been most generally used and is adequate for many, but not all, applications.

mass transfer coefficient, using an aerobic fermentation system as our generalized example.

5.2.3 Effect of diffusion

Fick's law of diffusion forms the basis for current theories on mass transfer. All postulate the existence of laminar or stagnant fluid films at the phase boundary for interfacial transfer and lead to relationships with molecular diffusion coefficients, D_L.

Values for D_L for binary liquid systems usually fall in the range 0.5 to 2.0×10^{-5} cm^2 s^{-1} for low molecular weight solutes in non-viscous liquids. For oxygen, D_L has the value of 2.10×10^{-5} cm^2 s^{-1} at 25°C in water. In media with high viscosity, as in polysaccharide fermentations or fungal broths, values of D_L are only slightly lower than those for water. For a stagnant medium (as within a multicellular aggregate) or for a stagnant or laminar-flow external film at a particle surface, it can be shown that:

$$k_L \propto D_L \tag{5.2}$$

However, the assumption of a stagnant or laminar-flow film next to the boundary in which the mass transfer resistance is highest is not appropriate in many practical flow conditions which require the application of Fick's law for unsteady-state diffusion. To do this simplifying assumptions must be made, especially with regard to the liquid film behaviour, and these lead to solutions of the form:

$$k_L \propto D_L{}^n \tag{5.3}$$

where the exponent n varies between 0.5 and 1.0 according to the hydrodynamic conditions.

5.2.4 Effect of interfacial phenomena

If we consider a fluid particle (gas or liquid, and for convenience spherical), moving relative to a continuous liquid phase, there are two possible extremes of interfacial movement as classified below:

(a) There is no internal circulation within the particle; particles behave essentially as if they are solid with rigid surfaces. We will refer to these as particles with 'rigid surfaces'.

(b) There is a fully developed internal circulation flow within the particle due to an interfacial velocity. The particle behaves as a part of an inviscid continuous phase with only a density difference. We will refer to these as particles with 'mobile surfaces'.

Illustrations of these concepts in relation to the relative velocity between the particle and the liquid are shown in Fig. 5.5, and they have been useful

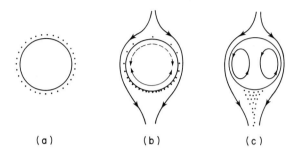

(a)   (b)   (c)

Fig. 5.5 Surfactant effects on bubble/drop surface-flow at (a) zero (b) low and (c) high relative particle velocities.

in explaining many drop and bubble phenomena. For example, trace amounts of surface-active materials can hinder the development of internal circulation by means of their differential surface pressure effect. Small bubbles rising slowly are then apt to behave like particles with rigid surfaces (Fig. 5.5a). Larger bubbles, rising more quickly, may sweep their front surface free of trace impurities and therefore escape the contamination effect of surfactants as illustrated (Fig. 5.5b and c). These effects lead to significant variations of k_L with changing bubble size and agitation power.

In fermentation practice, 'clean' bubble systems are probably rarely achieved and it is conservatively safe to base a design on contaminated rigid interface behaviour as discussed later (Section 5.6).

5.3   Interparticle transfer rates correlations for k_L

5.3.1   Basic parameters

Because of the complex hydrodynamics usually found in the multiphase system, a useful approach to physical transfer problems in a bioreactor is by the use of dimensional analysis. For the relatively simple cases where theoretical analyses from fundamental principles are possible, the solutions can still be conveniently expressed in terms of these dimensionless groups.

For external particle mass transfer, the following dimensionless groups are relevant:

$$\text{Sh (Sherwood number)} = \frac{\text{total mass transfer}}{\text{diffusive mass transfer}} = \frac{k_L d}{D_L} \qquad (5.4)$$

$$\text{Sc (Schmidt number)} = \frac{\text{momentum diffusivity}}{\text{mass diffusivity}} = \frac{\mu}{\rho D_L} \tag{5.5}$$

$$\text{Gr (Grashof number)} = \frac{\text{gravitation forces}}{\text{viscous forces}} = \frac{d^3 \rho g \Delta \rho}{\mu^2} \tag{5.6}$$

$$\text{Re (Reynolds number)} = \frac{\text{inertial forces}}{\text{viscous forces}} = \frac{d u \rho}{\mu} \tag{5.7}$$

$$\text{Pe (Peclet number)} = \text{Re} \times \text{Sc} = \frac{d u}{D_L} \tag{5.8}$$

By analogy, the first three groups for heat transfer are as follows:

$$\text{Nu (Nusselt number)} = \frac{h d}{k} \text{ analogous to Sh} \tag{5.9}$$

$$\text{Pr (Prandtl number)} = \frac{c_p \mu}{k} \text{ analogous to Sc} \tag{5.10}$$

$$\text{Gr}_H \text{ (Grashof number)} = \frac{d^3 g \rho \Delta T}{\mu^2} \beta \text{ analogous to Gr} \tag{5.11}$$

In the following summary of correlations for k_L, different expressions for Sh are given for different flow regimes, which themselves are usually characterized by the Re-number. This can partly be understood by the increasing influence of the momentum boundary layer. Figure 5.6 illustrates the increasing complexity of external flow conditions in which a particle in a bioreactor may find itself as agitation intensity varies.

5.3.2   Particles in stagnant environments

For non-moving submerged particles (with rigid or mobile surface) in a stagnant medium, mass transfer occurs only by radial diffusion. Here, Re $= \text{Gr} = 0$, whence it can be shown that:

$$\text{Sh} = \text{Nu} = 2 \tag{5.12}$$

As the lower limit for Sh, we will see that this value is usually irrelevant for bubble mass transfer, but it may become significant when applied to small light particles like microbial cells. Pseudo-stagnant liquid environments can exist in viscous fermentations or with well dispersed single cells (Fig. 5.6a).

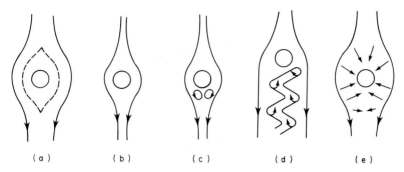

Fig. 5.6 Possible conditions of the momentum boundary layer around a submerged solid sphere with increasing relative velocity: (a) envelope of pseudo-stagnant fluid, (b) streamline flow, (c) flow separation and vortex formation, (d) vortex shedding, (e) localized turbulent eddy formations.

5.3.3   Moving particles with rigid surfaces

A range of these cases can occur in packed-bed, trickle-bed or free-rise or free-fall dispersed-phase reactor systems. For creeping flow, $Re < 1$, (as in certain packed-bed immobilised-enzyme reactors), it can be shown that:

$$Sh = 0.99 \, Re^{1/3} \, Sc^{1/3} = 0.99 \, Pe^{1/3} \qquad (5.13)$$

In the regime $10 < Re < 10^4$ (e.g. certain trickle-bed reactors):

$$Sh = 0.95 \, Re^{1/2} \, Sc^{1/3} \qquad (5.14)$$

For cases in which flow action just balances gravitational forces ('free-suspension'), Re can be expressed in terms of a bulk Gr, and the mass transfer coefficient is then given by the correlation:

$$Sh = 0.31 \, (Sc \, Gr)^{1/3} \qquad (5.15)$$

Thus,

$$k_L = 0.31 \, D_L^{2/3} \left[\frac{\Delta \rho g}{\mu} \right]^{1/3} \qquad (5.16)$$

revealing the strong effects of continuous-phase viscosity (e.g. polysaccharide fermentations) and low particle density (e.g. microbial cells) on mass transfer. Thus, *if oxygen demand at the cell interface is the limiting mass transfer step in a fermentation process, the performance of the fermenter may be outside the control of the operator in terms of aeration and agitation.* In these cases the particle velocity need not be known for design purposes.

At high agitation intensities, turbulence is expected to affect the mass transfer rates at solid particle surfaces (Fig. 5.6c and d). However, in these cases, since the actual particle velocity is unknown, conventional Reynolds numbers cannot be deduced. The concept of local isotropic turbulence may then be applied. According to this concept, large 'primary eddies' emerge, e.g. by impeller action, as waves in a turbulent fluid field. The scale of these primary eddies, L, is of the order of magnitude of the impeller diameter. Primary eddies are non-isotropic and result in a net undirectional velocity. These eddies are unstable and break up into smaller eddies of intermediate size, d, which may or may not be isotropic. Eventually, these 'intermediate eddies' break down into very small 'terminal eddies', of scale l, having completely lost their undirectional nature, and are therefore isotropic. Most of the energy dissipation takes place by these terminal eddies. Such a cascade of energy transfer is illustrated in Fig. 5.7. The scale of such terminal eddies is given as:

$$l = \frac{\mu^{3/4}}{\rho^{1/2}} \left(\frac{P}{V}\right)^{1/4} \qquad (5.17)$$

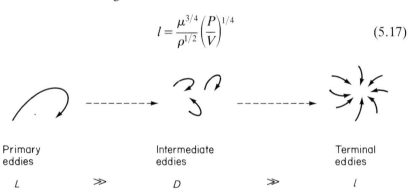

Primary eddies		Intermediate eddies		Terminal eddies
L	$\gg$	D	$\gg$	l

Fig. 5.7 Energy transfer from large primary non-isotropic eddies down to small terminal isotropic eddies according to the concept of local isotropic turbulence.

and for local isotropic turbulence to prevail it has been estimated that the ratio L/l should be $>10^3$. L being approximated by the impeller diameter for mechanically-stirred systems.

An appropriate Re-number expression characteristic of local isotropic turbulence can then be derived using the root-mean-square fluctuating velocity:

$$\mathrm{Re}_e = \frac{d^{3/4}\rho^{2/3}\,(P/V)^{1/3}}{\mu} \qquad (5.18)$$

This can then be used to develop a correlation for rigid-surface particle mass transfer in bioreactors in terms of the energy input to the system, as follows:

$$\text{Sh} = 0.13 \, \text{Re}_\text{e}^{3/4} \, \text{Sc}^{1/3} \qquad (5.19)$$

Combining expressions 5.18 and 5.19 for this case, k_L is seen to be dependent on $(P/V)^{1/4}$, a dependence which may be masked by the effect of power on interfacial area, as discussed later.

5.3.4 Moving particles with mobile surfaces

Mobile surface fluid particles show a behaviour which is less sphere-like than that of rigid-surface fluid particles. By viscous interaction with the continuous phase, oscillating shape variations of liquid drops and gas bubbles occur, and for $\text{Re} > 1$, mobile surface fluid particles in free-rising or falling conditions move in a wobbling or spiral manner, which has a marked influence on mass transfer rates. Overall, $\text{Sh} \propto \text{Sc}^{1/2}$ in the case of mobile interfaces, indicating greater influence of the velocity boundary layer than in the case of rigid interfaces where $\text{Sh} \propto \text{Sc}^{1/3}$ (Equation 5.19). As before, we can arrive at different correlations for different bulk flow regions. For creeping flow, $\text{Re} < 1$:

$$\text{Sh} = 0.65 \, \text{Pe}^{1/2} \qquad (5.20)$$

For higher Re numbers ($\text{Re} > 10\,000$), the equation takes the form:

$$\text{Sh} = 1.13 \, \text{Pe}^{1/2} \qquad (5.21)$$

Refinements between the two extremes are possible. In thick viscous liquids ($\mu > 70$ centipoise) large spherical-cap bubbles are frequently encountered and the relevant correlation is then:

$$\text{Sh} = 1.31 \, \text{Pe}^{1/2} \qquad (5.22)$$

As for small bubbles, large bubbles ($d_\text{B} > 2.5\,\text{mm}$) in non-viscous media appear to be in a state which approximates free suspension where gravitationally induced flow is responsible for the mass transfer; for these cases:

$$\text{Sh} = 0.42 \, \text{Sc}^{1/2} \, \text{Gr}^{1/3} \qquad (5.23)$$

As before, the absolute bubble velocity need not be known to evaluate Sh; note that the exponent of the Sc-number has changed from $1/3$ to $1/2$, consistent with transition from rigid to mobile interface behaviour.

In fluid–fluid dispersions, as discussed later, high agitation intensities which promote local isotropic turbulence will lead to particle disruption rather than increased mass transfer coefficients.

5.3.5 Interacting particles

In a swarm of bubbles where the gas hold-up is high, the relative proximity of the bubbles alters the fluid streamlines around them, and thus affects the mass transfer coefficient. Similar effects may be observed with spherical beads, etc. used as carriers for immobilized enzymes. The effect is that k_L increases with gas hold-up. For both circulating and non-circulating bubbles there is an increase in the Sherwood number at large Peclet numbers with increasing gas hold-up. Generally, Sherwood numbers for strong internal circulation are always higher than for no internal circulation.

5.3.6   Non-Newtonian flow effects

When the liquid phase exhibits non-Newtonian behaviour, the mass transfer coefficient will change due to alterations in the fluid velocity profile around the submerged particles. The trends for both mass transfer and drag coefficient are analogous. As for Newtonian fluids, two types of interfacial behaviour need to be considered, depending upon the mobility or rigidity of the particle surface.

5.3.7   Effect of bulk mixing patterns

In addition to the determination of the mass transfer coefficient, k_L, and the interfacial area, a, the development of gas and liquid phase mass balance equations for the species transferred depends on the flow behaviour of both gas and liquid phase.

In low viscosity liquids it is reasonably well established that in small stirred tanks the liquid phase can be considered to be 'perfectly mixed'. Under these conditions, the gas phase has also generally been assumed to be well mixed in tanks operating above critical impeller speed. In large tanks, however, the situation is less clear, and care must be taken to establish the behaviour of both phases. In cases where the degree of gas absorption is high, the contrasting assumptions of well mixed or plug-flow of the gas phase may predict significantly different gas absorption rates.

5.3.8   Measurements of $k_L a$

The volumetric liquid phase mass transfer coefficient, $k_L a$, can be determined experimentally by the methods illustrated below.

Static gassing out
The oxygen content of the liquid phase in a non-respiring system is reduced to zero by gassing out with oxygen-free nitrogen. The system is then aerated

and the value of the dissolved oxygen concentration, C_L, is measured as a function of time. The equation describing this is:

$$\frac{dC_L}{dt} = k_L a (C_g^* - C_L) \qquad (5.24)$$

which on integration, using the initial condition $C_L = 0$ at time t, becomes

$$\ln\left(1 - \frac{C_L}{C_g^*}\right) = -k_L a t \qquad (5.25)$$

The slope of the straight line obtained from a plot of $\ln[1 - (C_L/C_g^*)]$ against t is numerically equal to $k_L a$. It must be noted that this method positively requires a rapid response by the probe used for measuring the dissolved oxygen concentrations, and that a non-respiring system is employed to simulate the fermentation.

Dynamic gassing out
In this method, illustrated in Fig. 5.8, the air supplied to a respiring culture

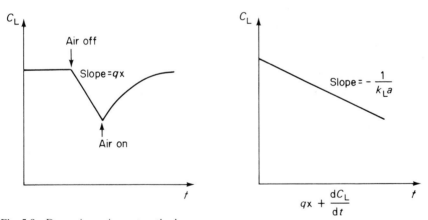

Fig. 5.8   Dynamic gassing out method.

is stopped and the resulting fall in dissolved oxygen is noted. Resuming the air supply (before the dissolved oxygen falls below the critical value), the increasing dissolved oxygen is recorded as a function of time. The changing dissolved oxygen concentration is given by:

$$\frac{dC_L}{dt} = k_L a (C_g^* - C_L) - qx \qquad (5.26)$$

which on rearranging becomes:

$$C_L = -\frac{1}{k_L a}\left(qx + \frac{dC_L}{dt}\right) + C_g^* \qquad (5.27)$$

The value of $k_L a$ is obtained from a plot of C_L against $[qx + (dC_L/dt)]$. This method determines $k_L a$ in the actual fermentation system and has been used extensively. However, it assumes that the culture is rapidly degassed when the air is switched off and this is not always the case, particularly for filamentous fermentation broth. It again relies heavily on the probe characteristics.

Oxygen balance method
This method employs the definition of the oxygen transfer coefficients, given in Equation 5.1:

$$N_a = k_L a (C_g^* - C_L)$$

The rate of oxygen transfer into the system is obtained by measuring the difference between the amount of oxygen in the streams to and from the fermenter and the respective flow rates.

In small fermenters, the gas behaves like the liquid in a well mixed vessel, hence a single point determination of C_L will be sufficient. It may be necessary to read the dissolved oxygen concentration at several points in large fermenters, however, as the gas flow through the system approximates plug flow. The saturation concentration is determined using Henry's law which gives the relationship of the equilibrium between dissolved oxygen and the partial pressure of oxygen in the gas:

$$p = H C_g^* \qquad (5.28)$$

In large fermenters, the partial pressure of oxygen in the gas will fall as it passes through the vessel. Thus a logarithmic mean oxygen concentration of the inlet and outlet gas streams should be used as given below:

$$p_{lm} = \frac{p_{in} - p_{out}}{\ln(p_{in}/p_{out})} \qquad (5.29)$$

The saturation concentration is then given by:

$$C_g^* = \frac{p_{lm}}{H} \qquad (5.30)$$

This is the most reliable method of measuring $k_L a$. It uses an actual fermentation system and is not dependent on the quality of the dissolved oxygen probe as are the methods described above. However, this method requires gaseous O_2 analysers and accurate measurements of temperature and pressure.

Sulphite oxidation method

This method is based on the oxidation of sodium sulphite to sulphate in the presence of a catalyst according to:

$$Na_2SO_3 + \tfrac{1}{2}O_2 \xrightarrow{Cu^{2+} \text{ (or } Co^{2+})} Na_2SO_4$$

The reaction rate is, within limits, independent of sodium sulphite or dissolved oxygen concentration. It is determined by filling the vessel with $1N$ Na_2SO_3, adding catalyst ($10^{-3}M$ Cu^{2+} or Co^{2+}) and commencing aeration, taking samples of the liquid at regular time intervals and determining the amount of unreacted sulphite. The rate of reaction is very fast; all the oxygen transferred from the gas will be consumed and $k_L a$ can be determined from:

$$\text{rate of sulphite oxidation} = k_L a \times C_g^* \qquad (5.31)$$

The sulphite oxidation method has been shown to depend on the equipment, pH, purity of Na_2SO_3 and the catalyst. The method employs a sodium sulphite solution to simulate a fermentation broth, which can lead to large errors in $k_L a$ estimations, and is, therefore, not suited for estimation of the oxygen transfer coefficients of individual fermenters. However, it can be helpful in scale-up studies, comparing the oxygen transfer characteristics in similar but different size vessels.

5.4 Intraparticle bioreaction rates

5.4.1 General concepts

In some biochemical systems the limiting mass transfer step is found not at a gas–liquid or solid–liquid interface (as discussed in Section 5.3), but in the interior of solid particles. Cases of increasing interest and importance include solid substrates and cell aggregates (microbial flocs, cellular tissues, etc.) and immobilized enzymes (gel-entrapped or supported in solid matrices). In the former, diffusion of oxygen (or other nutrients) through the particle limits metabolic rates while in the latter, substrate reactant or product diffusion into or out of the enzyme carrier often limits the overall bioreaction rates.

Generalized mass balances for these situations can be set up for specific cases where mass transfer resistances within the particle may be identified as an important factor. In the absence of physical diffusion effects, the 'intrinsic' biological reaction kinetics will prevail as defined by such expressions as the Monod and Michaelis–Menten equations (see Chapters 3 and 4).

5.4.2 Oxygen transfer in fungal pellets

Mycelial fungi growing in stirred tanks frequently modify their morphology so that the actual 'particle' may vary between a very open entangled colony and a very dense pellet. The practical importance is that at one extreme the 'open' colony confers greater homogeneity in the physiology of the biomass, but at the cost of high bulk viscosities, whereas the 'pellet' habit permits high biomass concentrations at relatively lower viscosity, but at the cost of considerable physiological differentiation between different zones of the pellet. For example, pellet suspensions do not grow exponentially, and this is ascribed both to intraparticle diffusion limitations (typically affecting oxygen transfer) and to the physical constraint that growth can only occur freely on the pellet surface. The effect of intraparticle diffusion on the reaction rate in microbial pellets or catalysts can be described by using an effectiveness factor:

$$E_f = \frac{\text{observed rate of substrate uptake}}{\text{uptake rate at pellet surface}}$$

The effectiveness factor depends on the properties of the organism or material and the order of reaction (normally between zero and one) for the substrate uptake. A typical solution for a material balance over a fungal pellet using the effectiveness factor is given in Fig. 5.9. Oxygen limitation

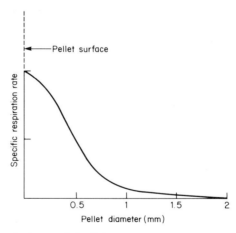

Fig. 5.9 Oxygen transfer in mycelial pellets.

in the pellet interior is often encountered in addition to the problems highlighted above, and this can also lead to pellet autolysis. An effective way of minimizing intraparticle diffusion limitations is to keep the pellet small.

In practice this may be achieved by judicious use of shear forces and/or surfactants in the fermentation broth.

5.4.3 Immobilized enzyme systems

In these systems the subtrates must diffuse into the support matrix to reach the enzyme. For fast enzymic reactions diffusional limitations will lead to concentration profiles in the immobilized particle as illustrated in the fungal pellet above (Fig. 5.9). A concentration gradient in the opposite direction will be established for reaction products suffering similar low diffusion rates. This can lead to product inhibition; the build-up of products near the reaction site will depress the rate of reaction. Conversely, if the rate of reaction rather than diffusion is limiting, substrate inhibition may be experienced. Thus, intraparticle diffusion can have a significant effect on the kinetic behaviour of enzymes immobilized on solid carriers or entrapped in gels.

5.4.4 Enzymic degradation of insoluble substrates

When the substrate in a bioreactor is insoluble (e.g. cellulose, oils), the effects of intraparticle mass transfer may again be important. In such systems, extracellular enzymes break down the substrate into water-soluble components which appear either as end-products or as intermediates for consumption by microorganisms. If a solid substrate is sufficiently porous, the enzyme can diffuse into it and degradation can proceed inside the material. The water-soluble substrate fragments must also diffuse out of the solid matrix through these same pores into the bulk solution. Much, therefore, depends upon pore sizes—typically, in the sub-micron range—relative to particle size; the reaction may concurrently proceed on the exterior surface of the particle, and for substrates of low porosity this is where much of the degradation takes place. This is also true for insoluble liquid particles, like oils, into which the enzymes cannot diffuse.

5.5 Physical properties of bioreactor media

5.5.1 Rheological properties

The rheological properties of the materials being processed in bioreactors will influence the power consumption and hence, the heat and mass transfer rates. These properties are particularly important in some antibiotic and similar fermentations involving filamentous organisms and semi-solid media because of the very viscous and frequently non-Newtonian behaviour of the materials.

Three classes of fluids have been encountered in bioreactors: Newtonian, non-Newtonian and (rarely) viscoelastic. The rheological characteristics of fluids can be described by the following general equation:

$$\tau = \tau_0 + K(\dot{\gamma})^n \tag{5.32}$$

(a) *Newtonian fluids* exhibit no yield stress, so $n = 1$, $\tau_0 = 0$, and K becomes the dynamic viscosity μ. Thus

$$\tau = \mu\dot{\gamma} \tag{5.32a}$$

Newtonian fluids with the viscosity being constant and not affected by shear rate will have constant rheological properties. This will apply to most bacterial and yeast fermentation fluids.

There are three types of time-dependent non-Newtonian fluids.

(b) *pseudoplastics* occur most often and these fluids follow a power law model

$$\tau = K(\dot{\gamma})^n \tag{5.32b}$$

where K and n change during the course of the fermentation. With no linear relationship between shear rate and shear stress, as for Newtonian fluids, the viscosity of pseudoplastics will not be constant and we can define an apparent viscosity by the equation:

$$\mu_a = \frac{\tau}{\dot{\gamma}} = K\dot{\gamma}^{n-1} \tag{5.33}$$

in which the viscosity decreases with increasing shear rate (i.e. $n < 1$).
(c) *dilatants* will also obey the power law model. In this case $n > 1$.
(d) *viscoplastic*, or, *Bingham plastic fluids* are recognized by their relatively high yield stress which must be exceeded to make the fluid flow. The fluid will then exhibit inherently Newtonian behaviour.
(e) *Casson behaviour* is exhibited by non-Newtonian fluids which demonstrate an apparent yield stress followed by pseudoplastic behaviour.

Filamentous fermentation broths and complex media with a high solids concentration will invariably exhibit non-Newtonian behaviour. The rheological characteristics of different fluids are demonstrated on shear stress–shear rate diagrams, as shown in Fig. 5.10.

5.5.2 Bubble dispersion

For oxygen transfer into microbial cultures, it is clear that since the maximum value of the concentration driving force for mass transfer is limited

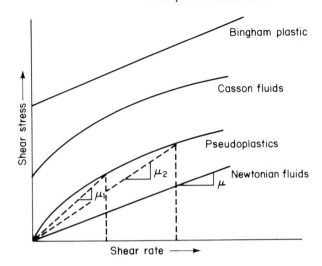

Fig. 5.10 Flow behaviour of biological fluids. μ_1 and μ_2 are the apparent viscosities of the pseudo-plastic fluid determined at different shear rates.

(due to the low solubility of oxygen), the oxygen transfer rate from air bubbles to the medium is largely determined by k_L (discussed in Section 5.3) and the interfacial area a, both of which are dependent on bubble diameter. The main variables which influence a are bubble diameter, d_B, the terminal velocity of the bubble, u_B, and the hold-up, φ.

Dispersion of bubbles in fermentation broths is greatly influenced by shear (e.g. by mechanical agitation) which is used to create high interfacial areas and also good mixing. In strongly agitated fermenters used for aerobic processes, the eventual bubble size is mainly determined by the bulk level of turbulence in the continuous phase, rather than by the primary action of the sparger system. Equally at low turbulence the small bubbles which are formed from tiny orifices, such as sintered glass, may coalesce to form large bubbles in the medium if there is inadequate mixing and/or ineffective surfactants present. Sparger design is then less critical than the prevailing bulk regime.

In viscous liquids, large free-rising bubbles will coalesce with each other if their surfaces approach to within a distance of about one bubble diameter. Thus, in such broths rapid coalescence may take place. This coalescence mechanism is self-accelerating and leads to formation of fast-rising spherical-cap bubbles, bringing about deterioration of the gas dispersion.

Typical patterns of rise velocities for bubbles in aqueous and viscous liquids are shown in Fig. 5.11. It should be noted that the aqueous solutions

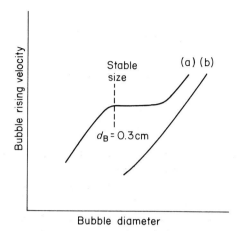

Fig. 5.11   Bubble rising velocity at various bubble sizes for (a) non-viscous and (b) viscous liquids. The plateau region on curve (a) stabilizes bubble size. Absence of a plateau on curve (b) results in continual coalescence as the bubble rises through the fluid.

show a unique plateau region in rise velocity which partly explains bubble size stability in these systems.

5.5.3   Gas flow effects on bubble swarms

There is a fairly extensive literature available on the behaviour of gas-sparged towers for height-to-diameter ratios ranging from unity to twenty or more. Due to the density difference between the gas and liquid phases, the rate of mass transfer is primarily determined by the force of gravity. Two regions in any gas-sparged reactor need to be considered: an orifice region and a bulk liquid region, where break-up and coalescence may occur.

For bubbles near sparger orifices in liquids with viscosities similar to that of water, it has been shown that the orifice diameter (d_0) influences the bubble size strongly at very low gas flow rates. In such cases the bubble size results from the balance between surface tension and buoyancy forces and we obtain

$$d_B = \left(\frac{6\sigma d_0}{g\Delta\rho}\right)^{1/3} \tag{5.34}$$

Above the gas flow rates for which Equation 5.34 is valid, and over a wide range of apparent viscosities, the bubble size is given by

$$\frac{d_B}{d_0} = 3.23 \, \text{Re}_0^{-0.1} \, \text{Fr}_0^{0.21} \tag{5.35}$$

where Re_0 and Fr_0 are based on liquid properties:

$$\text{Re}_0 = \frac{4\rho Q}{\pi \mu d_0}, \qquad \text{Fr}_0 = \frac{Q^2}{d_0^5 g}$$

In the region away from the orifice the bubble size will also be influenced by the liquid motion induced by the rising gas stream. If the power input from the gas phase is insufficient to generate turbulence in the liquid phase, the bubble size in the tank will be basically that of bubbles formed at the orifice (but may increase with liquid height in the reactor due to bubble coalescence, particularly in viscous solutions). Once the liquid is in turbulent motion, however, an equilibrium between coalescence and breakdown will determine the bubble size. Little information is available on coalescence of bubbles in highly agitated liquids, where fluid may separate the colliding bubbles before coalescence can occur. In electrolyte solutions, coalescence is greatly reduced and higher gas hold-ups have been reported. Bubble break-up is caused by the shear forces exerted on the bubble by the turbulent field. The ratio of this shear to surface tension forces is given by the Weber number:

$$\text{We} = \frac{\tau d_B}{\sigma} \tag{5.36}$$

The shear forces can be found from isotropic turbulence theory

$$\tau \propto \rho \left(\frac{P}{V} \frac{d_B}{\rho} \right)^{2/3} \tag{5.37}$$

so that at dynamic equilibrium, the Weber number is constant and a maximum stable bubble size can be predicted as

$$d_B \propto \frac{\sigma^{0.6}}{(P/V)^{0.4} \rho^{0.2}} \tag{5.38}$$

which relates to the interfacial area and hold-up:

$$a = \frac{6\varphi}{d_B}(1 - \phi) \tag{5.39}$$

5.6 Bioreactor performance

5.6.1 Bubble columns

Because of the very different bulk flow patterns that are induced, pneumatically agitated gas–liquid reactors may show wide variations in performance with the height-to-diameter ratio. In the production of baker's yeast, a tank-type configuration with a ratio of 3 to 1 is common industrially, whereas tower-type systems have height to diameter ratios of 6 to 1 or more (Fig. 5.12). As would be expected, the behaviour of both gas and liquid phases

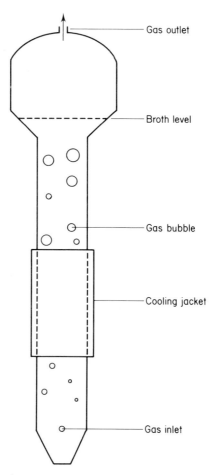

Fig. 5.12 Tower bioreactor.

may be quite different in these cases. In general, the gas phase rises through the liquid phase in plug-flow, under the action of gravity, in both types of system.

A variety of correlations for $k_L a$ are available for bubble columns. Generally they are of the form

$$k_L a = \text{constant} \times V_s^n \qquad (5.40)$$

where n is usually in the range of 0.9–1.0 in the bubble flow regime.

Many of the available correlations for k_L have been obtained on small-scale equipment, and have not taken cognizance of the underlying liquid hydrodynamics. As described earlier, provided the liquid is turbulent, the equilibrium bubble size in the bulk will be independent of the size at formation. The height of the region around the orifice where the bubble formation process occurs is a function of sparger geometry and gas flow rate and in laboratory scale equipment this height may be a significant fraction of the total liquid height (up to 30%), whereas in plant scale equipment this generally represents less than 5% of the total. Thus, correlations developed on small-scale apparatus need to be reviewed in light of the varying interfacial area with column height.

5.6.2   Systems with stationary internals

Several reactors which include internal elements to enhance mass transfer rates have appeared. These include draught tubes, multiple sieve plates staged along the length of the column, and static mixing elements (see Fig. 5.13).

A considerable literature exists on draught tube columns, where liquid is circulated due to a bulk density difference between the inner core and the surrounding annular space. The downcoming liquid in the annular space entraps air bubbles, and thus hold-up in the central core and annular region will be different. Several reports on small-scale air-lift columns of varied design detail have appeared, while industrial-scale air-lift devices have been used mainly for SCP production and waste treatment.

The overall mass transfer correlations for air-lift devices are usually expressed as:

$$k_L a = \text{constant} \times V_s^n \qquad (5.41)$$

Static mixing elements have been incorporated into air-lift devices in order to provide additional mixing and hence greater mass transfer capabilities. Static mixers are becoming increasingly more common in oxidation ponds for biological waste water treatment. Here, fine bubbles may be produced as the gas–liquid mixture passes through the mixing elements. These are

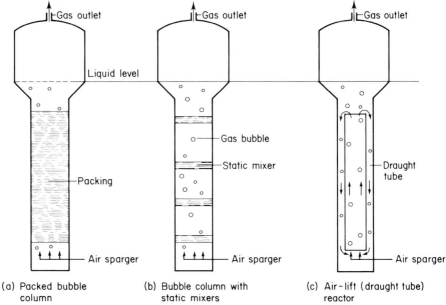

(a) Packed bubble   (b) Bubble column with   (c) Air-lift (draught tube)
    column   static mixers   reactor

Fig. 5.13   Bioreactors with stationary internals.

usually 45–60 cm in diameter, and are placed over sparger pipes. A fairly intense liquid circulation can be developed by such mixers due to entrainment by the gas–liquid jet rising from the mixing element.

In the ICI pressure cycle fermenter (see Chapter 10) several effects are combined. Overall circulation is on the air-lift principle, but mass transfer is considerably assisted by fixed horizontal perforated plates throughout the riser section. In addition, oxygen transfer into the bulk liquid is greatly assisted by the high hydrostatic pressure at the base of the very tall reactor, while gas desorption occurs in the lower pressure region at the top. The system is extremely effective from the oxygen transfer viewpoint but the element of plug-flow as broth passes round the reactor means that substrate additions can cause problems; since the design substrate (methanol) has adverse effects at very low excess concentrations, it was necessary to equip the reactor with a very large number of separate substrate addition injector points.

Tubular systems are still in the experimental stage as regards design details and a considerable variety of designs have been introduced with some potential advantages over conventional stirred tanks. They are essentially simple devices, in which the flow patterns of liquid and gas phases are well characterized. In a pipeline reactor, material can be transported while reaction takes place. This concept has been exploited in sewage treatment processes with multiple aeration points in a pipeline. Tubular systems may have potential use in algal cultivation, having a large surface area to volume ratio for maximum exposure to light. Dead spaces are minimized in tubular systems; on the other hand for most aerobic applications the problem of ensuring a sufficient oxygen supply on a large scale is essentially unsolved.

5.6.3 Stirred-tank reactors

Non-viscous systems
To obtain better gas–liquid contact, mechanical agitation is often required. The supply side of the overall mass transfer of oxygen from the air bubbles to the cells (and not the cell demand side) is considered in this section. The discussion is confined to non-viscous aqueous media in fully baffled sparged stirred tanks with submerged impellers (as shown in Fig. 5.14). The design details of a stirred-tank fermentor are given in Chapter 16.

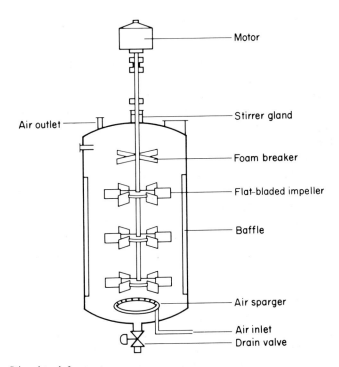

Fig. 5.14 Stirred-tank fermenter.

In Section 5.3, basic correlations for the individual mass transfer coefficient k_L were described for design purposes, but the overall mass transfer coefficient, $k_L a$, is ultimately required. Evaluation of the interfacial area, a, is therefore necessary. For particulates such as cells, insoluble substrates, or immobilized enzymes, the interfacial area can be determined from direct analyses, e.g. by microscopic examination. For gas bubbles and liquid drops, a can be evaluated from semiempirical correlations. Two cases are identified:

(a) for 'coalescing' air–water dispersions,

$$a = 0.55 \left(\frac{P_g}{V}\right)^{0.4} V^{0.5} \tag{5.42}$$

and

$$d_B = 0.27 \left(\frac{P_g}{V}\right)^{-0.17} V_s^{0.27} + (9 \times 10^{-4}) \tag{5.43}$$

(b) For 'non-coalescing' air–electrolyte solution dispersions,

$$a = 0.15 \left(\frac{P_g}{V}\right)^{0.7} V_s^{0.3} \tag{5.44}$$

and

$$d_B = 0.89 \left(\frac{P_g}{V}\right)^{-0.17} V_s^{0.17} \tag{5.45}$$

It is seen that the effect of electrolytes is significant. Electrolytes and surfactants inhibit bubble coalescence, resulting in the persistence of smaller bubbles with increased interfacial area compared to that in the clean water systems. A weakness common to both expressions is that neither accounts for any effects of cells in the aqueous phase. In addition, the above equations are only applicable if the impeller is not flooded by too high a gas flow rate and there are no gross surface aeration effects.

Several investigators have developed overall correlations. They usually take the form

$$k_L a = \left(\frac{P_g}{V}\right)^m V_s^n \tag{5.46}$$

and their ranges of applicability for bubble columns, air-lift systems, and stirred tanks are shown in comparable form in Fig. 5.15.

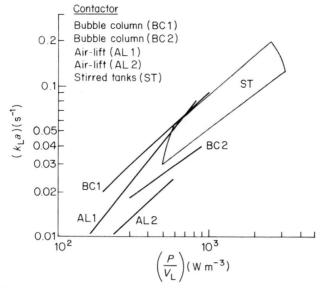

Fig. 5.15 Oxygen transfer coefficients (cell-free systems) for various bioreactors.

Overall correlations for $k_L a$ may be derived by combining the individual correlations for k_L and a developed previously. The equations for geometrically similar systems are:

(a) For 'coalescing' clean air–water dispersions

$$k_L a = 0.025 \left(\frac{P_g}{V}\right)^{0.4} V_s^{0.5} \tag{5.47}$$

(b) For 'non-coalescing' air–electrolyte solution dispersions

$$k_L a = 0.0018 \left(\frac{P_g}{V}\right)^{0.7} V_s^{0.3} \tag{5.48}$$

It should also be noted that the value of the exponents in Equations 5.47 and 5.48 depends on the size of the reactor (see section on scale-up) and to some extent on the method used for their evaluation.

The above equations suggest that the effect on the overall $k_L a$ of contaminants, such as electrolytes, is not constant for all aeration–agitation conditions.

From Fig. 5.15 it is seen that for a given power input, the magnitude of $k_L a$ obtained is *about* the same whether mixing is done mechanically in stirred tanks or pneumatically in bubble columns or air-lift devices. However, mechanically agitated systems offer the most commonly used means of providing the power inputs needed to secure really high $k_L a$ values such as are required for some antibiotic fermentations and the activated sludge method of treating waste water.

In large installations the power demand in stirred-tank fermenters can be excessive. An alternative is shown in Fig. 5.16. This fermenter combines

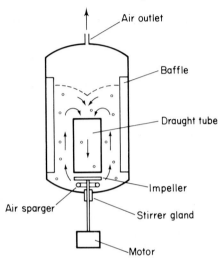

Fig. 5.16   Stirred-loop bioreactor.

the advantages offered by the air-lift and stirred-tank fermenter, employing an impeller between the air sparger and inner draught tube. The main function of the impeller is to break up gas bubbles, promoting a high interfacial area while the bulk mixing in the reactor is still promoted by the air-lift principle. Such fermenters may offer a solution to the problem of high oxygen transfer rates on large-scale installations.

As discussed later, scale-up procedures based on $k_L a$ values are often propounded. However, none of the overall correlations for $k_L a$ has universal applicability. Moreover, any scale-up procedure based on equalizing $k_L a$ at both scales according to a given correlation will usually lead to violations of other physical criteria which (depending on biological demands and tolerances) may be more important. For example, increasing the $k_L a$ by higher power input can sometimes result in damage to the organisms in a highly turbulent fermentation broth.

Viscous systems
Two types of viscous fermentations need to be distinguished: (a) mycelial fermentations (e.g. molds, actinomycetes) where the viscosity is due to

the microbial network structure dispersed in continuous aqueous phase; and (b) polysaccharide fermentations where the viscosity is due to polymers in the continuous aqueous phase resulting in an essentially homogeneous, viscous liquid.

The first type of fermentation broth can to some extent be simulated by material such as paper pulp, which has a macroscopic structure analogous to fungal hyphae, suspended in water. The second type may be simulated by aqueous polymer solutions of known properties. The simulations are not precise, and neither are the $k_L a$ correlations that have been developed from their observations.

An often quoted $k_L a$ correlation for non-Newtonian fluids is given in Equation 5.49:

$$k_L a = \text{constant} \times \left(\frac{P_g}{V}\right)^{0.33} V_s^{0.56} \qquad (5.49)$$

It has been demonstrated that this equation was valid for a non-Newtonian filamentous fermentation. By comparing equations (5.48 and 5.49) it is noted that the oxygen transfer coefficients of non-Newtonian fluids are less sensitive to increases in power input than those of Newtonian fluids. In general, therefore, more power input is required into the former fluids to attain the same oxygen transfer rates as in the latter.

One criticism of the two general equations (5.46 and 5.49) is that no account is taken of the properties of the fluid. The rheology of non-Newtonian fluids is time dependent and it would be desirable to include rheological properties into the correlation for $k_L a$. There have been attempts to develop more complete relationships between oxygen transfer coefficient and system parameter, using dimensional analysis and model solutions. However, the complex correlations which have emerged have found few applications in fermentation systems.

5.7 Power requirements

5.7.1 General concepts

Mixing is used to promote or enhance mass and heat transfer rates in a bioreactor. The productivity of most fermentation processes is limited by the aeration capacity of the fermenter. Power is required, therefore, for both of the mixing functions, which are to ensure good bulk mixing of the reactor content and to achieve high mass transfer rates. When the mixing is induced pneumatically or by pumping it is fairly easy to evaluate the power consumption from pressure-drop considerations. For mechanically-induced mixing, power consumption is more difficult to evaluate from operating variables. Since 100–$400 \, \mathrm{kW \, m^{-1}}$ (0.5–$2.0 \, \mathrm{HP}/100 \, \mathrm{USG}$) is normally required, and the usual range of capacities of industrial fermenters is 50–$200 \, \mathrm{m^3}$ power requirement is often 40–$600 \, \mathrm{kW}$ (50–$800 \, \mathrm{HP}$) per unit which is an important consideration in the process economics.

5.7.2 Air-lift systems

In these systems, gas sparging must supply all the energy for the required bulk mixing and mass transfer. The power input from sparging air through a reactor can be quite substantial in large bioreactors. Assuming the kinetic energy change of the gas (due to velocity differences between inlet pipeline and reactor) can be ignored and isothermal expansion of the gas is the source of all power, the power input can be estimated as:

$$P_g = GRT \ln \frac{P_i}{P_0} \qquad (5.50)$$

5.7.3 Agitated ungassed systems

Standard impeller configurations are given in Fig. 5.17. In general, the agitator shaft is positioned centrally in an upright vertical cylindrical tank fitted with baffles. The flat-bladed turbine (Rushton turbine) is used exten-

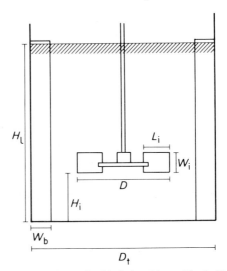

Fig. 5.17   Standard geometries for a flat-bladed turbine with six blades and four baffles. The design parameters are as follows: $D_t/D = 3$, $H_l/D = 3$, $H_i/H_l = 1/3$, $L_i/D = 1/4$, $W_i/D = 1/5$, $W_b/D_t = 1/10$.

sively and it provides adequate mixing for most bioreactors, and most published results are based on this configuration. Applying dimension analysis, it can be shown for Newtonian liquids that

$$\frac{P}{\rho N^3 D^5} = f\left(\frac{D^2 N \rho}{\mu}, \frac{D N^2}{g}, \text{geometric factors}\right) \tag{5.51}$$

The first term involves the inertia forces and is called the Power number, Po, the second term is Re, the Reynolds number, the third term is the Froude number, Fr, which takes into account gravity forces.

Although the complete functional relationship in Equation 5.51 is complex and can only be represented graphically, simple analytical expressions can be derived for certain ranges:

(a) In the turbulent flow regime,

$$\frac{P}{\rho N^3 D^5} = \text{Po} = \text{constant}$$

thus

$$P \propto \rho N^3 D^5 \tag{5.52a}$$

Thus, P is strongly dependent on impeller diameter but is independent of liquid viscosity.

(b) In the laminar flow regime,

$$\text{Po} \propto \frac{1}{\text{Re}_I}$$

thus

$$P \propto \mu N^2 D^3 \tag{5.52b}$$

Here, P is proportional to viscosity but independent of density.

Prediction of power consumption in non-Newtonian fluids can be more complex. The time-dependent rheological properties makes it difficult to estimate the Reynolds number. For power law fluids the average shear rate around the impeller is related to its speed and an apparent viscosity can be defined as shown in Section 5.6. A generalized impeller Reynolds number for power-law fluids can thus be derived:

$$\text{Re}_I = \frac{D^2 N \rho}{\mu_a} \tag{5.53}$$

This procedure allows the use of conventional power curve formerly developed for Newtonian fluids.

5.7.4   Gassed systems

The power required to agitate gassed liquid systems is *less* than for ungassed liquids since the apparent density and viscosity of the liquid phase decrease upon gassing. For Newtonian liquids this decrease may be as much as two-thirds of the ungassed power. The reduction in gassed power, P_g/P, is generally given as a function of the ratio of the superficial gas velocity to the impeller tip speed (the aeration number, N_A):

$$P_g/P = f(N_A) \tag{5.45}$$

where

$$N_A = \frac{Q}{ND^3}$$

The precision of this equation is poor and the following correlation

$$P_g = \text{constant} \times \left(\frac{P^2 \mu D^3}{Q^{0.56}} \right)^{0.45} \tag{5.55}$$

has been applied to both Newtonian and non-Newtonian systems. It should be noted that the equation is not dimensionally sound and predicts unrealistic results for very small gas flow rates.

A problem in the experimental determination of the effect of pseudo-plasticity on gassed power consumption results when the size of the bubbles formed at the sparger is comparable to that of the impeller blade. The impeller may then spin within a 'gas doughnut' leaving the bulk of the fluid unmixed. This can often occur in laboratory scale equipment, but not on an industrial scale. Another problem may be vibration caused when large bubbles are shed from the impeller.

5.8   Scale-up

5.8.1   Scale-up parameters

When the optimum process conditions are found at the laboratory scale there is a need to translate these findings for use in large bioreactors.

The scale-up methods which have been most often proposed are as follows:

(1) scale-up based on fixed power input.
(2) scale-up based on fixed mixing time.
(3) scale-up based on fixed oxygen transfer coefficient.
(4) scale-up based on fixed environment (e.g. dissolved oxygen).

(5) scale-up based on fixed impeller tip speed.

The list is not necessarily exclusive. The principle involved in using one of the parameters listed above in scaling up bioconversion processes is illustrated in Fig. 5.18. Assume it has been established that for a certain

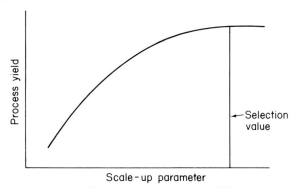

Fig. 5.18   Principle of scale-up and the selection of the value of the scale-up parameter.

value of (say) the mixing time in a bioreactor the production of (say) enzyme is high and not affected by small alterations in the value of the mixing time. The process can then be scaled up, using this particular value of the mixing time as a scale-up parameter. This assumes that everything else is constant or of little importance for process yield, which puts undue demand on the need for geometric similarities between bioreactors of different scale.

5.8.2   Scale-up at fixed $k_L a$

We have already implied that if physical mass transfer rates drop below certain values, growth is hampered or destroyed. Thus, $k_L a$ is frequently used as a basis for scaling-up, especially in aerobic systems. Table 5.3 gives an example of scale-up based on constant $k_L a$ in both reactors. From this table, it is seen that it may not be possible to maintain equal volumetric gas flow rates (vvm) since the linear gas velocity (V_s) through the vessel will increase differently with the scale and in fact would be impractical if the upper limit of liquid blow-out action is reached. However, it may be possible to reduce the volume of gas per volume of liquid per minute (vvm) on scale-up while increasing power input by changing the reactor geometry and/or power input per unit volume (P/V) as shown in the table. $k_L a$ will remain constant in both cases.

It appears that $k_L a$ is often, but not always, a reasonable design approach. An increase in $k_L a$ can sometimes have an adverse effect because of damage

Table 5.3 Scaling-up based on constant $k_L a$ for gas–liquid contacting in a sparged stirred-tank reactor. Effect of scale-up parameter on common operating parameters.

Operating parameter	Laboratory reactor 751		Plant reactor 10 m³	
H_L/T	1	1	1	2.8
P/V	1	1	>	1
VVM	1	1	0.2	0.1
V_s	0.1	0.5^a	0.1	0.1
$k_L a$	1	—	1	1

[a] Impractical liquid 'blow-out' conditions.

to organisms in highly turbulent fermentation broth and/or oxygen poisoning. Other problems, such as gross bubble coalescence, are also important in non-mechanically stirred fermenters.

For scale-up maintaining geometric similarities, Equation 5.46 may be used to predict the oxygen transfer coefficient. It must be noted, however, that the values for the exponents m and n are affected by scale as illustrated in Table 5.4 below.

Table 5.4

Vessel size (1)	Exponent in Equation 5.46	
	m	n
5	0.95	0.7
50	0.6	0.7
50 000	0.4	0.5

It is important to realize that $k_L a$ is not necessarily constant during a fermentation. In highly viscous fermentations, or in fermentations producing surfactants, large variations in $k_L a$ can occur. Thus, using the oxygen uptake rate throughout the fermentation as a scale-up parameter is often more successful. In this case, the oxygen uptake rate obtained in the small reactor is mirrored in the larger vessel by constantly manipulating the agitation and aeration. This allows for a much greater degree of freedom in geometric dissimilarities between the two vessels.

5.8.3   Scale-up on flow basis (Constant Power Input)

Another common design approach is based on constant input of agitator power per unit reactor volume, i.e.

$$\frac{P_1}{V_1} = \frac{P_2}{V_2}$$

where subscripts 1 and 2 refer to small and large scale vessel respectively.

Under turbulent conditions, the agitator power input is given by Equation 5.52:

$$P \propto \rho N^3 D^5$$

For constant power input in geometrically similar vessels we can therefore write

$$\frac{\rho N_1^3 D_1^5}{V_1} = \frac{\rho N_2^3 D_2^5}{V_2} \tag{5.56}$$

And this becomes

$$N_2 = N_1 \left(\frac{V_2}{V_1}\right)^{1/3} \left(\frac{D_1}{D_2}\right)^{5/3} \tag{5.57}$$

giving the impeller speed in the large fermenter for successful scale-up.

It must be noted that this scale-up method relies on correlation for power input in ungassed systems and the demand for geometric similarity also limits its applicability. For constant power per unit volume in turbulent flow it is to be noted that impeller tip speed and hence, shear, increases with the cube root of the diameter, so that many other flow parameters cannot be maintained constant on scale-up (see Table 5.5). This table also

Table 5.5 Examples of incompatible flow parameters on scaling-up a geometrically similar ungassed stirred-tank reactor

Flow parameter	Laboratory reactor 201			Plant reactor 2.5 m³	
P/V	1	1	25	0.2	0.0016
N	1	0.34	1	0.2	0.04
ND	1	1.7	5	1	0.2
Re	1	8.5	25	5	1

demonstrates how different flow parameters will affect flow conditions in the reactor when used as a scale-up parameter.

5.8.4 Scale-down

It is very common for the fate of a new process to depend on successful scale-down, rather than scale-up. Large fermenters are expensive and it is very rare that fermentation technologists get an opportunity to design a large fermenter according to specifications developed in a pilot fermenter. With given large production vessels, the question is 'how can we mirror these conditions in a small fermenter?' It is much more expensive to modify a large vessel than a small one, so a successful scale-up based on oxygen transfer coefficient is achieved when high product yields in the small fermenter are obtained at the value of $k_L a$ which is obtained in the large vessel.

In general, therefore, the real problem in scale-up is normally to reproduce on a smaller scale the non-ideal flow conditions which exist in plant scale.

Reading list

Aiba, S., Humphrey, A. E. and Millis, N. F. (1974). 'Biochemical Engineering', 2nd Edn. Academic Press: New York and London. [Aeration; scale-up and design of equipment.]

Bailey, J. E. and Ollis, D. F. (1977). 'Biochemical Engineering Fundamentals'. McGraw-Hill: New York. [Principles of mass and heat transfer.]

Banks, G. T. (1979). Scale-up of fermentation processes. *In* 'Topics in Enzyme and Fermentation Biotechnology' (Ed. A. Wiseman), Ellis Horwood Westergate, Sussex, Vol. 3, pp. 170–266. [Aeration and agitation; scale-up.]

Faust, U. and Sittig, W. (1980). Methanol as a carbon source for biomass production in a loop reactor. *Adv. Biochem. Engng* **17**, 63–99. [Fermenter design; mechanism of oxygen transfer; behaviour of air-lift reactors.]

Kristiansen, B. and Chamberlain, H. E. (1983). Fermenter design. *In* 'Filamentous Fungi' (Eds J. E. Smith, D. R. Berry and B. Kristiansen), Vol. 4, pp. 1–19. Edward Arnold: London. [Fermenter designs; fluid rheology.]

Sinclair, C. G. (1984). Formulation of the equations for oxygen transfer in fermenters. *Biotechnol. Lett.* **5**, 111.

Sinclair, C. G. and Mavituna, F. (1983). Mass and energy transfer. *In* 'Filamentous Fungi' (Eds J. E. Smith, D. R. Berry and B. Kristiansen), Vol. 4, pp. 20–76. Edward Arnold: London. [Principles and mechanism of oxygen transfer; measurements of oxygen transfer rates; heat transfer.]

Sweeney, E. T. (1978). 'An Introduction and Literature Guide to Mixing'. BHRA Fluid Engineering Series, Vol. 5. [Introduction to mechanism of mixing; performance and design of impellers.]

Wang, D. I. C., Cooney, C. L., Demain, A., Dunnill, P., Humphrey, A. E. and Lilly, M. D. (1979). 'Fermentation and Enzyme Technology', Wiley, New York. [Oxygen transfer; aeration and agitation; scale-up.]

6 | Downstream Processing in Biotechnology

G. SCHMIDT-KASTNER and
C. F. GÖLKER

6.1 Introduction

'Downstream processing' is a useful collective term for all the steps which are required in order actually to recover useful products from any kind of industrial process. It is particularly important in biotechnology where the desired final forms of the products are usually quite far removed from the state in which they are first obtained in the bioreactor. For example, a typical fermentation process gives a mixture of a dispersed solid (the cell mass, perhaps with some components from the nutrient medium, etc.) and a dilute water solution; the desired product may be within the cells, as one constituent of a very complex mixture, or in the dilute aqueous medium, or even distributed between the two. In any case its recovery, concentration, and purification will require careful and effective operations which are also constrained by manufacturing economics. Any special requirements, such as a need to exclude contaminants or to contain the process organism, will only add to the difficulties.

Many operations which are standard in the laboratory will become impractical, or uneconomic, on the process scale. Moreover, bioproducts are often very labile or sensitive compounds, whose active structures can survive only under defined and limited conditions of pH, temperature, ionic strength, etc. Bearing in mind such restrictions, much ingenuity is called for if the available repertoire of scientific methods is to be used to best effect. It will also be apparent that there is no one unique, ideal, or universal operation, or even sequence of operations, which can be recommended; individual unit operations must be combined in the most suitable way for a particular problem.

6.2   Separation of particles

At the end of a fermentation, the first step in many cases is to separate the solids (usually cells, but alternatively cells or enzymes on a particulate support, and not excluding solid components of the reaction medium) from the liquid continuum which is almost always aqueous. Some properties of cells which are relevant to such separations are listed in Table 6.1; note that the specific gravity of the cells is not much greater than

Table 6.1   Properties of cells with reference to separation

Properties	Bacteria	Yeast	Fungi
Shape	Rods, spheres chains	Spheres, ellipsoids filaments	Filaments
Size	0.5–3 μm	1–50 μm	5–15 μm diameter; 50–500 μm length
Specific weight	1.05–1.1	1.05–1.1	1.05–1.1
Cell weight	10^{-12} g	10^{-11} g	—

that of the fermentation broth. The cell size can also cause difficulties with bacteria, but larger cells are more easily separated, sometimes even by simple settling in decanters. Ease of separation will also be dependent upon the nature of the fermentation broth, its pH and temperature, etc., and in many cases it must be improved by the addition of filter aids, flocculating agents, etc. (see later). Table 6.2 gives a general classification of separation methods.

6.2.1   Filtration

This method is most widely and typically used to separate filamentous fungi, and filamentous bacteria (i.e. streptomycetes) from fermentation broths. It can also be used for yeast flocs. According to the mechanism, filtration can be performed as surface filtration, or depth filtration, or centrifugal filtration; in all cases, the driving force is pressure, whether created by overpressure or by a vacuum.

The rate of filtration, i.e. the volume of filtrate which can be collected in a given time, is a function of filter area, the viscosity of the fluid, and the pressure drop across the filter medium and the deposited filter cake. The resistance of the filter medium and the filter cake together is therefore critical, and this resistance depends upon its compressibility. For non-compressible cakes the filtration rate becomes independent of pressure, but most biological materials are compressible, so the resistance of the

Table 6.2 Classification of separation methods

Separation principle	Separation method	Particle size (μm)
Particle size	Fibre filtration	>200–10
	Microfiltration	20–0.5
Molecule size	Ultrafiltration	2–0.005
	Hyperfiltration	0.008–0.00025
	Gel chromatography	2–0.0003
	Dialysis	0.002–0.00025
Temperature	Crystallization	<0.002
Solubility	Adsorption	<0.002
	Extraction	
Electric charge	Electrophoresis	2–0.02
	Electrodialysis	0.02–0.00025
	Ion exchange	0.02–0.00025
Density	Sedimentation	>1000
	Decantation	1000–5
	Centrifugation	1000–0.5
	Ultracentrifugation	2–0.02

cake increases with time quite independently of the increased overall resistance as the filtration proceeds and the cake builds up.

A considerable improvement in filter flow is attained by cross-flow filtration, in which the solids which do not pass through the filter are kept in suspension by a turbulent flow across the membrane. This can be brought about by arranging for the suspension to flow across the membrane, or by fitting moving blades or an impellor inside the filter. Simple plate filters are widely used for the clarification of liquids, and can be used for the filtration of small amounts of suspensions, but their loading capacity is limited; filter-press assemblies are sometimes used, especially for batch operations.

Rotary drum vacuum filters are perhaps the most widely used devices for the separation of microorganisms from the fermentation broth; in these, the filtration element is a rotating drum maintained under reduced internal pressure. The drum rotates into the liquor to be filtered, and its continuing rotation allows essential subsequent operations on the filter cake, as shown schematically in Fig. 6.1. To avoid build-up of biomass on the filter surface, which will increase filtration resistance, such filters are frequently fitted with a knife discharge, as illustrated; if a mycelium which forms a coherent 'carpet' is being separated, it can be lifted from the filter by strings. The drum is often precoated with a filter aid which helps to prevent blocking

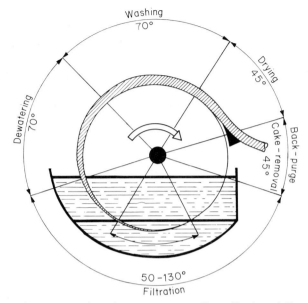

Fig. 6.1   Schematic representation of a rotary vacuum filter. (Section of filtration depends on immersion depth of the filter drum.)

and to allow a constant pressure drop to be maintained. Major advantages are the effectiveness of filtration, with minimal temperature rise, low power consumption, and the integration of filtration with washing and partial dewatering; contamination of the filtered-off material with filter aid can be a serious drawback. Rotary filters operating under positive pressure can also be used, while belt filters are an obvious modification of the principle, and are very suitable for readily-filtered precipitates requiring extensive washing. Belt filters can be combined with a press to promote dewatering.

6.2.2   Centrifugation

Bacteria are usually too small to separate on simple filter media, but their separation by centrifugation is also difficult because of the small difference in density between the particles and the suspension. Protein precipitates must very often be separated by centrifugation, with similar difficulties. The performance of a centrifuge can be characterized by the expressions:

$$Q = d^2 \Delta \rho g Z F / 18 \eta$$

and

$$Z = R \omega^2 / g \simeq R n^2 / 900$$

in which Q = volumetric feed rate, d = particle diameter, $\Delta\rho$ = density difference, g = acceleration due to gravity, F = volume held in centrifuge, η = viscosity, R = radius, ω = angular velocity and N = number of revolutions per minute.

The function $\Sigma = FZ$ is useful in comparing different centrifuges. Because F increases with the axial length of the rotor, and Z with its diameter and speed, performance improves by using longer rotors (F) or faster, wider rotors (Z); attainable values of Z are limited by the materials of construction. Figure 6.2 shows some rotor arrangements schematically, and

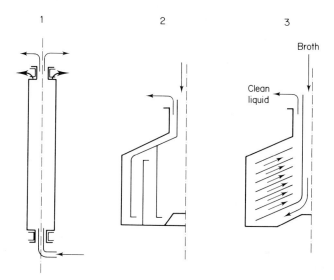

Fig. 6.2 Examples of solid centrifuges. 1, tubular bowl; 2, multichamber solid bowl; 3, disc bowl centrifuge.

the critical characteristics of different arrangements are summarized in Table 6.3. All designs have individual disadvantages, to which must be added the general ones of cost (including maintenance), power consumption, and (except where refrigeration is incorporated) temperature rise.

6.2.3 Flocculation and flotation

Where the small size of bacterial cells makes their recovery from fermentation broth very difficult, even by centrifugation, an improvement can often be achieved by flocculation; the sedimentation rate of a particle increases with its diameter (Stokes's law). Flocculation can occur reversibly if charges on the cell surface can be neutralized by oppositely charged

Table 6.3   Properties of different centrifuges

Type of centrifuge	Advantages	Disadvantages	Technical details
Perforated basket	Good dewatering; easy to clean; washing of cake possible	Limited solids capacity; solids recovery laborious; low centrifugal force; discontinuous operation	$n = 500\text{–}2500 \text{ min}^{-1}$ $Z = 300\text{–}1500$ $\Sigma = 900\text{–}1800$
Sieve-scroll-centrifuge	Possibility for classification of different solids; washing of cake possible; continuous operation	Low centrifugal force	$n = 500\text{–}1000 \text{ min}^{-1}$ $Z = 300\text{–}1500$
Tubular bowl	Good dewatering; high centrifugal force; easy to clean; bowl easy to remove	Limited solids capacity; solids recovery laborious; discontinuous operation; low Σ	$n = 13\,000\text{–}18\,000 \text{ min}^{-1}$ $d = 75\text{–}150 \text{ mm}$ $Z = 13\,000\text{–}17\,000$ $\Sigma = 1500\text{–}4000$
Multichamber solid bowl	Increase in solids capacity; no loss of efficiency up to complete filling of chambers	Solids recovery laborious; discontinuous operation	$n = 5000\text{–}10\,000 \text{ min}^{-1}$ $d = 125\text{–}530 \text{ mm}$ $Z = 6000\text{–}11\,400$
Scroll discharge centrifuge	Suitable for slurry with high solids concentration; high input of slurry; continuous operation	Low centrifugal force	$n = 700\text{–}2500 \text{ min}^{-1}$ $Z = 350\text{–}1400$ $\Sigma = 900\text{–}2300$
Disc bowl centrifuge	Partly continuous operation; high centrifugal force; liquid discharge under pressure; CIP cleaning	Poor dewatering; works only with low solids content in suspension	$n = 3000\text{–}10\,000 \text{ min}^{-1}$ $Z = 4000\text{–}7500$ $\Sigma = \text{up to } 270\,000$
Nozzle bowl centrifuge	Continuous operation; high centrifugal force	Poor dewatering	$n = \text{up to } 10\,000 \text{ min}^{-1}$ $Z = \text{up to } 14\,000$ $\Sigma = \text{up to } 80\,000$

ions, and irreversibly if charged polymer molecules form bridges between the cells. Thus flocculating agents include inorganic salts, mineral hydrocolloids, and organic polyelectrolytes. As well as choice of flocculating agent, flocculation also depends on such other factors as the nature (and physiological age) of the cells, the ionic environment, temperature, and surface shear stress.

If a sufficiently dense floc does not form, flotation can be used, in which small gas bubbles adsorb and entrain the organisms. The separation depends on the size of the gas bubbles; gas can be sparged in, or (better) very fine bubbles can be created from dissolved gases by releasing the over-pressure or by electrolysis. Formation of a stable foam is promoted by insoluble 'collector substances' such as long-chain fatty acids or amines, and the particles collected in the foam layer can be removed. In some SCP processes (see Chapter 10) a combination of flocculation and flotation is most efficiently applied for biomass recovery.

6.3   Disintegration of cells

6.3.1   Microorganisms

Disruption of microorganisms is usually difficult because of the strength of the cell wall and the high osmotic pressure inside; the particles are too small to be subjected to simple mechanical means, such as milling, whereby the necessary strong forces can be obtained. At the same time the disinteg-ration must be effected without damaging the desired cell components, and often the requirements are contradictory. Methods used for breaking microorganisms are summarized in Fig. 6.3. Their effect can often be assessed in terms of the level of activity of a cellular enzyme recovered

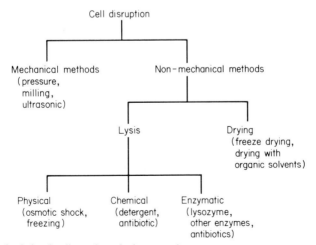

Fig. 6.3 Methods for the disruption of microorganisms.

in the disrupted suspension, combining a measure of the efficiency of disruption with an assessment of the degree of damage.

Mechanical methods
Mechanical methods may use shear (grinding in a ball mill, colloid mill, etc.), pressure and pressure-release (homogenizer), and ultrasound. A widely used process method uses high pressure followed by sudden decompression, resulting from the flow of the cell suspension through a fine nozzle. One such arrangement is shown in Fig. 6.4; here disintegration is due to

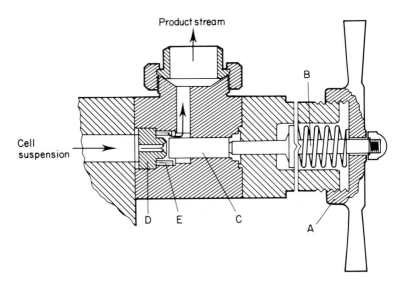

Fig. 6.4   Pressure valve for disintegration of microorganisms on suspension. A, handwheel; B, rod for valve adjustment; C, valve; D, valve seat; E, impact ring.

hydrodynamic shear and to cavitation. Ultrasound works mainly through cavitation, but is mainly used in the laboratory, since on a larger scale heat removal is difficult.

Non-mechanical methods
Cells can also be broken by thermal, chemical, or enzymic means. One widely used method is drying, which causes changes in the cell-wall structure and often allows a subsequent extraction of the cell contents by buffer or salt solutions; a well-known procedure is to prepare an 'acetone powder' by introducing the cells into a large excess of cold acetone. Lysis of cells can be caused by chemical means, such as salts or surfactants, or by osmotic shock, or by the application of appropriate lytic enzymes, which are now

available in considerable variety and can be matched to the specific nature of the microorganism in question. Where the full variety of methods can be used, and compared, the recovery of active cellular enzymes following disruption is usually best after enzyme treatment or ultrasound; thermal or osmotic methods are less good, and mechanical disruption is commonly the least effective method.

6.3.2 Plant and animal cells

Cells from animal tissue or from plants are usually less stable and can be more easily disrupted by mechanical means. Additional shear can be applied if the material is first frozen, while appropriate enzymes can assist extraction from either animal or plant materials. Technically, considerable attention has been given to increasing the yield of plant extractives of all kinds by treatment with appropriate lytic enzymes (cellulases and hemicellulases) prior to conventional extraction methods.

6.4 Extraction methods

Extraction as applied to bioproducts has the function both of separation and also of concentration. It is especially useful for the recovery of lipophilic substances, whether these are initially extracellular or are liberated by suitable treatment of the cells. The solution, or even suspension, containing the desired product is mixed with an immiscible solvent in which it is preferentially dissolved and from which it can be more easily and specifically recovered. The distribution of a substance between the two phases is governed by its characteristic partition coefficient, K:

$$K = \frac{\text{concentration of substance in phase A}}{\text{concentration of substance in phase B}}$$

However, the efficiency of any practical extraction process also depends upon the relative volumes of the two phases which are brought to equilibrium. Simple calculation confirms that where the substance is to be extracted into a given volume of solvent the extraction is more efficient if it is carried out with successive smaller aliquots, and where differential extraction of one component rather than others is sought, back-extraction will increase any selectivity. Thus liquid–liquid extraction can be performed in a single step (if the partition coefficient is very favourable), by multistage parallel-flow extraction, or by counter-current extraction (the most complex arrangement but one which is also capable of fairly good resolution of mixtures).

In the recovery of antibiotics, solvent extraction is often an early step after the removal of cells or cell debris. The extraction of whole broth can also be achieved using extraction-decanters, extraction columns, or mixer-settlers using centrifuges. Figure 6.5 shows a process for whole broth

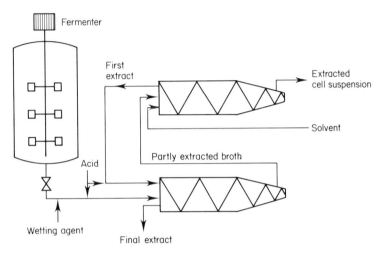

Fig. 6.5   Whole-batch recovery process for antibiotics extraction.

extraction for recovery of antibiotics direct from the fermentation broth.

Whole broth recovery methods clearly lead to a reduction in the number of recovery stages, and should therefore reduce stage-by-stage losses even though the extraction may be impaired by the presence of cells. Reducing the time in which products are left in the aqueous phase, in which degradation is usually most rapid, will also be advantageous.

Separation of enzymes from cells or cell debris can be achieved using extraction with aqueous multiphase systems. Cell debris and enzymes are distributed in aqueous polyethyleneglycol–dextran mixtures, which form separate phases and can thus be separated easily. This method avoids some of the difficulties which arise in centrifugation (small particle dimensions) and gives high yields.

6.5   Concentration methods

The products formed at low concentrations in bioprocesses may well be more concentrated as a result of some initial separation step, as in extraction (see above), but in general they must be further concentrated before economic purification is possible. Again, because of the lability of many products, methods for their concentration must be very gentle. Some suitable methods will include:

- extraction (already considered)
- evaporation
- membrane processes
- ion exchange methods
- adsorption methods.

6.5.1 Evaporation

Evaporation is usually applied to solutions obtained by solvent extraction, in which case careful arrangements for recovery of the evaporated solvent are obligatory; because of the higher specific heat of evaporation, it requires more critical arrangements when it is applied to an aqueous preparation. Direct evaporation of whole culture broth is often used for relatively low-grade products, usually by the use of some sort of spray-dryer.

Quite generally, evaporators with a short residence time are used to minimize losses due to thermal lability. The continuous flow evaporator allows concentration of labile products in one rapid throughput; in falling film evaporators the heat transfer is better, based on the large heat transfer surface, which allows a minimal temperature difference between the heating medium and the solution. Residence times are in the order of a minute. In thin-film evaporators the thin layer is created mechanically and is very turbulent; heat transfer is very effective and these evaporators can be used for the concentration of viscous liquids and even for concentration up to dry products. Centrifugal thin-film evaporators, in which evaporation takes place on heated conical walls inside a rotating bowl, give even shorter residence times.

6.5.2 Membrane filtration

Membrane filtration is a versatile procedure which can be used for both the enrichment and the separation of different molecules or particles. Under hydrostatic pressure, small particles will pass through a suitable membrane if the applied pressure exceeds the osmotic pressure of the solution, which tends to drive solvent from the concentrate into the more dilute solution. In ultrafiltration the membrane can be considered as a sieve (microporous membrane) whose pore size governs the separation, but this simple model is not applicable in the case of reverse osmosis in which molecules of similar

sizes may still be separated. This depends upon diffusion and solution interactions between the molecules and the membrane material. Examples of
the application of membrane filtration processes are given in Table 6.4.

Table 6.4   Use of membrane processes

Type of membrane	Application	Relative molecular mass
Microfiltration	Concentration of bacteria and viruses	>1 000 000 (or particles)
Ultrafiltration	Fractionation; dialysis; production of enzymes; production of protein; processing of whey	>10 000 (macromolecules)
Reverse osmosis	Concentration of pharmaceutical substances, production of lactose; part-desalination of solutions	>200 (organic compounds)
Electrodialysis	Purification and fractionation of charged substances; desalination	>100 (organic compounds)

   To obtain high flux rates it is usual to use asymmetric membranes, with
a dense thin layer of membrane material supported mechanically by a
coarser sponge-like structure. One serious problem in ultrafiltration is the
phenomenon of 'concentration polarization'. On the concentrate side of
the membrane the concentration of solutes increases with approach to the
membrane, and is zero on the other side of the membrane. This results
in the formation of what is, in effect, a 'secondary membrane', with different
flux and separation characteristics. As in cross-flow filtration (see above,
Section 6.2.1), the effect can be minimized, though not completely abol-
ished, by high flux rates across the membrane surface, achieved either
by high pumping speeds, inducing turbulent flow, or by equipment with
thin channels for high-velocity laminar flow. Typical arrangements are illustrated in Figs 6.6 and 6.7.

   Microfiltration can replace filtration for the separation of cells, or even
specific metabolites, from fermentation broths; the closed system facilitates
sterile work and also containment, where necessary.

6.5.3   Ion exchange and adsorption resins

The use of adsorbents to recover substances from fermentation broths is
a method of long standing, originally based on the use of active charcoal.

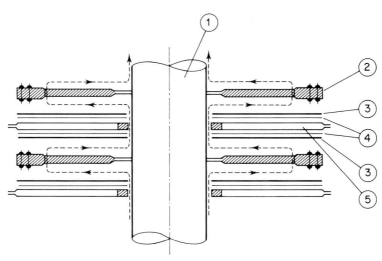

Fig. 6.6 Sectional drawing of 'plate-and-frame' ultrafiltration module (The Danish Sugar Co.). 1, centre bolt; 2, frame; 3, membrane; 4, filter paper; 5, membrane support plate.

Such methods can give a very large concentration effect in a single step, and with the availability of a wider range of more specific adsorption agents they are being used increasingly. For example, their exploration is usually one of the first steps in researching recovery methods for any new antibiotic.

Ion exchange resins
Ion exchange resins are polymers carrying firmly attached ionizable groups, which may be anions or cations according to choice, and which will be in the ionized or non-ionized forms according to the environment. Available types are listed in Table 6.5; the solid ion exchangers are used either by batchwise addition, followed by removing the resin by decantation, or else by column procedures. Chromatographic separations involving such ion exchange columns are dealt with below, Section 6.6.2.

Liquid ion exchangers operate through similar principles, but with ionizable substances which dissolve only in a non-aqueous solvent carrier; the separation is then by liquid–liquid extraction. The desired substance, in either case, is recovered from the ion exchanger by ion displacement, and the ion exchanger is then regenerated.

In suitable cases, antibiotics may be recovered from whole broth, as an alternative to clarified broth, by the use of ion exchange resins in the batch mode.

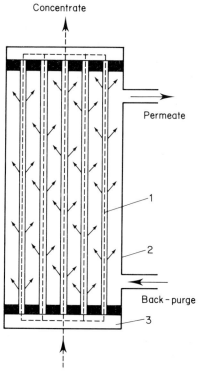

Concentrate

Permeate

1

2

Back‑purge

3

Fig. 6.7 Schematic drawing of a hollow-fibre ultrafiltration module (Romicon). 1, hollow fibre; 2, housing; 3, mounting of fibres.

Table 6.5 Chemical nature of solid and liquid ion exchangers

Ion exchanger	Matrix	Functional Group
Solid exchange resin	Styrene-divinyl-benzene Acrylate Methacrylate Polyamine Cellulose Dextran	Carboxyl Sulphonic acid Primary, secondary and tertiary amines Quaternary amines

Liquid ion exchange
  compounds
	Solvent is used as carrier
	  for functional group
		Primary, secondary and
		  tertiary amines
		Phosphoric acid-
		  monoester
		Phosphoric acid-diester

Adsorption resins

These are resins which, in effect, provide a solid phase with similar properties to an immiscible liquid solvent, and so may be used in place of extraction procedures. Today a wide range of such resins is available, and their important properties are summarized in Table 6.6. The resins are porous poly-

Table 6.6 Properties of adsorption resins

Physical properties
surface 20–800 m^2 g^{-1}
volume of pores 0.5–1.2 ml g^{-1}
average pore size 5–130 nm

Chemical nature
apolar: styrene-divinyl benzene
semipolar: acrylic ester
polar: sulfoxide, amide, N—O-groups

meric matrices whose functional groups, if any, modify the overall polarity of the matrix, without ionization. Most compounds are adsorbed in their non-dissociated state, usually from aqueous solutions; they are recovered by extraction into organic solvent or by changing the pH or ionic strength of the aqueous phase. The porosity of the adsorption resins is important, because this determines the available surface on which adsorption can occur, and thus the total capacity of the resin for the adsorbed substance.

6.6 Purification and resolution of mixtures

In a properly conceived process sequence, product recovery and concentration will already have been accompanied by some substantial degree of purification. However, the need for further steps to purify the product will usually remain, particularly when the desired product is accompanied by essentially similar cometabolites, which will usually accompany it through the earlier stages of recovery. Thus the situation to be faced in purification may or may not call for a high degree of resolution. The two methods most generally applied are crystallization and various kinds of chromatographic procedure; crystallization is more readily scaled up, but chromatographic methods lend themselves to a better resolution of mixtures.

6.6.1 Crystallization

Crystallization is mainly used for the purification of low molecular weight compounds, such as antibiotics. For example, penicillin G is usually extracted from the fermentation broth with butyl acetate and crystallized by the addition of potassium acetate in ethanolic solution. A similar sequence for the isolation and purification of actinomycin is set out in Fig. 6.8. The crystallization is from a concentrated solution in ethyl acetate

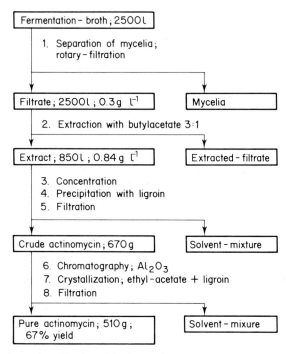

Fig. 6.8 Isolation and purification of Actinomycin.

at 50°C, by adding light petroleum until the solution becomes turbid; the solution is heated to 65°C and then cooled.

On a much larger scale, crystallization is the final stage in the purification of such products as citric acid and sodium glutamate, etc.

6.6.2   Chromatographic methods

Chromatographic methods are used for low molecular weight compounds where resolution of mixtures is needful (e.g. homologous antibiotics) and for macromolecules, particularly enzymes, where accompanying products are of rather similar nature. The necessary equipment is an arrangement of columns, packed with the carrier material which is usually particulate, reservoirs and pumping systems for the eluting liquids, and a means of selectively collecting specific eluate fractions.

According to the specific separation problem, the column proportions may vary, from relatively tall narrow columns to short wide ones; an example is shown in Fig. 6.9, where the construction is in stainless steel. For optimal resolution, column particles with an even size distribution are used; packing the column is usually effected by mixing the adsorbent with the solvent system in a separate tank, degassing under reduced pressure if necessary, and transferring to the column as a thick slurry. Some materials, especially those used in molecular-sieve chromatography, change their volume when equilibrated with mixtures of buffer and organic solvents, or with buffer solutions of varied ionic strength (salt gradient elution), and must then be fixed into the column by adjustable end-plates. Columns may be run with upwards or downwards flow, usually with a flow meter for the pumped liquid and an entry pressure control device. Installation of a sterile filter at each end of the column, to eliminate bacterial contamination and possible destruction of products, may be desirable.

It is normally necessary to monitor what is being eluted from the column at every stage, e.g. by following the ultraviolet absorption of the eluate at 254 nm or 280 nm for nucleic acids or proteins respectively; buffer gradients can be monitored by measurements of conductivity, or in extreme cases a by-pass leading to an autoanalyser can be fitted. To isolate the separated products some form of fraction collector is needed, and for large-scale separations using organic solvents this must be a flame-proof installation in an explosion-proof room.

Different chromatographic procedures can be distinguished, as follows:

- adsorption
- ion exchange
- gel filtration
- hydrophobic
- affinity
- covalent
- partition

Particular features of these are described below.

Adsorption chromatography separates products according to their different affinities for the surface of a solid matrix, either an inorganic carrier such as silica gel, alumina or hydroxyapatite, or an organic polymer.

Ion exchange chromatography uses resins of the kind already described (see Section 6.5.3 and Table 6.5), or insoluble or cross-linked polysaccharides (cellulose, sepharose) to which ionized functional groups have been attached. It can give high resolution for macromolecules; an example of

Fig. 6.9 C = PHARMACIA 1501-stainless-steel segment column; P = Particle filter; S_1, S_2 = Sterile filters; Fl = Flow meter; UV = UV-recorder

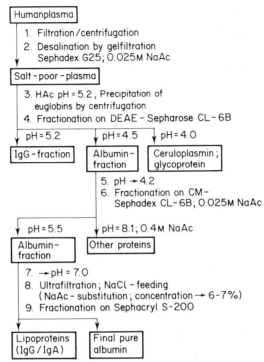

Fig. 6.10 Human plasma fractionation by ion exchange and molecular sieve chromatography (Curling, 1980)

the resolution of a complex mixture of proteins is given in Fig. 6.10. This shows the fractionation of human plasma proteins, on a production scale, by a combination of ion exchange chromatography, Steps 4 and 6, with gel filtration, Steps 2 and 9.

Gel filtration uses a 'molecular sieve', that is, a neutral cross-linked carrier with a definite pore size for molecular fractionation. Molecules larger than the largest pores cannot enter the matrix and pass directly through the column; smaller molecules enter the carrier and are retained; diffusion into the pores is then a function of the molecular size, so that the retained molecules are eluted in order of decreasing size.

Corresponding to the range of molecular weights, from a few hundred to a million or more, a wide range of carriers with controlled pore sizes is called for; they are generally produced from linear polymers (dextrans, agarose) subjected to varied degrees of cross-linking. These are generally used in aqueous systems, and gel filtration can also be used for desalting of peptide or protein solutions. Molecular sieves into which relatively hydrophobic groups have also been introduced can be used in organic solvents to separate lipophilic substances in the same manner.

Hydrophobic carriers. Molecules with different hydrophobicity can be separated on a carrier containing hydrophobic groups. Thus many enzymes and other proteins have hydrophobic regions, where there is an accumulation of neutral aminoacids (e.g. valine, phenylalanine), and will interact to varying degrees with a carrier on to which alkyl residues of varying lengths have been attached.

Affinity chromatography uses more specific interactions to obtain separations; a specific structure is attached to a solid support carrier and interacts specifically with the component to be isolated. The principle is shown in Fig. 6.11. The specific effector, such as an enzyme inhibitor (I) is immobilized on a water-insoluble carrier (C) which is packed into a column. A

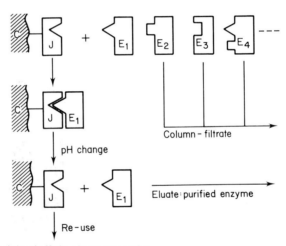

Fig. 6.11 Principle of affinity chromatography.

complex mixture of different enzymes (E_1, E_2, E_3, E_4 ...) in a neutral buffer solution is introduced, and while the enzyme E_1 is complexed by the inhibitor I, all the remaining enzymes pass through the column. Using a buffer solution at a different pH the enzyme E_1 can now be selectively eluted from the column. Typical specific effectors, and the complementary molecules that can be separated, include:

enzymes/enzyme inhibitors (or vice versa)
antibodies/antigens or haptens (or vice versa)
lectins/glycoproteins or polysaccharides

nucleic acids/complementary base sequences
hormones/receptors
vitamins/carrier proteins
etc.

This type of separation, therefore, depends upon the same high specific interactions that mediate biological processes. For example, dehydrogenase enzymes can be separated by virtue of their specific coenzyme interactions using a sepharose matrix carrying bound NAD^+. Antigen/antibody reactions can be used; for example human leucocyte interferon is recovered in high yield and in highly purified form after chromatography on sepharose carrying an immobilized monoclonal antibody. Obviously such specific immoblized effectors can be very costly, and have to be used for many cycles.

To prepare the specific immobilizate, a carrier such as agarose gel in bead form is activated by reaction with, for example, cyanogen bromide; it can then be coupled directly when small molecules, such as coenzymes, are to be bound, or through a bifunctional 'spacer' molecule when a larger effector molecule is to be attached.

Group-specific affinity chromatography uses somewhat less specific 'chemical' interactions in a similar way; for example polymers carrying dihydroxyboryl groups can be used to separate both glycoproteins and nucleotides because of the reversible complex formation between boric acids and 1,2-diols.

Covalent chromatography is a minor variant using reversible covalent bond formation between an immobilized functional group and the material being purified; in particular, thiolated polymers can be used to separate SH-containing proteins because of reversible disulphide formation.

6.7 Drying

Drying of bioproducts is in many cases the eventual method by which the products are brought to a stable form suitable for handling and storage; the heat sensitivity of most biological products means that the only methods which can be used are ones leading to water removal with minimal temperature rise. In some cases, the thermostability of products such as enzymes or pharmaceutical preparations is improved by the addition of sugars or other inert stabilizers.

To remove water as vapour, heat energy must be transferred and strictly controlled conditions are required to ensure that the temperature rise,

Table 6.7   Methods for drying of bioproducts

	Convection dryer	Contact dryer	Freeze dryer (radiation)
Addition of energy	gas current	heated surface	heated surface
			radient heat
Removal of water	convection by heated gas current		
		drain by pumping	sublimation
Mode of operation		batchwise	batchwise
	continuous	continuous	continuous
Status of drying material	resting layer		resting layer
	pneumatic movement	mechanical movement	mechanical movement
Apparatus	chamber dryer		
	shelf dryer	drum dryer	
	spray dryer	tumble dryer	
	belt dryer	film dryer	
	fluidized bed dryer		chamber freeze dryer
	pneumatic-conveyor dryer		belt freeze dryer
			plate freeze dryer
			tunnel freeze dryer

which results from the balance between the rate of heat input and the latent heat equivalent of the evaporation rate, is within the tolerable limit. The heat transfer may be effected by contact (conduction), convection, or radiation, or by combinations of these; Table 6.7 shows examples and operating conditions for the three most important drying methods. For fuller information the reader is referred to the Reading-list; here we describe the most important techniques in outline.

Vacuum drying is applied in batch mode in chamber dryers, or continuously, as in rotating drum vacuum driers. Heat transfer occurs mainly by contact with heated surfaces, and changes in the characteristics of the liquid phase as it becomes more concentrated must be taken into account.

Spray drying provides the most important example of a convective method, in which heat transfer, movement of product, and vapour removal are all effected by a gas current. Large quantities can be dealt with in continuous operation. The liquor to be dried is applied as a solution or slurry and is atomized by a nozzle or a rotating disc. A current of hot gas (150–250°C) causes such rapid evaporation that the temperature of the particles remains very low. Spray drying can be used for drying of enzymes or antibiotics, and when the presence of other materials is not deleterious it can be used for the drying of whole fermenter broth, e.g. for detergent-grade enzymes or feed-grade antibiotics. The method is also widely used in food industries.

Freeze drying is the most gentle drying method because water is sublimed from a frozen mass. For the sublimation of water vapour approximately 680 kcal per kg water have to be transferred to the sublimation surface by conduction from heated plates and by radiation on to the surface; to promote rapid sublimation, a very low pressure is maintained and the vapour must be removed by low-temperature condensation. The solid temperature is regulated through control of the pressure in the drying chamber, and measurement of the electrical conductivity of the material being dried provides a very sensitive check for the presence of any liquid water in the mass.

Many pharmaceutical products are freeze-dried, e.g. viruses, vaccines, plasma fractions, hormones and enzyme preparations, as well as labile and costly ingredients in diagnostics; at the same time, very large-scale applications of the technology are important in food industries.

Reading list

Aiba, S., Humphrey, A. E. and Millis, N. F. (1973). 'Biochemical Engineering', 2nd Edn, Ch. 13. Academic Press: New York and London.

Albertson, P. A. (1971). 'Partition of Cell Particles and Macromolecules', 2nd Edn. Wiley-Interscience: New York.

Atkinson, B. and Mavituna, F. (1983). 'Biochemical Engineering and Biotechnology Handbook', Chs 12, 13. Macmillan: London.

Curling, J. H. (1980). Albumin purification by ion exchange chromatography. *In* 'Methods of Plasma Protein Fractionation', pp. 77–91. Academic Press: London and New York.

Epton, R. (Ed.) (1978). 'Chromatography of Synthetic and Biological Polymers' (2 vols). Ellis Horwood: Chichester.

May, S. W. and Landgraff, L. M. (1979). Separation techniques based on biological specificity. *In* 'Recent Developments in Separation Science' (Ed. N. N. Li), pp. 227–255. CRC Press Inc: Boca Raton, Fl.

Mellor, J. D. (1978). 'Fundamentals of Freeze Drying'. Academic Press: London and New York.

Schmidt-Kastner, G. and Gölker, C. F. (1982). Aufarbeitung in der Biotechnologie. *In* 'Handbuch der Biotechnologie' (Eds P. Prave, U. Faust, W. Sittig and D. A. Sukatsch, Ch. 8. Akademische Verlagsgesellschaft: Wiesbaden.

Wang, D. I. C., Cooney, C. L., Demain, A. L., Dunnill, P., Humphrey, A. E. and Lilly, M. D. (1979). 'Fermentation and Enzyme Technology', Ch. 12. Wiley: New York.

7 | Sterilization and Sterility

L. B. QUESNEL

7.1 Introduction

Sterilization is the process of achieving sterility, of which there are no degrees; an object, surface or substance is either sterile or not. If sterile there are no viable organisms or cells present and if protected against re-contamination the sterile condition will be maintained indefinitely. *Disinfection* implies that the material has been treated so as to remove or reduce the risk from pathogenic organisms, but not that *all* viable organisms have been inactivated.

Sterilization procedures are applied:

(1) to ensure that a process or experiment is carried out only with the desired organism,
(2) to permit the safe use of products,
(3) to avoid environmental contamination,
(4) to prevent deterioration of a product.

Sterilization is carried out either by *removing* viable organisms, as in filtration, or by *killing* them by one of the following means:

(1) *heating* in the presence or absence of water,
(2) *irradiation* by ultraviolet, gamma or X-rays,
(3) *chemicals*, in solution or as gases.

Choice of agent depends on circumstances, available services, the nature of the material or equipment to be sterilized and cost.

7.2 Resistance to sterilization

Bacterial spores are the most heat resistant viable cells known and sterilization must be capable of eliminating the most resistant spores of the most resistant species. The spores of thermophilic bacteria have been reported to survive steam under pressure (200 kPa or 30 p.s.i.) at 134°C for between 1 and 10 minutes, and dry heat at 180°C for as much as 15 minutes. The resistance of individual spores in a population varies and the larger the population the greater the number of more resistant individuals. The resistances of spores will also be affected by sporulation conditions, storage conditions and age, as well as by the conditions during and after heating. The most radiation resistant organism is *Deinococcus (= Micrococcus) radiodurans* which has been found to survive doses as high as 6000 krad.

7.3   Mechanisms of killing

7.3.1   Heat

To kill a spore it is necessary only to denature irreversibly all the molecules of any one enzyme that is essential for germination, or outgrowth, or alternatively to damage irreversibly the gene for one essential enzyme. Heating causes breaks in DNA but the protective association between DNA and calcium dipicolinate in bacterial spores makes it more likely that the lethal damage in heat sterilization involves protein inactivation. The heat resistance of proteins is a function of hydration, and the greater the amount of water the more easily will it enter the hydrophobic, inner domains of protein molecules causing irreversible conformational changes. Estimates of the water contents of spores have ranged from about 5–20%.

In steam sterilization the steam under pressure has two important roles:

(1) by condensing on the material to be sterilized heat is transferred rapidly causing a rapid rise in temperature;
(2) the water molecules themselves increase, or at least maintain, the level of hydration within the spores.

In sterilization with dry heat the heat is transferred very slowly and the tendency is to reduce further the level of hydration and so to protect the spore proteins; spores are considerably more resistant to dry heat than to wet heat.

7.3.2   Radiation

Irradiation by ultraviolet light of wavelength 250–280 nm leads to DNA damage in proportion to the radiation dose. The main damage is the formation of pyrimidine dimers between adjacent bases; repair mechanisms are capable of restoring the integrity of the DNA but are unlikely to function in a dormant spore. Ultraviolet irradiation is not very penetrating and cannot be relied upon as a sterilizing agent unless direct exposure of the contaminating organisms can be guaranteed.

Gamma rays and X-rays are more useful because of their high penetration. There are two main classes of effect:

(1) a large number of single-strand and double-strand DNA breaks are caused;
(2) many molecules within the cell are ionized giving rise to highly reactive toxic species such as peroxides and free radicals to which -SH enzymes are particularly susceptible.

7.3.3 Chemicals

Chemical sterilizing agents may kill as a result of oxidizing or alkylating ability. Many are also highly toxic for man, or carcinogenic, and special equipment may be required for their use. For example ethylene oxide is violently explosive in mixtures with air, is chronically toxic at concentrations not detected by smell and may produce skin hypersensitization, erythema and oedema.

7.4 Measurement of killing

To kill organisms, by whatever means, a lethal dose will be required, which will depend on the duration of the injurious treatment. Thus in heat processes the temperature/time relationship will be paramount, in radiation the energy transmission/time and in chemical sterilization, the concentration/time relationships. In all cases other factors will alter the relationships, e.g. the presence of water, of oxygen or other gases, or protecting menstrua such as protein, temperature etc. Whenever data on killing are given the conditions of the experiment should be stated and, in particular, since the size of the population will determine the achievement of sterility the process should always relate to the initial contamination level.

Because of experimental difficulties in determination of killing times a wide variety of values can be found in the literature for purportedly the same strains or organism (see Fig. 7.1).

7.4.1 Definitions

The *thermal death time* (shortest time taken to destroy the organisms at a stated temperature) and the *thermal death point* (lowest temperature required to kill the organisms in 10 min) are often quoted but unless the

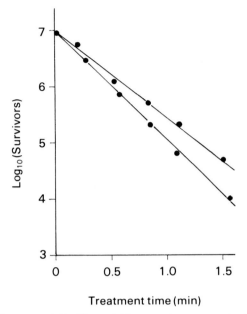

Fig. 7.1   Survival curves of *Bacillus subtilis* var. *niger* spores dry-heated at 160°C and recovered on two different agars.

initial population size and precise conditions are known they are not particularly useful. A more useful parameter is the *decimal reduction time* or *D-value*, the time (in minutes, at a given temperature) required to reduce the viable population to 10% of its previous value (Fig. 7.2). The *z-value* is the temperature change (in °C) required to change the *D*-value by a factor of 10 (Fig. 7.3).

7.4.2   The kinetics of killing

Under lethal conditions, organisms in a population do not die synchronously. Statistically the inactivation over any finite period (i.e. as the damaging dose increases) is proportional to the numbers viable at the beginning of a period, i.e. the population dies exponentially. Thus a plot of the logarithm of the numbers surviving at any time (dose) during the treatment against elapsed time (dose) of treatment, yields a straight line (Fig. 7.1). When the logarithmic decline is constant from time zero, as in Fig. 7.1, the curve is a 'single-hit' kinetic form (i.e. one irreversible lesion is enough to kill one cell) and is described by the equation:

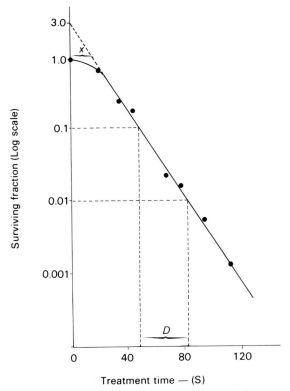

Fig. 7.2 Survival curves of *B. subtilis* MD2 steam-heated at 105°C, plotted as surviving fraction to show extrapolation number (3.0), *D*-value (0.55 min) and shoulder time, *x* (15 s), recovered on tryptone glucose extract agar.

$$N/N_0 = e^{-kd} \tag{7.1}$$

where N_0 = initial cell population, N = number of survivors after dose or treatment time, d = dose or treatment time, and k = specific death rate constant.

Not all populations exhibit 'single-hit' kinetics and Fig. 7.2 shows a curve with a shoulder before the onset of logarithmic inactivation. When a shoulder is found, the survival kinetics are of the form

$$N/N_0 = 1 - (1 - e^{-kd})^n \tag{7.2}$$

where n is an extrapolation number equal to the intercept on the N/N_0 axis said to give the number of 'hits' required for lethality. Either there is a multi-hit requirement for lethality, or, alternatively, each viable 'unit'

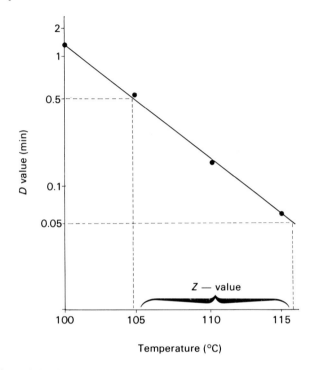

Fig. 7.3 Thermal death time curve for steam-heated *B. subtilis* spores showing derivation of z-value (11°C). Recovered on tryptone glucose extract agar.

may consist of several separate cells or spores and inactivation of the unit is not observed until every component is inactivated. (Non-linear survivor curves are also found in practice.)

The *D*-value is derived by interpolation as the time elapsing during any $\log_{10}$ unit of survivor reduction on the straight part of the graph, as shown in Fig. 7.2. When the logarithms of the *D*-values at different temperatures are plotted against temperature the relationship is usually found to be a straight line from which the z-value can be directly derived by interpolation (Fig. 7.3).

Whereas *D*-values are very dependent on the conditions of both treatment and recovery (Fig. 7.1), z-values are far less dependent on these factors (Fig. 7.3). The z-value may be important in determining the limits of a process since materials may be unable to withstand certain temperatures. However the z-value cannot be truly constant over a wide temperature range, since below a particular temperature no damage will occur and the *D*-value will be infinity.

7.5 Determination of sterilizing conditions

The sterilizing dose required will depend upon the initial level of contamination. For a population exhibiting single-hit curves (Fig. 7.1) at various temperatures or radiation doses, for any number of spores in the population initially the calculation of process conditions for any required degree of probability of success in sterilization is easily done. For example, for a contaminating population of 10^8 organisms the application of 8 D-values of process time at a fixed temperature would reduce the log number of survivors to zero, i.e. one organism, and assuming continuous logarithmic decline in population a further D-value of process time would represent a 1 in 10 probability of finding a survivor, that is of failure to sterilize.

7.5.1 The unit of lethality

To compare the relative sterilizing capacities of different heat processes a unit of lethality is required. The unit chosen is the lethal effect of 1 min of heating at a temperature of 121°C. Relative lethality can be expressed in terms of F-values based on the relationship:

$$F = t \times 10^{(T-121)/z} \tag{7.3}$$

where t = time of application of killing treatment, T = temperature in °C, and z = increase in temperature required to reduce the period of heating by 90% (i.e. the z-value).

In the standard case $t = 1$ min of heating and $T = 121°C$ so that $F = 1$. In practical cases, since the relative resistance of organisms to different temperatures varies, the z-value of the most resistant contaminant at the temperature to be used should be employed. The most heat-resistant spores commonly found have a z-value of c. 10°C and this value may be used if specific experiments cannot be carried out. Using $z = 10$ the standard value of F is designated F_0 at a temperature of 121°C.

If a process is assigned a value of $F = 4$, then this process would achieve a kill equivalent to 4 min of the standard treatment. If such a process was applied to a standard organism whose D-value was 2.0 min, then the reduction in viability would be 100-fold. Expected kill for any initial contaminating population where D for the organism and F for the process are known can be obtained from the equation:

$$F_s = D(\log N_0 - \log N) \tag{7.4}$$

where F_s is the integrated lethal capacity of heat received by all points in the heated material during the heating process. Conversely if the initial and final viable counts are known, then F_s for the process can be defined.

For fermentation media, an acceptable probability of failure to sterilize might be 0.001 (1 batch failure in 1000). If, for example, sterilization of 100 litres of corn steep liquor is to be carried out and the contamination level is 10^6 viable spores per ml, then the total 'bioburden' will be 10^{11} spores. If the process temperature is 121°C and the *D*-value of the contaminating *Bacillus* spore at this temperature is 3 min, then with single-hit kinetics the process time required (using Equation 7.4) for a 0.001 probability of failure is a total of $14 \times D = 42$ mins. (If the kill-curve is multi-hit then the shoulder time (Fig. 7.2) must be added).

If, for some reason, e.g. destruction of a nutrient component, it is not possible to use this process time/temperature combination it should be possible to find a new process time from the relationship:

$$t = 42 \times 10^{(121-T)/z} \qquad \text{(c.f. Equation 7.3)}$$

If the standard organism is being used and the new temperature *T* is 131°C then with $F = 1.0$ and $z = 10$, the new process time $t = 42 \times 10^{-1} = 4.2$ min. In other words if the sterilizing temperature is raised by 10°C then the heating time can be decreased ten-fold. If, however, the most resistant contaminant is much less resistant than the standard resistant organism, with a *D*-value of, say, 1.2 min at 121°C, then to achieve equivalent lethality at 121°C, from Equation 7.4

$$F_s = 1.2\,(\log 10^{11} - \log 10^{-3}) = 16.8\,\text{min}$$

If the 100 litres is to be sterilized by a continuous-flow heat exchanger holding 10 litres and maintaining a temperature of 131°C, then the bioburden per 10 litres is 10^{10}. The F_s equivalent for the process becomes $1.2 \times 13 = 15.6$ min, and the residence time (t_r) for which the temperature of 131°C is to be maintained will be:

$$t_r = 1.56\,\text{min}$$

Assuming perfect 'piston flow' and instantaneous heat-up on entry and cool-down on exit the required flow rate of medium through the sterilizer will be $10/1.56 = 6.4\,\text{l/min}^{-1}$. The practical advantage in using fast high temperature processes for sterilization is that the activation energies for thermal destruction of many compounds of concern range from 10–25 kcal mol^{-1} and the temperature coefficients for such reactions are smaller than that for sterilization.

In the food industry the most serious health hazard is the presence of *Clostridium botulinum*, the most heat-resistant pathogenic spore-former as well as the most toxic agent. The *D*-values of the spores at 121°C are of the order of 0.2 min. Because it was thought that unprocessed food used in the canning industry might contain the equivalent of one *Cl. botulinum* spore per gram and that a large factory might process say 10^{11} g per annum, a protection of the order of $12 \times D$ per batch treatment was considered reasonable. Rounding off upwards a process of 3 min at 121°C was considered the standard treatment and is familiarly called 'the bot. cook'. In practice the level of contamination is more likely to be 10^{-3} to 10^{-4} *botulinum* spores per gram and, with the exception of one or two notable episodes, the canning industry has had a remarkably good record.

In clinical sterilization processes where the bioburden in a single load of, say, contaminated dressings may be very considerable, the recommended time–temperature relationships for sterilization are:

For moist heat:	*For dry heat:*
steam at 134°C (207 kPa, 30 p.s.i.) for 3 min	160°C for 45 min
steam at 126°C (138 kPa, 20 p.s.i.) for 10 min	170°C for 18 min
steam at 121°C (103 kPa, 15 p.s.i.) for 15 min	180°C for $7\frac{1}{2}$ min
steam at 115°C (69 kPa, 10 p.s.i.) for 20 min	

On this basis a 'hold-time' of 15 min steam at 121°C is equivalent to $75 \times D$, making the probability of failure, in relation to the survival of *Cl. botulinum*, almost impossibly remote.

7.5.2   'Hold-time' and 'cycle-time'

Whenever large volumes of liquid (and even more so, solid) have to be sterilized, heating-up and cooling-down times become appreciable and occupy the major proportion of the cycle time. This is so even for laboratory-scale batches of media as shown by Table 7.1. These data clearly show

Table 7.1   Effect of liquid volume and number of containers on time required for liquid to reach 121°C in an autoclave

Liquid volume/ container (l)	No. of containers/ load	Liquid temperature at start (°C)	Time for liquid to reach 121°C (min)	Total time of cycle (min)
0.5	30	29	19	29
1.0	20	26	34	44
2.0	10	27	37	47
3.0	8	26	43	53
4.0	5	26	52	62
5.0	5	26	60	70
6.0	4	26	62	72

that with a set hold-time of 10 min the heating-up time was nearly twice as long for the smallest unit volumes and 6 times as long for the largest of these unit volumes.

Batch heating of the volumes needed on an industrial scale will involve considerably longer heating-up and cooling-down times before inoculation can be carried out. During such cycles the lethal effect extends over a very much wider range of the cycle than the actual hold-time and it is necessary to integrate the lethal effect over the whole of the time–temperature programme if serious degradation of the liquor is to be avoided (Fig. 7.4). The determination of integrated lethal effect throughout a large con-

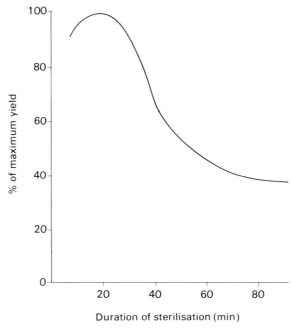

Fig. 7.4   Decrease of end-product yield from a fermentation, from thermal degradation of nutrients at increasing sterilization times.

tainer whose contents are not homogeneously heated is a complex procedure (see reading list). A simple practice is to measure the temperature (by thermocouple) at the most heat-impenetrable point, to record its variation with time, plot this relation on 'lethality' graph paper, and obtain the total integrated lethality (F-value) of the cycle from the area under the curve. Lethality paper has a time abscissa, and the ordinate is marked in temperature but scaled in relative lethality units. Taking the lethal effect of 1 min at 121°C as the unit, then 1 min at 111°C is 0.1 units, since D is 10 times as long ($1/10$ killing effect) for a fall of 10°C where $z = 10$°C (the standard value usually applied); similarly 1 min at 101°C is 0.01 unit.

7.6   Practical methods

7.6.1   Heating

Sterilization may be either batchwise or continuous. For steam sterilization the steam should always be 'saturated' so that a small drop in temperature (by contact with colder material) causes release of heat of condensation.

Batch process
(a) Autoclaves. These are relatively small pressure vessels for steam sterilizing volumes up to some tens of litres for laboratory-scale fermentation and for sterilizing small fermenters, and (extensively) for certain sterile products in industry. Autoclaves are normally controlled by reference to pressure and temperature gauges and both should be checked. All air must be removed before the heating cycle since mixtures of air and steam at any given pressure will achieve a lower temperature than pure steam at the same pressure.

At the end of the hold-time, an adequate cooling-down time must be allowed, during which the pressure must decrease slowly. Abrupt release of pressure will either cause breakages or violent frothing and boiling-over. When loading assemblies such as fermenters, care must be taken over the settings of valves and closures to allow access of steam to containers.

(b) Direct steam injection. A large fermenter is usually fitted with the necessary piping and valves for sterilization of the vessel and associated feed lines by direct steam injection. In some instances separately sterilized liquor may be added to the already sterile fermenter but it is also possible to sterilize the batch of medium *in situ*. If direct steam injection is used allowance must be made for the fact that some 10–20% final volume will be due to condensate.

While the thermal efficiency of this process is high, severe foaming during sparging may limit heat transfer. Also many fermentation media go through a high viscosity phase during heat-up, particularly in high-starch liquors. Build-up of 'mash' into solid lumps also occurs and the centre of such lumps may never sustain an adequate total lethal dose. To minimize these problems adequate care must be taken in batching, mixing and agitating constituents during preparation and sterilization of the 'liquor'.

Other causes of failure to sterilize may be due to bad engineering design where the thermometer pocket wells, joints, sample ports, etc. produce crevices or pockets in which aggregated solids resist subsequent cleaning and sterilization procedures (see reading list).

(c) Indirect heating. This is usually achieved by passing steam through heat-exchange coils or a jacket. For jacketed vessels the available heat transfer

area will depend on the tank diameter and the larger the vessel the slower will be the heat-up rate. For internal coil systems the surface area/volume ratio can remain more or less constant whatever the reactor size. The process is less heat efficient than direct injection, and for both types the problem of cooling, usually by coils, remains.

Undesirable reactions due to heating must always be considered. Common among these are oxidation of phenols and the Maillard reaction between reducing sugars and amino groups (amino acids and proteins) leading to condensation products, which may inactivate much of the carbohydrate and amino-nitrogen. These reactions are favoured by alkaline pH. However, a certain amount of 'cooking' can also enhance the nutritional value of a medium (see Fig. 7.4). Depending on available processes it may be possible to sterilize ingredients separately and mix aseptically, and this is often done in laboratory-scale work. Sterilization will generally cause a pH drift, usually toward more acid pH, and adjustment after sterilization is common.

Continuous flow sterilization
These processes have the advantage that they are rapid, they save plant space (compared to an independent batch sterilizer) and because of the short process time they give improved quality media. The three stages of the cycle should be considered in designing equipment to achieve:

(1) rapid heating-up,
(2) the hold-time at the sterilizing temperature,
(3) rapid cooling.

Possible systems are outlined in Fig. 7.5.

Direct steam injection
As in the batch process this offers high heat efficiency and lower capital cost. Steam is injected directly into the flowing medium through a venturi-type mixer, and the flow is held at the sterilizing temperature (say 140°C) in a highly lagged coil where the residence time determines the F-value of the treatment. It then passes directly to a rapid cooler. The most rapid cooling is attained by flash evaporation: fluid from the hold coil passes through a throttling valve to an expansion chamber, where the rapid drop in pressure causes vaporization of steam and the temperature drops to *c.* 100°C. One disadvantage is the excessive foaming which may result. Alternatively (or additionally) cooling may be carried out in a heat exchanger using incoming cold medium as the heat absorber, with a considerable energy saving. Cooling to inoculation temperature may be completed by a water-cooled heat-exchanger.

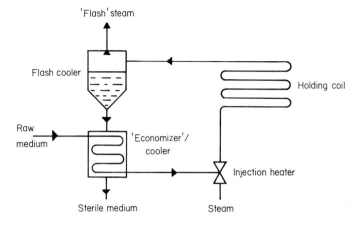

(a) Direct steam injection

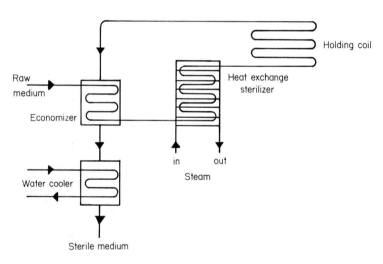

(b) Indirect steam heat-exchange system

Fig. 7.5   Schematic diagrams of continuous flow sterilization.

Indirect steam heating

For rapid heat-up a plate heat exchanger may be used through which steam at 500–690 kPa (80–100 p.s.i.) is passed. The heat exchanger area can be considerable compared to the hold volume, and the use of materials such as titanium has enabled strong, light, efficient anticorrosive equipment to be made. Cleaning and reassembling can also be easily performed with these units; no condensate is introduced and 'impurities' in the steam cannot enter the medium. Holding and cooling can be achieved as before (Fig. 7.5).

With both categories of sterilizer, heating will cause proteins in the medium to be coagulated and the insoluble material may cause problems by adhering to surfaces. Piped steam is also used for the 'in-line' sterilization of valves, seals and sampling ports, as well as air filters. Design requirements for these are critical, particularly in full-scale plant (see Reading list).

Dry heat sterilization
Dry heat sterilization is carried out in ovens fitted with fans to circulate the hot air. The method of packing the oven is of great importance since heat transfer from dry air is very slow. It can only be used for solid materials capable of withstanding the high temperatures needed, such as laboratory glassware, talc, etc. Thermocouples should be employed to follow heat-up rates.

7.6.2   Radiation sterilization

Radiation sterilization is usually carried out with a cobalt-60 or caesium-137 source. In either case very strict safety precautions must be applied and such installations are unlikely to be found in the average laboratory.

The short wavelength (high energy) of X-rays makes them highly penetrating. The unit of measurement of radiation dose is the rad, equivalent to an energy absorption of 100 ergs g^{-1} air, doses usually being measured in megarad (Mrad). The röntgen (R), based on the ionizations produced per unit volume of air, is 83 ergs g^{-1} air. In SI units the Gray is used equivalent to 1 J kg^{-1} energy in tissue (1 Gy = 100 rad).

Vegetative cells are less resistant than bacterial spores but considerable variation in resistance levels is found, depending on the efficiency of repair systems. The most resistant vegetative cells belong to the species *Deinococcus radiodurans* (originally isolated from irradiated meat) which can withstand a dose of 6 Mrad (6×10^4 Gy). Among spores those of *Clostridium botulinum* are highly resistant, the lethal dose being in the range from 0.5–2.2 Mrad ($5 \times 10^3 - 2.2 \times 10^4$ Gy). Anoxic atmospheres and low relative humidity protect against radiation, the lethal effect always being greater in the presence of oxygen and high water content.

Viruses are of equivalent resistance to bacterial spores but some have been reported to survive much higher than average lethal doses; encephalitis viruses required as much as 4.5 Mrad (Jordan & Kempe, 1956). In practice doses of 2.5 Mrad (2.5×10^4 Gy) are used for sterilizing pharmaceutical and medical products. As with heat sterilization chemical effects on culture media are likely to be complex and difficult to control, and changes may continue after irradiation has been completed.

7.6.3   Chemical sterilization

Formaldehyde gas is only effective if it can be guaranteed to contact the contaminating organisms. It has very poor penetration and diffusibility, and a pungent and lasting smell; it is used only in special circumstances.

Hydrogen peroxide has found increasing use in the food industry. It is a powerful oxidizing agent, killing both vegetative cells and spores, with activity increasing with concentration, temperature and, usually, pH. It has the advantage that its ultimate breakdown products, water and oxygen, are harmless and it has been used in concentrations from 10–25% w/v in the sterilization of milk and for the containers used for food products.

Ethylene oxide (b.p. 10.7°C) and propylene oxide (b.p. 33.9°C) can be used in gaseous form in an appropriate installation. Ethylene oxide is much more effective but is highly toxic, irritating and violently explosive in mixtures with air. Its lethal effect on bacteria is dependent on humidity; optimum conditions include 40–80% relative humidity and a temperature of 60°C for 3–4 h at a gas concentration of 800–1000 mg l^{-1}. The process is used where equipment would be damaged by higher temperatures.

7.6.4   Filtration sterilization

Filtration differs from other sterilization methods because organisms are not killed, but physically removed; the filter assembly must itself be sterilized by some other means before it can be used to sterilize a flowing gas or liquid. There are two main types: (a) depth filters and (b) screen filters.

(a) Depth filters are made from fibrous or powdered materials in a relatively thick layer pressed or bonded together, to form a weft of multidirectional interconnected channels of varying size. They are made from fibreglass, cotton, mineral wool, cellulose or asbestos fibres and moulded into mats, wads, stacks or cylinders depending on design. In use, particles penetrate within the filter and are trapped as a result of a number of factors including direct interception by fibres, inertial impaction, Brownian movement, convection, and electrostatic effects. The efficiency of the filter is in general terms related to the depth of the bed, the proportion of particles passing through a given depth being given by the relationship

$$\frac{N}{N_0} = e^{-Kx} \tag{7.5}$$

where N_0 = concentration of particles entering the filter, N = concentration of particles passing through a given depth of filter, x = depth of filter, K = is a constant which will depend on numerous factors such as the size of the particles, the diameter of the fibres, the porosity, etc.

This equation, of exactly the same form as that describing the kinetics of killing, allows calculation of the depth of filter for any desired probability of failure to retain particles. The deeper the filter bed the more efficient, but also the greater the resistance to flow, measured by ΔP, the pressure drop caused by it. Figure 7.6 shows the dependence of both penetration and pressure drop in glass fibre depth filters on air flow rate.

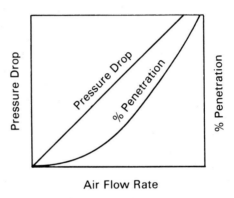

Fig. 7.6   The relationship between sodium chloride penetration and air flow rate, and between flow rate and pressure drop, for Microflow filters.

Modern designs of depth filter are cylinders made from bonded borosilicate microfibres in which the layers increase in fineness and density from in-to-out, so that the greatest filtration area is the final stage filter layer, and the whole cylinder is sheathed in reinforcing polypropylene mesh. Gradings are provided down to 98% (nominal) capture at 0.1 μm, which in appropriate aggregate assemblies allow flow rates of water as high as $6000\,l^{-1}$ at 0.4 bar (5.5 p.s.i.) ΔP. Gas filtration rates and efficiencies are invariably much higher, and assemblies delivering over $3\,m^3\,sec^{-1}$ at 0.1 bar ΔP (from a 7 bar supply) are readily obtainable. The efficiency of gas filtration is greatly diminished if liquid condenses on the filter, increasing resistance and allowing 'channelling' to occur, so that organisms 'grow through' the filter bed leading to downstream contamination.

Very high sterile air supply rates are often necessary since an industrial aerobic 'fermentation' may require 1 vol air (vol liquor)$^{-1}$ min^{-1}, so that a 30 000 gallon reactor may 'consume' nearly 200 000 m³ of air per day.

(b) Screen filters('absolute' filters) are based on a simple sieve principle. They are membranes made from cellulose esters or other polymers having more or less rectilineal pores of fairly uniform diameter which exclude contaminating particles, trapping them on the surface of the filter. Since the smallest bacterial contaminants are about 0.5 μm or more in diameter a membrane filter of 0.45 μm pore size will remove bacteria (but not viruses). To achieve high flow rates a large membrane surface is needed and the membranes must be given a rigid support to withstand the pressures used. Modern air filter units achieve both aims by forming the membranes into cylinders on a rigid support core, such as polypropylene, which is slotted to permit passage of the filtered air to the lumen. The membrane is usually further protected by fabric mesh on the outer layer of which is a polypropylene mesh sheathed in a glass microfibre prefilter and nylon sock. Multiple 'Millipore aerotubes' of this sort assembled in aggregates of 20 allow gas flows of $70 \, m^3 \, min^{-1}$ with a pressure drop of 0.7 bar across a filter area of $2.4 \, m^2$. Such filters require a prefilter, which is usually built in, to remove gross contamination before it can occlude the membrane.

Whatever type of filter is used it must be sterilized, and many are designed for in-line sterilization either with steam or ethylene oxide. Small units for sterilizing air to a laboratory fermenter can be autoclaved *in situ*.

Filters are also used to sterilize the effluent gases from fermenters which will almost invariably contain aerosolized microbes. In the culture of pathogenic viruses the escaping particles may be as small as 0.05 μm in diameter; more generally it should be remembered that particles below about 5 μm diameter will readily penetrate into the lung, and that most pathogenic organisms are considerably more infectious by inhalation than by ingestion.

7.7 Evaluation of sterilization efficiency

(1) Thermocouples with associated chart recorders allow reliable assessment of a thermal process on the principles outlined earlier for derivation of total F-values. This is the method of choice for large-scale processes.

(2) Browne's Tubes: sealed glass tubes containing about 0.15 ml of a red fluid which gradually changes to green as the heat dose is increased. They are available for testing of both wet and dry heat processes and give an accurate indication of the 'clinically recommended' temperature–time treatments listed earlier.

(3) Spore strips. These are bio-indicators for heat and chemical or radiation methods. Filter paper strips are impregnated with *c.* 10^5 spores of *Bacillus stearothermophilus* for wet heat and *Bacillus subtilis* var *niger* for dry heat

treatments. To check for growth of survivors after treatment, the former are incubated at 55°C for 5–7 days in a tryptone glucose yeast extract broth, the latter at 37°C. Clearly the indication of sterility is retrospective. Various types of spore ampoule are also available.

*(4) Thermalog S indicator strips.*These devices respond to wet heat in the range 115–123°C, the heat dose being indicated by the distance moved (in mm) by a blue dye.

*(5) Autoclave tape.*Indicates that steam at a minimum 120°C has reached the tape, whose stripes turn from white to black. It does not give assurance of sterility but merely indicates that an item has been steam processed.

7.7.1   Tests of filtration efficiency

(1) Methylene blue test. Described in British Standard 2831, the test cloud has a particle size distribution of 0.02–2.0 μm and 99% of particles are below 0.6 μm diameter.

*(2) Sodium chloride test.*Described in British Standard 3928 the test cloud particle size distribution is similar to the methylene blue cloud. For sterilizing grade the filter must have a sodium test efficiency of 99.997% or greater, i.e. a NaCl penetration of less than 0.003% at maximum design flow-rate.

*(3) Bacillus subtilis spores.*A test devised at MRE (now CAMR), Porton, uses monodispersed *subtilis* spores (size range 0.7–1.0 μm) injected into the airstream delivered to the filter; samples are collected for 5 min from either side of the filter and from the counts obtained a percentage penetration value is calculated.

*(4) T3 bacteriophage test.*The test cloud is monodispersed T3 bacteriophage (0.03 μm) and the samples are collected by electrostatic precipitator impaction, counted by phage plaque formation and a percentage penetration calculated.

A filter having a sodium flame penetration of 0.001% has an equivalent *B. subtilis* spore penetration of *c.* 0.000005% and a T3 penetration of *c.* 0.00005%.

Reading list

Ashley, M. H. J. (1982). Continuous sterilization of media. *The Chemical Engineer*, February 1982, 54–58.

Block, S. S. (Ed.) (1977). 'Disinfection, Sterilization and Preservation', 2nd Edn. Lea and Febiger: Philadelphia.

Gaughran, E. R. L. and Goudie, A. J. (Eds) (1977). 'Sterilization of Medical Products by Ionizing Radiation'. Multiscience Publications: Montreal.

Peppler, H. J. (Ed.) (1967). 'Microbial Technology'. Rheinhold Publishing Corporation: New York.

Pirt, S. J. (1975). 'Principles of Microbe and Cell Cultivation'. Blackwell Scientific Publications: Oxford.

Process Biochemistry. Wheatland Journals: Watford, UK. A monthly journal frequently publishing articles on the principles and practice of sterilization.

Rivière, J. (1977). 'Industrial Applications of Microbiology' (Ed. and trans. M. O. Moss and J. E. Smith). Surrey University Press: London.

Russell, A. D. (1982). 'The Destruction of Bacterial Spores'. Academic Press: London and New York.

Russell, A. D., Hugo, W. B. and Ayliffe, G. A. J. (Eds) (1982). 'Principles and Practice of Disinfection, Preservation and Sterilization'. Blackwell Scientific Publications: Oxford.

Stumbo, C. R. (1973). 'Thermobiology in Food Processing', 2nd Edn. Academic Press: New York and London.

8 | Microbial Screening, Selection and Strain Improvement

R. P. ELANDER

8.1 Introduction

The screening of fastidious microorganisms for novel microbial metabolites requires highly selective procedures that allow for the isolation of the rare organisms which will afford new metabolites of potential interest from an extremely large background population of common microorganisms producing known or unwanted metabolites. In most large-scale screening programmes, rare and fastidious microorganisms are selected since the probability of discovering new, interesting activities is maximized; highly sensitive assays are used, designed to detect a specific activity from a microorganism which is difficult to isolate from nature. The enrichment for, and eventual selection of, unusual organisms is difficult and involves considerable technique and patience. Interesting cultures must often be repurified, because of slower growing commensal organisms, and as isolated from nature are inherently variable in their growth characteristics and metabolic activities. Hence, the first task of the microbiologist is to minimize genetic variability by selecting stable, genetically uniform isolates producing the minimum of unwanted metabolites and large amounts of the desirable component.

After selection of the producer microorganism, genetic improvement and long-term preservation of the strain are essential for the eventual industrial process. Most of this chapter is devoted to directed microbiological screening for desirable metabolites, and ways of securing the essential genetic improvement of the producing microorganism.

8.2   Sources of industrially important microorganisms

8.2.1   Essential characteristics

It is axiomatic that the producer microorganism is the major determining factor for the success or failure of an industrial fermentation process. An economically important strain will possess certain general attributes regardless of the nature of the actual fermentation process, which can be summarized as follows:

- a pure culture, free from both microscopically visible organisms and also from phage
- readily producing both vegetative cells and spores or other propagation units (species which produce only a mycelium are rarely used)
- growing vigorously after inoculation into seed stage vessels
- producing a desirable product, preferably a single one, easily recovered and preferably with the absence of any toxic by-products
- producing the required product in a short time, preferably 3 days or less
- preferably protecting itself against contamination, e.g. by growing at acid pH or high temperature or by producing an acceptable microbial inhibitor
- amenable to change by mutagenic agents but genetically stable in their absence
- readily conserved for long periods.

8.2.2   Natural habitats and culture collections

Microorganisms can be isolated from nature or obtained as pure cultures from culture repositories; to obtain unusual cultures, one usually isolates the microorganism from an unusual ecological niche. Although microorganisms are ubiquitous, the most common sources of industrial microorganisms are soils and lake and river mud. However the vast majority of microbial genera remain to be explored (Table 8.1). Resourcefulness and ingenuity are needed to devise selective enrichment methods to isolate rare and fastidious microorganisms, and to preserve them when interesting.

Table 8.1   Soil microorganisms screened for biological activity (Berdy, 1980)

	Actinomycetes	Fungi	Bacteria
Known: Genera	34	2000	97
Species	600	49 000	300
Screened: Genera	16	210	18

The main isolation methods used in routine soil screening are outlined in Table 8.2. Soil treatment is often used to enrich for particular classes of microorganisms and novel approaches include:

- Ultraviolet radiation
- Air drying, or heating at 70–120°C
- Filtration, or continuous percolation
- Washings from root systems
- Treatment with detergents and alcohols
- Preincubation with toxic agents, antimetabolites, or other growth regulators

Table 8.2   General isolation methods and preservation procedures for industrial microorganisms

Isolation

Sponge (soil directly)	Aerosol dilution
Soil dilution	Flotation
Gradient plate	Differential centrifugation

Selective enrichment

Soil treatment	Nutritional (C, N sources)
Selective inhibitors (antibiotics, antimetabolites, etc.) added to plating medium	Temperature, pH range, aeration, etc.

Long-term preservation

Soil desiccation	Microporous beads
Mineral oil immersion	Liquid nitrogen refrigeration (−196°C)
Low temperature refrigeration (−80°C)	

Specific enrichment procedures designed to select unusual classes of soil, water or marine microorganisms are commonly used in laboratories with large screening programmes. Examples of such procedures, with their specific target classes of microorganism, are summarized in Table 8.3.

Growth inhibitory substances (antibiotics, chemical agents, etc.) may be added to prevent growth of undesirable microorganisms; the use of aminoglycoside, anthracycline and polyene antibiotics has been reported useful in the isolation of rare actinomycetes.

Microbial cultures can be obtained from permanent culture collections. Martin and Skerman (1972) list 349 collections in their 'World Directory of Culture Collections of Microorganisms' (Wiley-Interscience, New York). Examples of well-known culture collections available either as sources of

Table 8.3   Isolation procedures for specific types of microorganisms

Dilute organic media	*Micromonospora, Arthrobacter*
	Aquatic gram-negative bacteria
	Oligotrophic bacteria
Pollen-baiting; haematin	*Actinoplanaceae*
Deoxycholate; triphenyltetrazolium chloride	Gram-negative bacteria
Chitin (as nutrient source)	*Lysobacter* and other gliding bacteria
Bark, roots, animal faeces	*Myxobacter*
Plant debris (fresh water)	Motile, spore-forming actinomycetes
High temperature (42–55°C)	Thermotrophs
Low temperature (4–15°C)	Psychrotrophs
Extreme pH	Acidophiles
High NaCl concentrations	*Nocardia*, halophiles
Incubation in N_2 atmosphere	Anaerobes
Deep sea muds	Marine bacteria

reference strains or as repositories for cultures involved in patent applications include:

- American Type Culture Collection (ATCC), Rockville, Maryland, USA
- Commonwealth Mycological Institute (CMI), Kew, Surrey, England
- Centraal Bureau voor Schimmelcultures (CBS), Baarn, Netherlands
- Deutsche Sammlung von Mikroorganismen (DSM), D-3400 Göttingen, West Germany
- Institute for Fermentation (IFO), Osaka, Japan
- Fermentation Research Institute (FERM), Tokyo, Japan
- Northern Regional Research Center (NRRL), Peoria, Illinois, USA
- USSR Research Institute for Antibiotics (RIA), Moscow, USSR

Industrially important microbial cultures are permanently preserved if they are interesting or useful. Research cultures are organisms newly found to elaborate, by synthesis or transformation, a useful product—which is often evaluated biologically by a stable assay microorganism. If the research culture achieves significant commercial interest, it will be used in a fermentation improvement or development programme, allowing the isolation of sufficient material for clinical trial or field studies. If the product is successful and deemed marketable, the development culture becomes a production culture and is continuously improved by genetic manipulation. The successful preservation of research, assay, development and production cultures is essential for the industrial biotechnology laboratory, and valuable strains must be preserved for long periods of time, free from adverse phenotypic changes.

A number of basic procedures are used for the maintenance of industrially important microorganisms, each with slight variations depending on the peculiarities of the producer microorganism (Table 8.2). They include:

(1) drying on sterile loam sand, or other natural substrates (corn kernels, rice, bran, etc.);
(2) storage on nutrient agar slants or in liquid media, usually as frozen suspensions at $-20°$ to $-100°C$;
(3) removal of 'free' water from cells or spores by lyophilization and storage of the dried product in a vacuum under various conditions;
(4) storage as vegetative cells, arthrospores, or conidia in liquid nitrogen ($-196°C$) or in the vapour phase of liquid nitrogen ($-167°C$).

8.3 Microbial screening

An outline of a critical path for a microbial product is summarized in Fig. 8.1. Primary screening is designed to detect and isolate microorganisms with potentially interesting commercial applications, eliminating many unwanted microorganisms and/or products and allowing detection of the small percentage of potentially interesting isolates. An ideal primary screen must also be predictive, sensitive, rapid, inexpensive, adaptable to a large sample base, specific to target activities, but effective for a broad range of compounds!

8.3.1 Direct selection on solid media

A natural microbial source, for example a soil sample, is plated directly or diluted to give a microbial cell concentration such that aliquots [either applied directly or sprayed using a variety of procedures (Table 8.2)] yield individual isolated colonies. In the 'crowded plate' technique, less diluted soil samples are spread on agar so that antimicrobial activity can be observed directly by zones of inhibition of adjacent colonies. Rapid plate detection methods for specific properties (Table 8.4) are used; the devising of such direct methods of obtaining relevant information is often critical, and calls for skill and ingenuity.

8.3.2 Screening of fermentation broths

Though the use of rapid plate screens has been effective in discovering useful microbial metabolites, it has a limitation that plate activity cannot always be duplicated in submerged cultures. This has led to the more extensive use of shaker vessels for the initial tests. This allows a broader variety

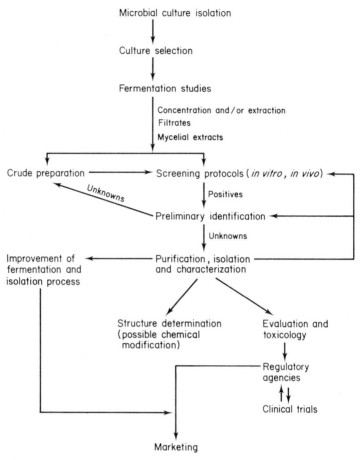

Fig. 8.1 A typical critical path analysis for the successful development of a new microbial product for clinical application.

of fermentation conditions (variable temperature and aeration level), and additional post-fermentation steps can be included in the screening procedure. Application of multiple fermentation samples to automated detection systems such as multi-channelled colorimetric assays (Autoanalyzer[R]), high performance liquid chromatography (HPLC) or gas–liquid chromatography (GLC) allows rapid screening of concentrated fermentation broth samples. The automated use of paper discs, wells, cylinders or multi-point plastic sheets in conjunction with multi-capillary pipettes has facilitated the handling of large numbers of broth samples. However in the last resort

Table 8.4   Typical rapid plate procedures for the detection of microbial metabolites

---

Extracellular enzymes

Amylase: hydrolysis of starch in agar revealed by clear zones after iodine staining.

Proteases: solubilization of target protein at specified pH in agar suspension revealed by cleared zones; liquefaction of gelatin gel (collagenases).

Lipases: digestion of target lipid emulsified in agar revealed by cleared zones or by precipitation of liberated fatty acids by Ca^{2+} in agar.

Pectinases: liquefaction of pectate gel; buffered pectin agar (pH 5 or 7) shows cleared zones with cetyltrimethylammonium bromide (CTAB) for endopolygalacturonidase or pectate lyase respectively.

Carboxymethylcellulase: cleared zones on CMC agar with CTAB (see pectinases).

Cellulase: dye release into clear agar from agar containing azure-dyed cellulose.

Xylanase: like carboxymethylcellulase.

Urease: pH indicator (phenol red) in urea agar.

Nucleases: precipitation of unhydrolysed nucleic acid (RNA, DNA) with added HCl; ultraviolet fluorescence of unhydrolysed nucleic acid with acridine orange.

Phosphatases: agar with phenolphthalein diphosphate as combined substrate and pH indicator.

Enzyme inhibitors

In general, by modifying procedures like the above using enzyme-containing agar.

Other metabolites (examples, excluding antimicrobials)

Citric acid: pH indicator on paper overlay; solubilization of $CaCO_3$ in agar.

Oestrogen (from steroids): red colour with *p*-nitrobenzenediazonium fluoroborate reagent.

NAD: bioautography using auxotrophic microorganism (auxanography).

Other products: similarly by specific colour, pH or bioassay reactions.

---

it remains true that the scientific insight, instinct and natural curiosity of the researcher are probably the most important criteria for success.

8.3.3   Antimicrobial screening

During the first two decades (1940–1950s) of the 'antibiotic era' the simple plate prescreens (crowded-plate technique, cross-streak technique, etc.) were sufficient to discover new useful antimicrobial agents. Modern screening for rare, useful antibiotics has reached such complexity that such simple plate tests are of limited value. Today, most antibiotic screens are applied to rare *Actinomycetes* (actinomycete genera other than *Streptomyces* are referred to as rare, and collectively account for less than 5% of the total population of actinomycetes in soil). The common characteristics of these rare microorganisms are slow growth rates, more exacting nutritional requirements, poor sporulation and low stability in soil populations.

These characteristics probably account for their rare occurrence; however, their study has been fruitful (Table 8.5).

Table 8.5   Secondary metabolites from rare genera of actinomycetes

Genus	Metabolite
Streptoalloteichus	Tallysomycins A, B
	Nebramycin, factors II, IV, V
Actinomadura	Rifamycin O
	Carminomycin
	Maduromycin
	Luzopeptin
Streptosporangium	Bleomycin-Phleomycin
	Sibiromycin
	Sporamycin
	Chloramphenicol
Chainia	Chainin
	Aburamycin
Spirillospora	Spirillomycin
Actinoplanes	Gardimycin
	Purpuromycin
	Teichomycin
	Lipiariamycin
Saccharopolyspora	Sporacin A and B
Planomonospora	Sporangiomycin
Actinosporangium	Marcellomycin
	Musettamycin

Primary screens

A typical primary antimicrobial screen contains a broad range of specific resistant or hypersensitive microorganisms (Table 8.6). *Penicillium avella-*

Table 8.6   A primary antimicrobial screen

Staphylococcus aureus	*Penicillium avellaneum*
Streptococcus faecalis	*Acholeplasma laidlawii*
Proteus vulgaris	*Bacillus subtilis* (rec⁻) on minimal agar
Escherichia coli	*E. coli* (poly A⁻) on minimal agar
Candida albicans	Bacteria resistant to aminoglycosides, etc.
Trichomonas vaginalis	Bacteria hypersensitive to β-lactams,
Bacteroides fragilis	aminoglycosides and macrolides

neum is included as it is particularly sensitive to antitumour agents, as are recombination-deficient strains of *Bacillus subtilis* and *Escherichia coli*. *Acholeplasma laidlawaii* is particularly sensitive to polyether antibiotics which are commercially important in agriculture. Anaerobic bacteria (*Bacteroides fragilis*) are included because of their importance in the etiology of coccal pneumonia, brain abscess infection, periodontal disease and deep-seated skin infections. Naturally, company strategies will also be reflected in the selection of screening organisms at this stage.

Pure cultures of interesting soil microorganisms are prescreened (on agar plates, in fermentation tubes, or in flasks) on several kinds of media (differing in carbon and nitrogen source availability or ratio, lipid-based, or buffered for pH range). If liquid cultures are used, samples of culture fluid are withdrawn aseptically and analysed in the primary screen; sensitivity can be increased by concentrating the broth. Hypersensitive bacteria are frequently used, and selective resistant bacteria are used to exlude commonly known, less interesting metabolites from the screen. The early exclusion of known, and therefore uninteresting, compounds in a screen ('dereplication') is most important.

Secondary screen
Interesting activities on the primary screen are retested using more specialized media and conditions. This allows for the further sorting out of metabolites with real value for activities of special interest.

8.3.4   Enzyme inhibitor screens

Screens using the inhibition of enzyme systems for the discovery of antimicrobials of clinical or agricultural interest include the following:

- β-lactamase
- purine and pyrimidine synthesis
- aminoglycoside inactivation (phosphorylation, adenylation, hydroxylation)
- chitin or β-glucan synthesis
- fatty acid synthesis
- protein synthesis
- rumen activity
- viral neuraminidase (sialidase), etc.

Screens based on inhibition of β-lactamase have led to the discovery of many promising new β-lactam antibiotics (Table 8.7). These were all discovered by combined uses of β-lactam supersensitive mutants, β-lactamase inhibition and differential affinity for penicillin binding proteins; Fig. 8.2 shows one example of this approach.

Many medically useful compounds (examples include tetracyclines, rifamycins, erythromycins and anthracyclines) are secondary metabolites that are biosynthesized in part by the polyketide pathway. In this pathway,

Table 8.7   Representative β-lactam antibiotics produced in bacteria and actinomycetes

Compound	Structure	Organism
Cephamycin C		*S. clavuligerus*
Wildfire Toxin (Tabtoxin)		*Pseudomonas tabaci*
X-372A		*Streptomyces* sp. 372A
Clavulanic Acid		*S. clavuligerus*
Nocardicin A		*Nocardia uniformis*
Thienamycin		*S. cattleya*
SQ 26,180 (monobactam)		*Chromobacterium violaceum*

small organic acid units (acetate, propionate, etc.) are activated and combined in a manner quite analogous to fatty acid biosynthesis. Screens designed to discover metabolites which inhibit fatty acid biosynthesis have been used with success; conversely cerulenin, an antibiotic which specifically inhibits polyketide biosynthesis, has been used to select improved polyketide producers.

8.3.5 Antifungal screens

The design of cell wall inhibitor screens to discover more potent antifungal agents has been emphasized by several laboratories in recent years. Fungi

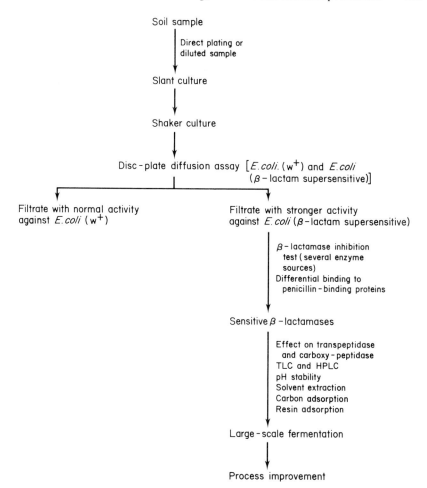

Fig. 8.2   Development of new β-lactam antibiotics by application of different selective screens in succession.

have long been overshadowed as agents of human disease by viruses, bacteria and protozoa, but fungal diseases have become more prevalent with the increased use of immunosuppressive drugs as therapy for cancer. Due to their eukaryotic nature, fungi possess fewer differences in their overall metabolism when compared to animal cells, so that they offer fewer selective targets than bacteria. One major difference between fungal and human tissue is the nature of the fungal cell wall, in which chitin, a $(1{\rightarrow}4)$-β-homopolymer of *N*-acetylglucosamine, is a major constituent. Inhibitors of chitin synthase have been shown to be potent antifungal agents, and non-toxic inhibitors represent important new antifungal therapy. Chitin is also a major component in the insect integument, so insecticidal activity can also be discovered by a search for chitin synthase inhibitors. Polyoxin D is a well-known inhibitor of this type, but its toxicity precludes its use in animal or human therapy. Another interesting microbial metabolite, nikkomycin, has much structural similarity to UDP-*N*-acetylglucosamine and may become a useful insecticidal agent.

8.3.6   Screening for pharmacologically-active agents

The development of screens for the inhibition of enzymes of specific pharmacological interest was initially reported by Umezawa. The ensuing activity in his laboratories in Japan, and the discovery of bestatin (a potent inhibitor of aminopeptidase B) as an agent for tumour therapy will encourage more emphasis on this type of screening. Examples of the kinds of pharmacological activity that might be searched for using suitably-designed screens are given in Table 8.8.

Table 8.8   Useful activities, other than antibiotic action, found in microbial metabolites: potential targets for directed screening

Activity	Example	Producing microorganism
Antitumour	Bestatin	*Streptomyces olivoreticuli*
Anticoagulant	Phialocin	*Phialocephala repens*
Antidepressant	1,3-Diphenethylurea	*Streptomyces* sp.
Antihelmentic	Avermectin	*Streptomyces avermitilis*
Antilipidemic	Ascofuranone	*Ascochyta vicrae*
Antipernicious anaemia	Vitamin B_{12}	*Streptomyces griseus*
Coronary vasodilator	Naematolin	*Naematoloma fasiculare*
Detoxicant	Detoxin	*Streptomyces caespitosus*
Estrogenic	Zearalenone	*Gibberella zeae*
Herbicide	Herbicidin	*Streptomyces sagamonensis*
Immune response enhancer	*N*-acetylmuramyltripeptide	*Bacillus cereus*
Immunosuppressor	Cyclosporin	*Trichoderma polysporum*
Insecticide	Piericidin	*Streptomyces mobaraensis*
Miticide	Tetranactin	*Streptomyces aureus*
Plant hormone	Gibberellic acid	*Gibberella fujikuroi*
Psychotropic	Lysergamide	*Claviceps purpurea*
Salivation inducer	Slaframine	*Rhizoctonia leguminicola*
Serotonin antagonist	HO_{2135}	*Streptomyces griseus*

8.3.7 Antiviral screening

An agar-diffusion tissue culture method for *in vitro* and *in vivo* screens using complex fermentation broths has been used in antiviral screening. In this way the microbial metabolites pyrazomycin and mycophenolic acid were discovered to be potent inhibitors of a number of important viruses. Virus diseases in need of effective therapy are numerous, and there is an outstanding need for more effective screens for new antivirals.

8.3.8 Screening for veterinary compounds

Screening for potential veterinary use has advantages over programmes designed exclusively for human use in that active compounds can be screened directly in target animals. Government requirements are less stringent and probabilities of success are therefore much greater.

Examples of successful microbial screening are the development of polyether antibiotics for treatment of poultry infected with the coccidium *Eimeria tenella*. Monensin, a streptomycete polyether originally detected in antibacterial screens, was reported to be a very active coccidiostat, and was also found to be effective in feed efficiency improvement in ruminants, and against swine dysentery and certain animal viruses. Avermectin, a new potent anthelmentic antibiotic from a strain of *Streptomyces avermitilis*, was found using a direct animal-test screen.

8.3.9 Screening for antitumour agents

Screening for antitumour agents is more difficult than screening for antimicrobials since the initial screens generally require *in vivo* models, needing longer periods of time (and more compound) for evaluation. Consequently, a number of *in vitro* 'prescreen' methods have been developed, providing rapid and sensitive *in vitro* screens for potential antitumour activity. One effective prescreen for antitumour compounds uses microorganisms which have been shown empirically to be particularly sensitive to antitumour antibiotics. Oxidative bacterial and yeast mutants have been used as test organisms, and the fungus *Penicillium avellaneum*, resistant to antibacterial antibiotics, is highly sensitive to maytansine alkaloids and bleomycin-type antibiotics. Mycoplasmas lack cell walls and have been shown to be highly sensitive to antitumour compounds; for example *Acholeplasma laidlawii* is highly sensitive to a number of antifungal and antitumour agents and reveals kinds of antitumour activity not detected in other screens. Bacterial mutants (*B. subtilis*, *E. coli*, *Salmonella typhimurium*) with defects in their DNA repair systems have also been used effectively. The use of inhibition screens based on carefully selected enzymes with specific roles in cancer etiology has already been noted, and has been particularly fruitful.

8.3.10   Identification of metabolites

Active fermentation broths selected by primary screens are next fermented under a variety of conditions on a larger scale (10–70 litre vessels). The broths are usually extracted by subjecting active aliquots to a variety of solvent-extration procedures over a wide pH range. Usually a cup-plate or paper disc assay is used to determine biological activity during chemical purification. Stability of the metabolite to pH and to heat, and broth stability in the absence or presence of vegetative cells, etc., are important characteristics which are determined early, together with adsorption or non-adsorption of the metabolite on a variety of ion exchange or exclusion chromatography resins (see Chapter 6). Such observations are crucial for subsequent development of recovery processes.

Early identification of bioactivity is usually established by paper chromatography or thin-layer chromatography. Recently, both preparative and analytical high performance liquid chromatography have been shown to be particularly useful. Given the sophistication of modern techniques, the chemical study of promising new products is not usually a rate-limiting step in their development, once the separation techniques to provide individual products in reasonable purity have been worked out.

One of the most important and time-consuming aspects of a screening programme is the early identification of potentially new metabolites and the elimination of known ones. With more than 3000 microbial metabolites now reasonably well characterized, and reports of approximately 100 new ones each year, rapid and accurate identification of new antibiotics is necessary to prevent wasteful duplication of effort.

Computer technology plays an important role in this task. Table 8.9 lists characteristics generally included in the data base for a known antibiotic compound. Much of the information comes from published reports in the *Journal of Antibiotics*, which in each new issue lists considerable information on newly discovered compounds. Also, Umezawa has published an 'Index of Antibiotics from Actinomycetes' which contains most of the information listed up to 1978.

Use of specific antibiotic-resistant bacteria, and knowledge of the spectrum of cross-resistance for antibiotics with many specific strains of bacteria, are invaluable in eliminating known and unwanted antibiotics.

Table 8.9 A typical antibiotic data base profile

Antibiotic names (synonyms)	Infrared spectra
Structural type	Solubility
Chemical formula	Qualitative chemical reactions
Elementary analysis	Chromatography (paper, TLC)
Antibiotic code number (Berdy)	Stability
Producing organisms	Antibiotic activity
Molecular/equivalent weight	Microbial cross-resistance
Appearance–physical characteristics	Toxicity data
Optical rotation	Antitumour/antiviral activity
Ultraviolet spectra	Isolation characteristics:
	Filtration, extraction, ion-exchange,
	absorption, chromatography,
	crystallization

8.3.11 Future directions

Given the well known diversity of microorganisms (Table 8.1), soil screening programmes continue to provide useful new microbial metabolites. However, for success, the special skills of the microbiologist, microbial geneticist, chemist, and biotechnologist must all be combined; screens must be based on novel nationales for marketing targets seven to ten years in advance.

To date, most soil screening activity has been devoted to the discovery of new antibiotics. However, the evolutionary diversity of microorganisms is not restricted to their production of mutual inhibitors. As living cells, they synthesize a vast array of natural substances, many of which are likely, on the principles of comparative biochemistry, to influence biological activities of more highly evolved organisms. Microbial products can be expected to influence various human and animal disease syndromes and conditions not associated with microbial infections.

8.4 Strain improvement

As soon as a useful organism emerges from a screening programme, it becomes necessary to improve its productivity for the metabolite under evaluation. The initial phases of fermentation development will often consist of modifications to the medium and fermentation conditions used, but the major source of progress will be to improve the performance of organisms by selection and genetic manipulation. Even at the research stage when a promising drug is under laboratory and clinical evaluation, a dependable supply is necessary; during an evaluation of an antitubercular antibiotic, a supply of 250 kg of pure drug was required for clinical trials alone.

8.4.1   Mutation and selection of improved strains

Large-scale programmes concerned with strain selection, maintenance, and improvement begin shortly after favourable clinical reports are obtained. The most effective method for increasing the yield of a fermentation product has undoubtedly been the use of induced mutation followed by selection of improved strains. The foremost difficulty of this approach stems from the extremely low frequencies at which desired mutations occur, so that any desired mutant must be selected from a large non-mutant population base. The major disadvantage then becomes the lack of a scientific rationale for selecting desirable mutants; at this stage, relatively little will be known of the chemical structure, biosynthetic pathway, or metabolic regulation of the desired fermentation product. Therefore, screening for the rare mutant is time-consuming and costly.

Initially, strain selection was dependent upon the spontaneous variability encountered in natural populations of microorganisms, but by the late 1940s a number of effective mutagenic agents were known, and some were finding their way into strain development laboratories. The general view that all mutations were based on loss of function was repeatedly challenged by the isolation of enhanced variants (gain mutants) from conidial populations exposed to radiation and chemical mutagens. However, there is no evidence that increased production is not a result of decreased function of some enzyme system. The mutation-selection process is still the most important method for obtaining improved strains; for further information about the nature of the mutagenic agents used, and their effects, consult the Reading list.

Selection of strains as spontaneous variants
Antibiotic-producing microorganisms exhibit great natural variation, and selection from natural spore populations is of great practical importance. It also plays a key role in the maintenance of improved cultures. Heterokaryosis, with subsequent culture run-down, is still an important problem in commercial culture laboratories despite improved preservation methods. Continued selection for the preferred colony type is one of the major requisites for maintaining production at constant high levels.

Selection from natural variants is also important in the early stages of development programmes. Most mycelial organisms are probably heterokaryotic when first isolated. Selection of single spores, hopefully uninucleate, often leads to diverse colony types, some of which may represent superior antibiotic-producing entities. In the early stages of development, one separates and selects highly conidiating, stable (non-sectoring) isolates and proceeds to evaluate them for desirable characteristics; notably, antibiotic species or titre. One example of such direct selection was the isolation of *Penicillium* variants which synthesized copious amounts of penicillin G in submerged fermentation.

Role of major mutation in strain development

Screening after subjection of a parent strain to physical or chemical mutagens greatly increases the probability of finding improved strains. As applied to industrially important microorganisms, mutation has two aspects. The first, major mutation, involves the selection of mutants with a pronounced change in a biochemical character of practical interest. Such variants are commonly used in genetic studies and are generally (though perhaps not exclusively) loss mutants. They are isolated routinely from populations surviving prolonged exposure to a mutagen. In contrast, minor mutants will show only a subtle change in a particular character; often the changes are so slight that the variants are not morphologically distinguishable from parent strains. Such mutants are common in all of our important antibiotic-producing organisms.

Examples of major mutation in strain development are numerous. An early and classic example was the selection of non-pigmented *P. chrysogenum* strains with high penicillin production. In the streptomycin-producing organism *Streptomyces griseus*, the initial strain synthesized small amounts of streptomycin and large amounts of a related substance with low activity, mannosidostreptomycin (streptomycin B), whose formation represents a loss of potential streptomycin and interferes with efficient isolation of the antibiotic. A variant was finally isolated which produced negligible amounts of the unwanted substance, thus allowing for greater synthesis and recovery of the desired antibiotic.

A careful study of variants with impaired productivity may elucidate biosynthetic pathways, contribute to the identification of precursors, and guide further steps in the improvement programme. In recent years, the use of major mutation has acquired particular significance, and has sometimes led to new and more efficacious products. The tetracycline-producing organisms appear to be particularly amenable to this approach. A modified tetracycline synthesized by a mutant strain of *Streptomyces aureofaciens* was shown to be changed at the C-5a position and was almost devoid of antibiotic activity. Another mutant, strain S-604, synthesized 6-demethyl-tetracycline, a new antibiotic material not elaborated by the parent strain, which proved to have several advantages over the methylated form and is today one of the leading commercial forms of tetracycline.

The use of major mutants coupled with some biosynthetic information can be fruitful. Modified phenylpyrroles with differing antifungal properties have been synthesized by metabolism of tryptophan analogues in *Pseudomonas fluorescens*, the producer of pyrrolnitrin. The addition of 6-fluoro-tryptophan to either wild-type strains or tryptophan analogue-resistant mutants led to the formation of 4′-fluoropyrrolnitrin, a modifed metabolite with significantly greater antifungal activity.

Role of minor mutation in strain improvement
Minor mutation usually plays the dominant role in strain development. By definition, such mutations affect only the amount of product synthesized, and such variants are usually phenotypically similar to the parent, with rapid and abundant mycelial and conidial development. As they produce only slightly more antibiotic than the preceding parent strain, quantitative definition of a 'significant' yield increase is somewhat relative, depending on the productivity of the parent and the statistical reliability of the quantification method. Usually, a 10 to 15% increase is implied. Such variants are usually selected from conidial populations exposed to moderate doses of a mutagen, repeatedly isolating 'minor' (positive) variants and using each succeeding strain for further mutation and selection. After several stages, a worthwhile yield increase is likely. Such increases have also been obtained by repeated selection without the introduction of mutagens. Minor variants with a slight quantitative change in one feature may also vary in other features owing to pleiotropic effects. Since such variation is slight, success depends on efficient and accurate selection techniques. The population to be tested must be large, and assay for the desired product must be accurate and specific.

Large-scale programmes concerned with the development of superior penicillin-producing variants of *P. chrysogenum* have been carried out in government, university, and industrial laboratories for over 30 years. From the cumulative screening of hundreds of thousands of strains, a series of very superior penicillin producers has been developed, certain of which are utilized today for commercial manufacture; for full accounts, see the Reading List. Over the years, 'mutation breeding' has been more extensive with strains of *P. chrysogenum* than with any other industrially important microorganism. The genealogy of the Wisconsin family of strains of *P. chrysogenum* and of two important modern industrial lineages is presented in Fig. 8.3. Initial screening was based on natural (non-mutated) colony populations, but subsequent natural populations failed to lead to major yield increases (for example, at Stanford University some 60 000 cultures

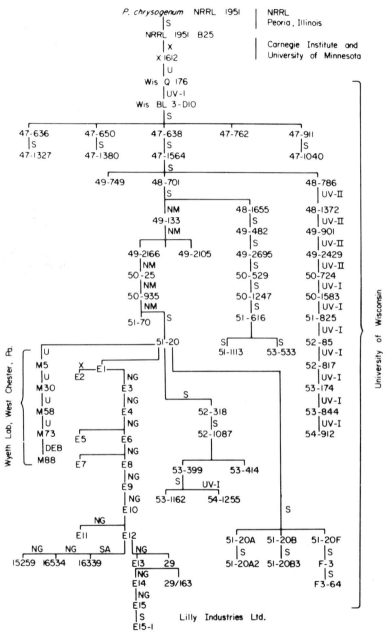

Fig. 8.3 The genealogy of 'Wisconsin' strains of *Penicillium chrysogenum* and of two industrial lines developed subsequently. S, mutations occurring spontaneously; X, isolated after X-ray irradiation; U, isolated after ultraviolet irradiation, wavelength unspecified; UV-I, as U, wavelength 275 nm; UV-II, as U, wavelength 253 nm; NM, isolated after nitrogen mustard treatment; NG, isolated after nitrosoguanidine treatment; DEB, isolated, after diepoxybutane treatment.

were screened without success). Key strains early in the ancestry of the Wisconsin series were the famous Q-176 culture, with significantly improved antibiotic titres, and strain BL3-D10, which did not produce the characteristic and troublesome chrysogenin pigment. All further mutant selections over the next decade were derived from Q-176.

Three distinct selection lines were established from the pigmentless mutant. One line was based on the selection of natural variants, another on ultraviolet-radiation survivors, and a third highly successful line was based on treatment with a nitrogen mustard mutagen. Industrial lines were later derived from the 51-20 strain using a variety of chemical mutagens. Throughout, it was noted that with ultraviolet radiation the highest morphological mutation rate did not necessarily coincide with the highest rate of kill; lower doses, yielding 25–30% survivors, were often more effective in inducing higher productivity mutants.

Examples of gradual step-wise improvement in antibiotic production are numerous, and some very typical data on penicillin demonstrate this well (Table 8.10).). Step-wise selection based on 10–12% greater penicillin productivity than the immediate parent resulted in the eventual selection of a variant which exceeded its original parent by 55%.

Table 8.10   Improvements in penicillin production in strains
of *Penicillium chrysogenum*

Strain[a]	Improvement in production over:	
	Previous strain	Fleming strain[b]
NRRL 1951	0.50	0.50
NRRL 1951 B25	2.70	4.05
X1612	0.06	4.90
Q176	0.52	9.00
47-1564	−0.04	8.55
48-701	−0.14	8.15
49-133	0.08	12.70
51-20	0.73	22.60
E15	1.05	50.64
E15-1	0.09	55.00
Cumulative total	5.55	176.09

[a] The strains concerned can be identified in the family lineage in Fig. 8.3.
[b] Using modern fermentation conditions.

In summarizing the mutation concept as applied to strain development, certain generalities appear appropriate.

   (a) Strains selected as obvious variants after exposure to mutagens are usually inferior in their capacity for accumulation of antibiotic;

improvements are extremely few, and their selection and evaluation is extremely important.

(b) Mutagen dose is important; mutants sought for major mutation roles are best isolated from populations surviving prolonged doses of mutagens, whereas variants for increased productivity are generally isolated from populations surviving intermediate dose levels.

(c) Strains with enhanced capacity for antibiotic synthesis generally exhibit wild-type morphology and growth habit; strains with altered morphology, etc., may be inherently better producers but may require considerable fermentation development. Since positive variants are so few, it is better to screen a smaller number of variants under a wider variety of environmental conditions, particularly when one seeks only quantitative changes.

(d) Step-wise selection implies small increments in antibiotic productivity, generally 10–15%, once the initial strains are obtained by natural selection, and as productivity increases, the probability of finding superior strains decreases. Hence, accurate and more sophisticated evaluation procedures are increasingly important. The development programme is only as effective as the mutation, selection and assay procedures used.

(e) Variant strains may require special propagation and preservation procedures, and actual production gains depend also on stability and reliability of performance. Maintenance through continued selection and purification plays an important role.

(f) Though a strain may meet the criteria of superiority in the laboratory, there is no guarantee that enhanced productivity will occur in production fermentors, and long-term pilot-plant studies are often necessary before any enhanced strain potential can be realized in actual production.

8.4.2   Isolation of mutant classes and their use in microbial processes

Though the application of a suitable mutagenic treatment procedure is essential, so is the development of rapid, selective isolation and testing procedures. Procedures for the isolation of mutant classes can be considered as variations on the proverbial problem of 'finding the needle in the haystack'. The simplest, and probably the most tedious, is to examine a large number of clones in liquid fermentations in hopes of finding one mutant which shows improvement in some particular biochemical character. This approach has often been used, but it has become increasingly important to consider new methods which allow selective isolation of particular classes of mutants, and particularly of overproducers.

The chance of discovering the desired mutant may be improved by increasing the frequency of mutation by a cyclical application of mutagens, either singly or in combination. An equally fruitful method is to design selective enrichment procedures which allow the mutants to grow more rapidly than the parent culture. Selection procedures using colour reactions to identify mutant colonies are very useful, while replica-plating procedures allow the efficient transfer of hundreds of colonies at a time, on to a variety of selective or test media.

The theoretical basis for all selective mutant isolation is to create conditions such that mutant cells, capable of producing larger quantities of enzymes or possessing greater permeability to substrates, will have distinct growth advantages and eventually outgrow and replace the less efficient parental strain. Nutrient media which restrict the growth of undesirable clones, but allow expression of particular mutant classes, are highly desirable. Selection for growth on novel substrates often restricts the growth of the parent but allows the mutants to grow, albeit at reduced rates. Often, leaky auxotrophs (bradytrophs) may be more sensitive to inhibition by metabolic analogues and this represents a way to discover mutants which are derepressed for the production of biosynthetic intermediates. The use of synchronized cells in chemostat culture is excellent for obtaining mutants resistant to high concentrations of toxic inhibitors, metabolic analogues, antibiotics, etc.

Auxotrophic mutants

Auxotrophic mutants are mutants which are defective in some metabolic step; they are recognized by their growth on a complex 'complete' medium and their lack of growth on a chemically defined 'minimal' medium. They are then classified into categories by propagation on minimal medium supplemented with amino acids (amino acid-requiring), vitamins (vitamin-requiring), or nucleic acid components (purine- or pyrimidine-requiring). By replication of a master-plate of mutant strains to a number of Petri dishes, each containing a pool of substances, arranged in such a manner that each substance is present in two different pools, only 12 plates are needed to test 36 different compounds. For example, auxotrophic mutants of the glutamic acid bacterium, *Corynebacterium glutamicum*, are the basis for the commercial production of amino acids in Japan; examples are listed in Table 8.11 (see Chapter 13). As a further example, Fig. 8.4 shows the remarkable progress in productivity of mutants producing high yields of the aromatic amino acids when subsequent analogue-resistance mutations are imposed upon the initial auxotrophs.

Table 8.11   Examples of amino acid accumulation by auxotrophic strains of
Corynebacterium glutamicum

Auxotrophic markers	Amino acid accumulated
hom; thr; leu; ile; met + thr; ile + leu	lysine
ile; leu; thr; arg	valine
lys	diaminopimelate
tyr	phenylalanine
phe	tyrosine
phe + tyr	tryptophan

Analogue-resistant mutants

The first step in isolating mutants resistant to structural analogues of meta-bolites is to determine, by plating on a graded series of inhibitor levels, the minimal concentration that prevents growth of the wild-type strain. Resistant mutants are then selected by plating out large numbers of muta-genized cells on plates containing increasing concentrations of the inhibitor.

Mutants resistant to amino acid analogues are likely to owe their resis-tance to a mutation (a) in a regulatory or operator gene, resulting in depres-sion of biosynthetic enzymes, thereby overproducing the antagonized amino acid and thus overcoming competitive inhibition by the analogue; (b) in the structural gene of a permease for the antagonized amino acid, resulting in reduced capability to incorporate the analogue; (c) in the structural gene of an amino acid acyl-tRNA synthetase, resulting in discrimination between the antagonized amino acid and the analogue during protein syn-thesis; and (d) in the structural gene of a biosynthetic enzyme subject to end-product inhibition, which has resulted in loss of feedback control.

Mutants resistant to analogues of primary metabolites often overproduce since the mutants are no longer subject to feedback repression. Such a concept was for example applied to the pyrrolnitrin fermentation. This antifungal phenylpyrrole is derived from tryptophan in strains of *Pseudomo-nas*. Tryptophan is a direct precursor, but impractical to use in large-scale fermentation because of its high cost. Efforts were made to select fluoro- and methyltryptophan-resistant mutants no longer subject to feedback inhi-bition by tryptophan. Resistant mutants were ultimately selected which no longer required tryptophan for pyrrolnitrin formation and which pro-duced nearly three-fold more antibiotic than the sensitive parent culture (Fig. 8.5).

Reversion mutants

(a) Reversion of non-producing strains. To increase the efficiency of selec-tion methods utilizing tetracycline-producing strains of *Streptomyces rimosus*

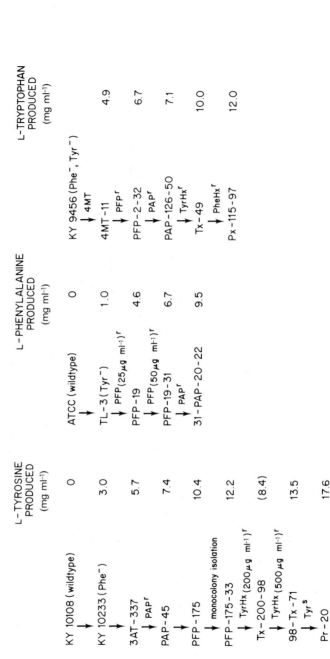

Fig. 8.4 Overproduction of tyrosine, phenylalanine and tryptophan by mutants of *Corynebacterium glutamicum*. Agents used to select resistant strains were: 3AT, 3-aminotyrosine; PAP, *p*-aminophenylalanine; PFP, *p*-fluorophenylalanine; TyrHx, tyrosine hydroxamate; 4MT, 4-methyltryptophan; PheHx, phenylalanine hydroxamate. After K. Arima (1977). *Dev. Ind. Microbiol.* **18**, 79–117.

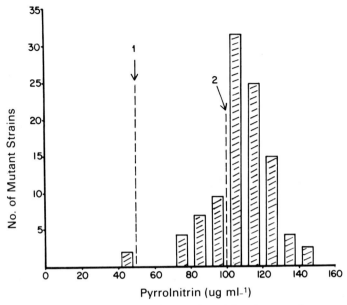

Fig. 8.5 Levels of pyrrolnitrin produced by strains of *Pseudomonas fluorescens* selected for resistance to 1 mg ml^{-1} of 5-fluorotryptophan. Index (1) marks the yield given by the grandparent 5-FT-sensitive strain, and (2) the yield given by the intermediate 5-FT-resistant parent strain.

and *Streptomyces viridifaciens*, procedures were developed based on the selection of high-producing variants from mutant populations. The greatest number of these variants was obtained from revertants of non-producing mutants, i.e., mutants in which a previous mutagenic treatment had induced lesions in loci controlling tetracycline biosynthesis. Reversion mutants from non-producing parents are readily screened by an overlay assay using a sensitive assay organism. Revertants will produce clear zones, whereas non-producers will show no zones of inhibition. By this technique, revertants of a non-producer parent were obtained which produced more than twice as much chlortetracycline as the original (grandparent) culture; one such revertant showed more than a six-fold increase.

(b) Reversion of auxotrophic mutants. Overproduction of precursor is another possible way to increase antibiotic formation, and can be achieved by relieving feedback regulation of its biosynthetic pathway. Modification of the structure of a feedback-sensitive enzyme through auxotrophic mutation, followed by enzyme replacement via reversion mutation, is a common method. The technique was used to increase chlortetracycline

production in a low-producing strain of *S. viridifaciens*. Reverse mutation of a homocysteine auxotroph resulted in 88% of the revertants producing three-fold more tetracycline than the original prototrophic culture.

Mutation to non-production of undesirable metabolites and the use of blocked mutants

Antibiotic mixtures containing a variety of related or unrelated antibiotics are commonly encountered. The problem is to produce the desired component exclusively, or at least to produce more of the desired component and less of the others. Mutation and medium manipulation with the nebramycin-producing organism *Streptomyces tenebrarius*, a culture originally producing a number of nebramycin components, yielded a variety of strains, some of which elaborated a single nebramycin. A methionine auxotroph of *Streptomyces erythreus*, the producing culture of erythromycin, when fed varying concentrations of the required amino acid, synthesized increased levels of erythromycin C relative to the erythromycin A component. *Streptomyces noursei* produces both cycloheximide and the unrelated metabolite, nystatin. Mutants were selected which no longer synthesized cycloheximide and which concomitantly produced markedly increased levels of nystatin. Strains of *C. acremonium* produce a variety of different antibiotics; intensive development programmes culminated in strains producing dramatically increased levels of cephalosporin C with proportionately lower levels of the related but unwanted penicillin N. In further mutants, blocked in the synthesis of cephalosporin C, there were strains which synthesized enhanced levels of deacetoxycephalosporin C (as well as deacetylcephalosporin C). Such strains could be useful in producing an important intermediate for the manufacture of orally active cephalosporin derivatives.

Metabolite-resistant mutants

Metabolites are often toxic to growing cultures of the producer organism, though this toxicity is seldom important in production terms. Antibiotics which inhibit the growth of the producer organism include tetracycline, novobiocin, actinomycin, streptomycin, and nystatin. Differences in biosynthetic activity may be affected by genetically determined resistance to the producing culture's own antibiotic. Therefore, increasing autoresistance, either by adaption to gradually increasing concentrations of antibiotics or by similar selection after mutagenesis, is a useful procedure for improving production strains. A four-fold increase in chlortetracycline (CTC) productivity in *S. aureofaciens* was obtained by repeated transfers in media containing 200–400 µg CTC ml^{-1}. The resistance of the surviving strains to increasing concentrations of chlortetracycline was in direct proportion to their capacity to synthesize the antibiotic. Selection at low concentrations of tetracycline led to increased frequency of low-producing variants, while high concentrations resulted in an increased number of active mutants. Clearly the lethal effect of the antibiotic selectively deprived the natural population of low-producing sensitive clones. This method is most effective during the early phase of strain improvement or when a population of a high-producing strain must be repurified. The imperfect fungus *Geotrichum flaveobrunneum* NRRL-3682 produces an interesting series of azasterols which are inhibitory to fungi. Mutants selected for increased resistance to the metabolite AzB or to the polyene antibiotic pimaricin showed increased ability to synthesize azasterols.

Mutations affecting permeability
Production of large amounts of glutamic acid in bacteria can be attained by selecting mutants having an altered cell membrane with increased outward permeability of the amino acid (see Chapter 13). The increased permeability can be attained in wild-type strains of *C. glutamicum* by biotin deficiency or addition of penicillin or fatty acid derivatives; it can also be obtained, with some process advantages, by oleic acid limitation of oleic acid-requiring auxotrophs or by glycerol limitation of glycerol-requiring mutants. Permeability may also be important in the production of L-lysine. Mutants of *E. coli* resistant to S-(β-aminoethyl)cysteine excrete lysine and are defective in the lysine transport system. The overproduction of lysine was not due to mutations affecting biosynthetic enzymes, but rather to active transport systems which enable an efflux of lysine against a five-fold higher accumulation of extracellular (compared to intracellular) lysine.

Miscellaneous classes of mutants
The metabolic poison monofluoroacetate (MFA) inhibits aconitate hydratase, an enzyme which converts citrate to isocitrate. Mutants of *Candida lipolytica* sensitive to MFA produced less aconitate hydratase and, when used to produce citric acid on *n*-paraffin, synthesized citric acid and isocitric acid at a ratio of 97:3 compared to the usual ratio of 60:40. Selenomethionine-sensitive strains of *C. acremonium* have increased capacity to synthesize cephalosporin C due to enhanced methionine uptake.

   The production of glucose isomerase (GI) in strains of *Streptomyces*, used commercially for the conversion of glucose to fructose, was repressed by glucose and required the presence of D-xylose as an inducer. Mutants resistant to D-lyxose, an analogue of glucose, were found to produce GI constitutively (without xylose) but were still repressed by glucose. Additional selection for resistance to a glucose analogue, 3-*O*-methylglucose, resulted in the isolation of a unique mutant insensitive to catabolite repression, which produced GI in the presence of glucose.

Blocked mutants and product cosynthesis
Cosynthesis depends upon the use of two blocked mutants which, when propagated together, are able to synthesize the end-product. Intermediates which accumulate in the blocked mutants and diffuse out of the cell are used by mutants blocked earlier in the pathway. The phenomenon has been reported for a variety of antibiotics and is particularly useful in elucidating biosynthetic mechanisms.

Localized mutagenesis and comutation
Localized mutagenesis, affecting small selected regions of the chromosome, offers a promising new approach. Mutation programmes can be directed to maximize mutations in any marked area on the chromosome, especially areas known to affect the formation of end-products.

The technique has been effectively applied to the isolation of temperature-sensitive mutants and to select a large number of nutritional mutants within the histidine operon of *Streptomyces coelicolor*. Isolation of comutants in unknown loci linked to the revertant site can be done by a heterokaryon method or by the use of temperature-sensitive mutants. Comutation also appears promising as a relatively simple tool for gene linkage and mapping studies.

Mutational biosynthesis (new metabolite formation)
Mutation of microorganisms producing secondary metabolites has resulted in the selection of strains capable of producing new modified metabolites either directly or in response to some added precursor analogue. The modified metabolites usually possess structural features of the parent compound but often lack certain functional groups, or contain structurally modified functional groups which confer differing biological activities. This approach to generate new secondary metabolites has been especially useful with new aminoglycoside antibiotics. For example, wild-type strains of *S. fradiae* normally synthesize neomycins A, B, and C, all of which include a deoxystreptamine subunit. A mutant strain derived from an NG-treated population was unable to synthesize deoxystreptamine, and was therefore dependent on an outside source of diaminocyclitols for antibiotic synthesis. When the mutant was cultured with added streptamine instead of deoxystreptamine, two new antibiotic substances were produced which were called hybrimycin A_1 and A_2. Similarly the addition of epistreptamine gave two more antibiotics, hybrimycin B_1 and B_2.

By a similar approach, biosynthetic analogues of butirosin, another aminoglycoside antibiotic, were obtained using mutant strains of *Bacillus circulans*. Other examples of new biosynthetic products produced by mutant strains are listed in Table 8.12.

Table 8.12   Mutational biosynthesis and new metabolite formation

Organism	Original product	New product
Pseudomonas aureofaciens	Pyrrolnitrin	4'-Fluoropyrrolnitrin
Streptomyces peucetius	Daunomycin	Adriamycin
Nocardia mediterranei	Rifamycin B	Rifamycin W
Micromonospora inyoensis	Sisomycin	Mutamycins
Streptomyces griseus	Streptomycin	Streptomutin
Cephalosporium acremonium (ATCC 20389)	Penicillin N	6-(D)-[(2-amino-2-carboxy)-ethylthio]-acetamido-penicillanic acid
Micromonospora purpurea	Gentamicin	Hydroxy- and deoxy-gentamicins

8.4.3   Rational selection procedures

All of the methods and mutant classes so far described can be made part of a rational selection programme, and usually a number of the procedures will be called upon. One of the most important guidelines to follow in making a procedure truly rational is an understanding of the biosynthetic pathways leading to the desired product, including its precursors, and of the regulatory mechanisms those pathways involve. Even rather general information of this kind can be fruitful, and individual examples have already been cited. The operating advantages of rationally directed selected procedures are also illustrated by the data in Table 8.13, which quantifies their benefits in the cases of penicillin and cephalosporin strain improvement programmes. Here the consistently higher yields of mutants deserving retention for further study represent very considerable saving in effort, time, and skilled manpower requirements.

8.4.4   Sexual and parasexual breeding as a basis for strain improvement

In contrast to eubacteria (including actinomycetes), the yeasts and moulds possess true nuclei and many species have regular sexual cycles comparable

Table 8.13 Camparison of random vs. directed selection procedures in strains of *P. chrysogenum* and *C. acremonium*

Type	Muta-genesis	Organism	Selected method	% Retained[a]
Random	UV, NG	*P. chrysogenum* and *C. acremonium*	None	0.81
Directed	UV, NG	*P. chrysogenum* and	1. Colony plate	1.36
			2. Auxotrophs	5.71
	X-ray	*C. acremonium*	3. Haploidization agents	1.59
			4. Mitotic inhibitors	1.33
			5. Mercury resistance	1.23
			6. Amino acid analogues	1.41
			7. Sulphur analogues	3.98
		C. acremonium	1. Methionine analogues	3.64
			2. Increased methionine sensitivity	
			(a) growth	1.62
			(b) β-lactam synthesis	3.85

[a] Superior on both primary and secondary screening tests.

to those in higher plants and animals, the major difference being the prolonged and generally predominant haploid phase in fungi. *Neurospora crassa* is typical, and is a fungus which has been thoroughly exploited by geneticists. The vegetative mycelium is haploid and can be propagated almost indefinitely by serial transfer of hyphal fragments or asexual spores (conidia). Two strains of opposite mating types (*A* or *a*) are required to initiate the sexual cycle, and are mated by mixing the conidia of mating type *a* with mycelia of mating type *A* on appropriate media. After a period of nuclear division and nuclear migration, fusion between *A* and *a* nuclei takes place. Each fused nucleus (diploid) immediately undergoes meiosis to form four haploid nuclei which, in turn, undergo two successive mitotic division to form eight ascospores, contained linearly in a sac (ascus). Analysis of the linear segregation of each genetic marker as well as the recombination between markers in each ascus provides a unique way of mapping genes in *N. crassa*.

Few of the industrially important fungi, however, display the conventional sexual cycle as found in *N. crassa*. However, they often form heterokaryons in which rare diploid nuclei result from the fusion of two haploid nuclei. This process is called parasexuality (Fig. 8.6). Although recombination (mitotic recombination) is much less frequent in the parasexual cycle compared to the meiotic process, it can occur by mitotic crossing over or by other mechanisms. The practical importance of mitotic recombination is that it makes possible genetic analysis and controlled breeding in organisms with no sexual cycle. The parasexual cycle has been demonstrated

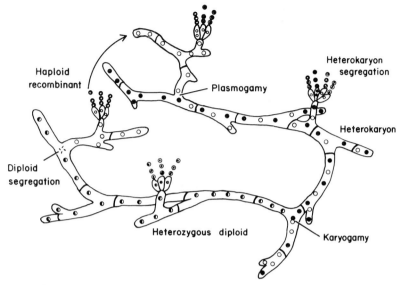

Fig. 8.6 The parasexual cycle in a fungus.

in many industrial fungi including *P. chrysogenum*, *C. acremonium* and *Aspergillus oryzae*. At least one successful attempt to apply parasexual genetics to strain improvement in *P. chrysogenum* has been reported and in one study, a homozygous diploid representing genome duplication of the haploid parent was an efficient producer of penicillin V (Fig. 8.7).

Yeasts have a life cycle similar to that of *N. crassa* but morphologically simpler. The haploid phase of *Saccharomyces cerevisiae* consists of free cells with single nuclei which propagate by budding. The diploid phase is initiated by fusion of haploid cells of opposite mating type (*a* or *α*), followed by meiosis leading to the formation of four haploid cells (ascospores). Unlike the situation in *Neurospora* and other fungi, diploid nuclei in yeast can divide by mitosis and a diploid culture can multiply indefinitely by budding.

8.4.5 Protoplast fusion technology and strain improvement

Gene transfer following the fusion of microbial protoplasts represents one of the newer approaches. Protoplasts are generated by the treatment of whole cells with a variety of lytic enzymes, and complementary biochemical mutants can be used to 'force' recombinants. An osmotic stabilizer is essential to provide osmotic support for the protoplasts, and fusion is enhanced by the addition of polyethyleneglycol (PEG).

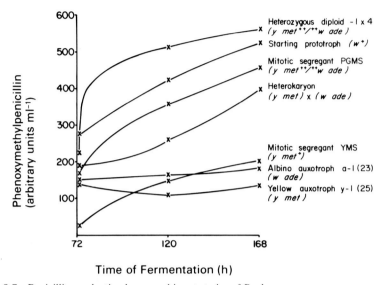

Fig. 8.7 Penicillin production by recombinant strains of *P. chrysogenum*.

Protoplast fusion technology has been used with success with commercial strains of *C. acremonium*, in which conventional genetic manipulations have proved to be difficult. Recombinant strains which produced significantly more cephalosporin C than the biochemically deficient parents, and strains which produce cephalosporins efficiently from inorganic sulphate, have been found.

Protoplast fusion technology has also been applied for the development of fast-growing, low *p*-hydroxypenicillin V strains of *P. chrysogenum*. The development of low *p*-hydroxypenicillin V fermentations was important because the hydroxylated product interferes with the chemical ring expansion of penicillins to oral cephalosporin products. Table 8.14 summarizes some results obtained in this example of what promises to be a technique of wide application.

The protoplast fusion technology has recently resulted in the production of indolizomycin, a new indolizine antibiotic. The antibiotic was produced following protoplast fusion between two non-antibiotic-producing strains of *Streptomyces griseus* and *Streptomyces tenjimariensis*.

8.4.6   Potentials for R-DNA technology in microbial improvement

In contrast to conventional mutation-selection methodology, which often involves random trial and error approaches to generate superior microbial

Table 8.14   Application of protoplast fusion for the breeding of low *p*-hydroxypenicillin V-producing strains of *P. chrysogenum*

Strains	Phenotype	Progeny following fusion	Recombinant (R) or parental type (P)	No.
Pc A ×	$G^f\ OHV^h$	$G^f\ OHV^h$	P	13
Pc B	$G^s\ OHV^o$	$G^f\ OHV^L$	R	3
		$G^f\ OHV^o$	R	1
		$G^s\ OHV^o$	P	14
		$G^s\ OHV^L$	R	3
		$G^s\ OHV^h$	R	3

G^f = fast grower; G^s = slow grower; OHV^h = 8–10% *p*-PHV; OHV^L = 2–5% *p*-PHV; OHV^o = 0.05% *p*-OHV.

strains, recombinant DNA technology requires a directed effort to create desired, guided genetic changes. R-DNA technology uniquely enables one to *design* microbial strains with genetic capabilities to synthesize end-products which otherwise could not be obtained by microbial fermentation. The relevant principles and methods are considered in detail in Chapter 19; briefly, the steps for expression of foreign DNA in a host cell can be summarized as follows:

(1) A vector DNA molecule (plasmid, phage, 2 μm DNA, etc.) capable of entering and replicating within a host cell. The vector should be small, easily prepared and must contain one site where the integration of foreign DNA will not destroy an essential function.
(2) A method of splicing foreign DNA into the vector.
(3) A method of introducing the hybrid DNA recombinants into the host cell and a discriminating procedure for selecting the presence of the foreign DNA.
(4) A method for assaying for the foreign gene product.

Since its first demonstration in *E. coli*, R-DNA technology has been one of the most rapidly developing areas of science. Dramatic advances have appeared, and general prospects are presented in Chapter 19. However, the basic techniques were developed for the transfer of single fully-identified units of the genome, and their most spectacular applications have correspondingly been in the production of single-gene products. Nearly all antibiotics and similar metabolites, by contrast, correspond to rather large numbers of genes, and it is not clear how the full rationale of R-DNA technology is to be applied in such cases. Specific opportunities do arise when an identified enzyme activity can usefully be added to an established pathway—for example the possible value of introducing the acyl transferase capability from *Penicillium* into *Cephalosporium* to produce new cephalosporins is a clear target for the application of newly developed R-DNA techniques that can be used in fungi.

R-DNA technology is also undergoing rapid development in streptomycetes, which produce 60% of the known antibiotics. Development of DNA cloning systems for antibiotic-producing *Streptomyces* is under way, and should greatly facilitate detailed genetic analysis of specific antibiotic biosynthetic pathways and of molecular mechanisms involved in their control. DNA cloning can also result in new combinations of gene sequences which lead to novel antibiotics or to increased yields of existing or hybrid antibiotics. Gene cloning of a whole set of biosynthetic genes for the actinorhodin antibiotic pathway in *Streptomyces coelicolor* resulted in the formation of new hybrid isochromanequinone antibiotics. The new antibiotics are known as mederrhodins and dihydrogranatirhodin. The demonstration of novel compounds in genetically engineered strains shows the usefulness of gene cloning techniques as a new tool in antibiotic discovery.

Meanwhile, considerably less specific techniques of gene transfer such as protoplast fusion can be used to create new assortments of genetic diversity, to which the methods described earlier in this chapter can then be applied to search either for improved production capabilities or wholly new products; this is an important though less spectacular development which is already being widely investigated and has resulted in new antibiotic formation.

In conclusion, the applications of protoplast fusion, high efficiency DNA transformation and R-DNA technology to industrially important microorganisms appear numerous and will undoubtedly receive considerable attention over the next decade. Moreover, the new genetic technologies offer exciting prospects for the systematic genetic construction of improved strains.

Reading list

Genetics of Industrial Microorganisms
    Vol. 1 (1973) (Ed. Z. Vanek, Z. Hostalek and J. Cudlin). Elsevier: Amsterdam.
    Vol. 2 (1976) (Ed. K. D. Macdonald). Academic Press: London and New York.
    Vol. 3 (1979) (Ed. O. K. Sebek and A. I. Laskin). American Society of Microbiology: Washington DC.
    Vol. 4 (1982) (Ed. Y. Ikeda and T. Beppu). Kodansha Ltd: Tokyo.
Abelson, P. H. (1984). 'Biotechnology and Biological Frontiers'. American Association for the Advancement of Science: Washington DC [Collection of 36 recent papers on basic research and applications of biotechnology.]
Berdy, J. (1980). *Proc. Biochem.* **15**, 28–35. [Classification of known antibiotics.]

Bu'Lock, J. D., Nisbet, L. J. and Winstanley, D. J. (Eds) (1982). 'Bioactive Microbial Metabolites—Search and Discovery'. Academic Press: London and New York.

Elander, R. P. and Chang, L. T. (1979). *In* 'Microbial Technology', 2nd Edn, Vol. 2 (Eds H. J. Peppler and D. Perlman) pp. 243–302. Academic Press: New York and London. [Culture selection and industrial microorganisms.]

Gomi, S., Ikeda, D., Nakamura, H., Naganawa, H., Yamashita, I., Hotta, K., Kondo, S., Okami, Y. and Umezawa, H. (1984). *J. Antibiot.* **37,** 1491–1494. [First report of new antibiotic produced by interspecies protoplast fusion.]

Hopwood, D. A. and Chater, K. F. (1980). *Phil. Trans. Roy. Soc.* (*London*) **B290,** 313–328. [Genetic approaches for discovery and improved yields.]

Hopwood, D. A., Malpartida, F., Kieser, H. M., Ikeda, H., Duncan, J., Fujii, I., Rudd, B. A. M., Floss, F. G. and Omura, S. (1985). *Nature* **314,** 642–644. [First report of hybrid isochromanequinone antibiotics by genetic engineering.]

Iwai, Y. and Omura, S. (1982). *J. Antibiot.* **32,** 123–142. [Effect of culture conditions in screening programmes.]

Nakayama, K. (1981). *In* 'Biotechnology', Vol. 1 (Eds H. J. Rehm and G. Reed) pp. 355–410. [Comprehensive review of soil screening procedures.]

Peberdy, J. F. (1979). *Ann. Rev. Microbiol.* **33,** 21–39. [Protoplast fusion.]

Perlman, D. and Peruzotti, G. P. (1970). *Adv. Appl. Microbiol.* **12,** 277–294. [Classic review, especially for plate testing procedures.]

Umezawa, H. (1977). 'Enzyme Inhibitors of Microbial Origin'. University Park Press: Baltimore. [Definitive review of discovery of industrial antibiotics.]

Umezawa, H. (1978). 'Index of Antibiotics from Actinomycetes'. University Park Press: Baltimore.

Vandamme, E. J. (1984). *In* 'Biotechnology of Industrial Antibiotics', pp. 3–31. Marcel Dekker: New York.

Vournakis, J. N. and Elander, R. P. (1983). *Science* **219,** 703–709. [Genetic engineering approaches to strain improvement.]

9 | Instrumentation

B. KRISTIANSEN

9.1 Introduction

This chapter deals with instruments used in measurement and control of fermentation processes. It will be seen that they are by no means exclusive to fermentation technology, and most find widespread application in biotechnology and related technologies. In fact the majority of the analytical instruments used in biotechnology are not biotechnological developments, and until quite recently, biotechnologists usually employed instruments which were developed for use in the chemical industry. Though excellent instruments in their own right, they were not designed for the higher degree of complexity of biological reactions and with the onset of computer control, this became very obvious.

The development of fully automated computer controlled fermentation processes depends upon the availability of sensors producing meaningful signals which can be translated into control action. Considerable effort is now being devoted to the development of more suitable instruments which can provide more detailed information about biological processes, needed for continued improvements in process yields and productivity. Unfortunately, few ideas or developments have yet been realized on a commercial scale, but it is only a matter of time before this will change. However, this chapter concentrates on current instruments, their application (on all scales) and their present limitations.

Computer control is the current by-word in fermentation technology and will soon reach the point where you dare not admit that your fermenters are not coupled to a computer; unfortunately, amidst the improvements this brings, we are beginning to see cases of the old dogma 'instruments prevent creative thinking'. Computer control requires on-line instruments (or at least off-line instruments where the time lag between sampling and analysis is short) and this is reflected in this chapter.

9.2 Terminology

Given our total dependence on the instrumentation for understanding bio-technological processes, it is important that we understand the instruments we use, otherwise we can easily make incorrect assumptions about their suitability or performance. Some common characteristics are explained below.

Response time is normally defined as the time required for the output signal to change from the initial value to 90% of the final value following a step change. As a rule of thumb instruments used in biological systems should have a response time less than 10% (preferably much less) of the system doubling time. Thus, in a typical fermentation, with doubling time around 3 h, instruments with response time above 18 min are not suited to on-line control. Most instruments have much lower response times and it is often the time taken for other operations associated with the sampling (e.g. pumping fluid through tubing, sample preparation, etc.) which produces an excessive lag between the measurement and the control action.

Sensitivity is a measure of the change in output of an instrument resulting from a change in input. Generally, instruments with high sensitivity are preferred, making it possible to measure small input changes. However, other instrument parameters such as linearity (of output to input), accuracy and range should be taken into consideration when choosing an instrument.

Linearity between input and output is the simplest and most desired relationship between the two. The greatest simplification is in calibrating (or recalibrating the instruments) with a minimum number of calibration points required.

Accuracy indicates an uncertainty about the output signal. It is usually expressed as a percentage of the instrument range. The higher the accuracy, the smaller is the uncertainty about the output signal as an indicator of the true value of the input signal.

Drift is the variation in the output independent of input. Normally this is seen as an output signal which falls with time. This is an inherent property of the instrument and can only be overcome by recalibration at regular intervals. Given the high demand for asepsis over long runs, drift may be a severe problem with instruments applied to fermentation processes.

Resolution is the smallest measurable value of changes in the input and is normally expressed as a percentage of full scale deflection.

Offset (error, static error or residual error) is a deviation in the output from the true value of the input when the latter is held constant.

Reproducibility should never be assumed and instruments should be recalibrated whenever possible, particularly instruments involved with O_2 and CO_2 measurements. Often the absolute value of the measured variable is not as important as its relation to the input value. Constant recalibration will, therefore, compensate for changing instrument characteristics.

9.3 Process control

There are three possible aims for process control:

(1) maintain a variable constant through time
(2) force a variable to follow a prescribed path through time
(3) optimize some function of the system variables.

The first is achieved by regulation, the second by servomechanisms and the third by optimal controllers. All these items of hardware are generally termed automatic controllers.

In a process control system, we have four classes of variables:

(1) controlled variables
(2) disturbance variables
(3) manipulated variables
(4) reference variables.

The controlled variable is the output variable we want to control and the manipulated variable is the input variable with which we are controlling it. A disturbance variable is an input variable affecting the controlled variable through variation in other inputs than the manipulated variable and the reference variable is the desired value of the controlled variable. An illustration of the way in which these four classes of variables are linked is given in Fig. 9.1.

The three control structures commonly referred to, open loop, feed forward and feedback, are shown in Fig. 9.2. Open loop control can only be used if disturbances are absent and process conditions never change. (Feed forward control can be used as long as disturbances can be measured and a process model is available.) Manual control systems are normally described as feed forward. Feedback (or closed loop) control can accommodate any number of disturbances and no knowledge of the process is (in principle) necessary.

Typically, temperature control systems will be based on feedback control. A typical feed forward system is a continuous fermenter in which feed is pumped into and broth taken out of the fermenter continuously. Here, the dilution rate will be the controlled and the feed flow rate the manipulated variable.

The speed at which a system corrects a transient input or adjusts to a new set point is decided by its dynamic response. This may be adjusted

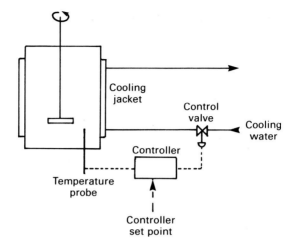

Fig. 9.1 A scheme for controlling bioreactor temperature illustrating how the four types of process control variables are linked. In this case the temperature of the reactor content is the controlled variable, the controller set point is the reference variable and the flow of cooling water is the manipulated variable with the temperature of the cooling water acting as a disturbance variable.

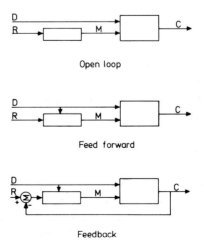

Fig. 9.2 Three common control structures.

for cyclic or damped response as illustrated in Fig. 9.3 by selecting suitable control strategies.

(a)   Time

(b)   Time

Fig. 9.3   Typical dynamic responses of a system to a set point change. (a) Cyclic response; (b) damped response; --- input — system response.

The simplest type of control is the on–off or two position action which is used extensively in control of fermentation processes. The controlled variable will cycle between the two set points but the dynamic response of fermenter broths are sufficiently poor to dampen the effect of the control action, thus avoiding overshooting. A typical relationship between controller output and response is illustrated in Fig. 9.4.

Process control technology is well advanced and typical fermentation processes will only need relatively simple controllers. The problems are encountered in measurements and actuation of control action demanded by the controller and this chapter has sought to highlight the present state of the technology.

9.4   Air flow monitoring

In aerobic fermentations, air is passed through the fermenter to provide oxygen and also to remove carbon dioxide which would otherwise affect metabolic activity. In tower and air-lift bioreactors, the air also promotes bulk mixing. Normally, the air flow rate lies in the range 0.5–1.5 vvm (volumes of air per reactor volume per minute). Higher flow rates should

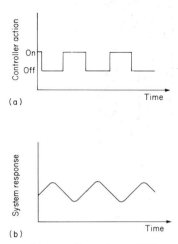

(a)

(b)

Fig. 9.4   Relationship between (a) controller action and (b) system response.

be avoided as this will lead to impractically high superficial gas velocities ($m\,s^{-1}$) in large reactors when scaling-up is attempted.

The air flow is normally monitored by manually controlled variable area flowmeters (rotameters) (usually calibrated for flow at NTP conditions). If the line pressure changes, a control valve should be used to keep the flow rate constant. The position of the rotameter float can be detected by a proximity sensor, but if a signal is required for process control purposes, it is more common to replace the flow meter with a thermal mass flow meter, as illustrated in Fig. 9.5. Here, two probes read the temperature

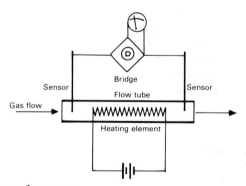

Fig. 9.5   Thermal mass flow meter.

difference between inlet and outlet gas as it passes through a heated tube. Incorporating the probes in a Wheatstone bridge allows the temperature difference to be translated into an air flow rate. Such flow meters are usually more accurate than rotameters. However, their flow rate range is more restricted, limited to laboratory and small pilot-scale operation.

9.5 Measurements of power input

The unit of work (or quantity of heat) has the unit joule (J). Power is the rate of work, or rate of flow of energy, and is measured in watts (W) where $W = J\,s^{-1}$. To assess fermentation processes in stirred tank reactors, it is important to determine how much energy is imparted to the liquid by the impeller. The power input per unit of fermenter volume has often been used as a scale-up parameter and the convention of $1\,HP/100\,gal$ is often quoted. In SI units, this is equal to $1.64\,kW\,m^{-3}$.

Total power uptake can be determined by employing a wattmeter measuring the rate of power consumption by the drive motor (also a convenient way of measuring money spent, which on large-scale fermenters can be of the order of 10% of the operating cost!). For industrial-scale vessels, this will also give the power input to the liquid (shaft power). On pilot and laboratory-scale fermenters, however, friction losses in stirrer glands and motor inefficiency cannot be neglected and other methods must be employed. Dynamometers and strain gauges are used in pilot plants; the latter can be incorporated in the stirrer shaft providing direct measurements of the shaft power input. A simple dynamometer for small fermenters can be made by mounting the motor in two load bearings on the motor shaft positioned vertically above the fermenter. The power input is determined by the torque required to prevent the motor from turning during operation.

Another scale-up criteria is constant tip speed (see Chapter 5), being defined as ND where N is the speed of rotation of the impeller and D is the impeller diameter. It is quite common for large reactors to have fixed speed motors, typically 100 r.p.m. in $100\,m^3$ vessels. The impeller speeds in smaller reactors can be determined using tip speed as a scaling-down factor, as shown in Table 9.1. In this case, the impeller diameter and the aspect ratio (liquid height to diameter) reflect typical units. The table indicates that small fermenters must be operated at excessive impeller speeds. Note that power per unit volume is not kept constant when tip speed is used as a scale-up parameter (see Table 5.5).

9.6 Temperature measurements

Most microorganisms have a very narrow temperature range for optimal biological activity (see Chapter 4). Heat energy is readily transported into cells and temperature control is, therefore, an important process parameter.

Table 9.1   Scale-down on constant impeller tip speed in a stirred tank reactor. Data based on impeller speed of 100 r.p.m. in a 100 m³ fermenter

Fermenter volume (m³)	Liquid height / Vessel diameter	Impeller diameter / Vessel diameter	Impeller speed (r.p.m.)
0.001	1	1/3	2775
0.01	1	1/3	1191
0.1	1.5	1/3	633
1	2	1/4	431
10	2	1/4	200
100	2.5	1/4	100

Moreover, temperature affects the performance of a range of instruments not usually associated with temperature control and it is important that these are fitted with manual or, preferably, automatic temperature compensation circuitry.

The temperature probes which have found widespread application are resistance thermometers and thermistors.

9.6.1   Resistance thermometers

These are based on the principle that metal resistance increases with temperature. They are constructed as a length of wire, often wound in a coil to reduce size. When current is passed through the coil, a voltage change is produced which can be related to the temperature. The thermometer itself will have a response time of the order of 1 s, but normally it is inserted in a sheath to protect it from the environment, which may increase the response time by up to 5–10 s.

Resistance thermometers have a considerable range (typically −100 to 650°C for a platinum thermometer). The relationship between temperature and thermometer resistance is not linear over the whole range, but can safely be assumed for the limited range required for biotechnological processes. In a few cases, particularly in fermentation systems with a high solids content and little mixing, it has been found that the current passing through the thermometer may heat the probe and its environment, thus producing erroneous readings.

9.6.2.   Thermistors

These are semiconductors exhibiting increasing conductivity with temperature. They have response times of the order of 1 s when the thermal contact is good. Encapsulating the thermistor (usually in teflon or stainless steel) may increase this to 10 s.

Thermistors are very sensitive, but the output is highly non-linear. The considerable response to small changes in temperature makes thermistors well suited to application over a limited temperature range (as in fermentation technology), over which the response to temperature changes may be assumed to be linear. Thermistors are also relatively inexpensive.

Temperature control based on temperature sensor outputs, is normally achieved by using steam or electric energy for heating and water for cooling. The technology is highly developed, giving control to $\pm 0.1°C$ for laboratory-scale and $\pm 0.5°C$ for large-scale. The main problem in temperature control is encountered in large fermenters where the surface area designed for heat transfer may not be sufficient to remove the metabolic heat and heat of mixing.

9.7 Rheological measurements

Viscosity, the resistance of a fluid to flow, may be used to indicate the rheological properties of a fluid, and is an important process parameter which is seldom given the attention it merits. The performance of bioreactors can be significantly affected by the flow characteristics of the broth, as shown in Chapters 5 and 17, and it is important to have instruments which can measure rheological properties quickly and reliably.

The viscosity of a fluid is determined in a shear stress–shear rate diagram as shown in Fig. 9.6 with a unique curve for different types of fluids. In the case of a Newtonian fluid, the viscosity is equal to the slope of the straight line, and such a fluid has constant rheological properties. This is not the case with the other fluids for which the viscosity changes with shear rate, making it important to state at which shear rate the viscosity is measured. In general, one tries to work at the same shear rate as that near the fermenter impeller, which can be difficult to determine as shear rate will change with distance from the impeller.

Unfortunately, rheological measurements can, with a few exceptions, only be carried out off-line. The most common viscometers are described below and shown in Fig. 9.7a–d.

9.7.1 Tube viscometers

There are two types of tube viscometers. In a capillary viscometer, the liquid flows under gravity through a tube of known diameter. The time taken for a given volume to pass can be related to viscosity, using the viscosity of pure water as a reference.

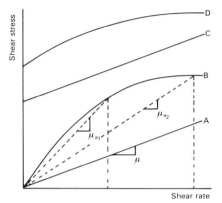

Fig. 9.6   The relationship between shear stress and shear rate for different classes of fluids. A, Newtonian fluids; B, pseudoplastic; C, Bingham plastic; D, Casson fluid. μ = viscosity and μ_{a1}, μ_{a2} = shear-dependent viscosity (also referred to as apparent viscosity) of pseudoplastic fluid determined at two different shear rates.

In the second type of tube viscometer, the liquid is pumped through a narrow tube under laminar flow conditions: the pressure drop over a fixed length of tube is measured and related to viscosity.

Tube viscometers are simple and can be used for quick and reliable off-line viscosity determinations of Newtonian fluids. A few reports on on-line applications have been published. In general, however, tube viscometers are not well adapted to fermentation broths, which are often non-Newtonian and will usually contain solids which can block the tube.

9.7.2   Cone and plate viscometers

The fluid is placed between a rotating cone and a stationary plate. Provided the angle between the cone and plate is less than 4° and edge effects can be neglected, the shear rate is uniform throughout the liquid and the shear stress and, hence viscosity, can be evaluated. This instrument is of a much more complex construction; a high degree of engineering skill is required in making the cone and matching it with the plate. It is not suited to fluids with solids such as mycelium or cell aggregates.

9.7.3   Concentric cylinders (cup and bob)

Concentric cylinders, also termed cup and bob viscometers, are used extensively. The fluid is placed in the annulus between the cylinders and the shearing action is produced by rotating one (normally the inner) while

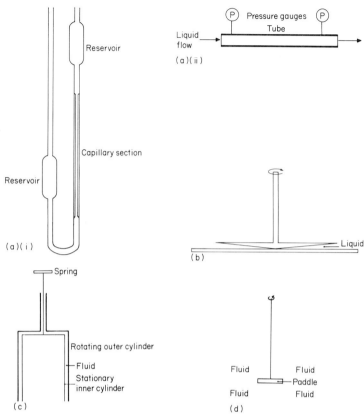

Fig. 9.7 (a) Tube viscometers (i) and (ii); (b) cone and plate viscometers; (c) concentric cylinders (cub and bob viscometer); (d) infinite sea viscometer.

the other is kept stationary. The torque exerted on the second cylinder is related to the viscosity of the fluid. Strict demands on the engineering of these viscometers are essential for reliable readings.

The cup and bob viscometers are excellent for Newtonian fermentation fluids and manufacturers have produced modified versions which can be used on fluids containing filamentous organisms or other high viscosity imparting substances. However, there may be the problems of solids blocking the annulus or wall slip which will lead to inaccurate readings.

9.7.4 Infinite sea viscometers

In these instruments, the torque is measured on an impeller or disc rotating in a large volume of the fluid; the fluid is free flowing and not affected by the walls of its container. Theoretically, these instruments are suitable for all kinds of fluids, and unaffected by suspended solids. However, shear rate changes with distance from the rotating disc and there will be a problem in determining at which shear rate the readings should be taken. Normally an average shear rate can be defined, allowing these relatively simple viscometers to be used.

9.8   Foam control

Foaming can be a serious problem in fermentation. If allowed to form it may block outlet filters, encourage contamination, or reduce yield. Unfortunately, most fermentation media encourage foam formation, and its depression can be achieved mechanically and/or chemically.

The simplest form of mechanical foam break-up for a stirred tank is to fix an additional impeller on the stirrer shaft above the liquid level. A number of manufacturers supply separate mechanical devices including specially designed impellers operating under their own power independent of the agitation system and sophisticated gas–liquid separation devices. They normally give good foam control, but their relatively high power consumption can lead to high operating costs.

Commonly, foam control is achieved by using a simple break-up device aided by chemical foam control, for which a range of antifoam agents are available. Automatic antifoam addition systems incorporate a conductivity probe immersed through the top of the fermenter vessel. The probe can simply be a stainless steel rod, coated with teflon, but exposing the tip. The probe will be activated by foam touching the tip of the probe. A signal will be sent to the control system leading to the addition of antifoam by a pump (small fermenters) or pneumatically (large-scale). Antifoam will enter the fermenter, break down the foam which will leave the tip of the probe, and the circuit is broken.

It is important to ensure that the correct amount of antifoam is added. Too much antifoam will reduce the gas hold-up, thus affecting the oxygen transfer rate, and may complicate product recovery. As there is always a lag between controller activation and foam breakdown, it is important to incorporate a lag in the control circuit to prevent addition of too much antifoam. It should also be possible to alter the sensitivity of the probe to avoid splashes from the broth surface activating the probe.

Some antifoam 'controllers' rely simply on addition of a fixed amount of antifoam at regular intervals. This can only be used satisfactorily when the necessary experience about a fermentation has been gained. Unfortunately, foaming often occurs in short, irregular bursts and there is little way of predicting these.

There a number of antifoams on the market including natural oils, higher alcohols, silicon oils and n-paraffins. Some antifoams are metabolized by the microorganisms and regular addition to the fermenter is necessary; indeed by careful design of the medium, the antifoam can be used as an additional carbon source, often more efficiently than the main carbon source.

9.9   pH probes

External pH has little direct influence on the internal pH of microbial cells, but the breakdown of substrates, their transport through the cell wall, and the excretion of cell products are all affected by the pH value of the environment. The effect of pH on specific growth rate was demonstrated in Chapter 4 and it is clear that pH represents an important process parameter.

pH is a measure of hydrogen ion activity, defined as $pH = -\log a_H^+$, and its determination is given by employing the Nernst equation, which is temperature dependent (see Section 9.12.2). Thus, probe output will change with temperature and it is important to have temperature compensation (preferably automatic) in the contoller/indicator circuitry.

Probes which combine a glass electrode with a reference electrode are now used extensively, making external reference probes obsolete. As the demand for sterility is normally strict in biotechnology, steam-sterilizable probes are gaining favour. However, one drawback is that repeated sterilization affects the probe performance (irrespective of manufacturers' claims, though it must be said that progress is being made in this respect). The pH-sensitive glass tip is very brittle (again improvements are made), but the major drawback is that the best probes are unnecessarily expensive. However, new manufacturers are entering the market producing adequate probes compatible with most pH meters, at reasonable prices. There are a number of excellent pH meters/controllers on the market which can be used for control of pH; it is important to be able to adjust the sensitivity and response time of the controller since the addition of acid or alkali to restore the set pH has to take effect over the whole fermenter contents.

The pH probe is referenced to the pH meter ground, but sometimes interference through a different ground may be experienced (indicated by erratic reading indicator off-scale etc.). This can be overcome by employing a common ground for the pH meter and the fermenter itself.

9.10   Redox probes

It has been suggested that redox potential may be a better indicator of the state of a fermentation than either pH or dissolved oxygen. There are a number of steam sterilizable redox probes available and most manufacturers of pH probes will also supply redox probes. One reason for the lack of application of redox probes in the fermentation industry is the difficulty in interpreting results as dissolved oxygen and the presence of oxidizing or reducing compounds will all affect the probe output.

9.11   Dissolved oxygen probes

The critical role played by dissolved oxygen (DO) in fermentation processes was discussed in Chapters 4 and 5. Thus, the advent of DO probes has offered a significant contribution to our understanding of the performance of bioreactors.

DO probes basically consist of a stainless steel or glass sheath containing two electrodes and a suitable electrolyte. To separate the electrodes and electrolyte from the fermentation broth, the probe is covered with a membrane. Oxygen diffuses through the membrane and is reduced at the cathode, which is negatively polarized with respect to the anode. This produces a current which can be translated into oxygen concentration. The DO probe construction is shown in Fig. 9.8.

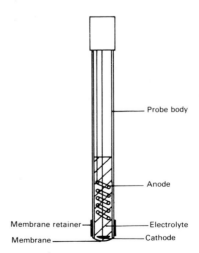

Fig. 9.8   The dissolved oxygen probe.

9.11.1 Polarographic DO probes

Polarographic probes have an external power source to set up an 0.6–0.8 V polarization voltage between the two electrodes. The cathode is typically platinum, with $Ag/AgCl$ as a reference anode. The electrode reactions are:

$$Cathode: O_2 + 2H_2O + 2e^- \rightarrow H_2O_2 + 2OH^-$$
$$H_2O_2 + 2e^- \rightarrow 2OH^-$$
$$Anode: Ag + Cl^- \rightarrow AgCl + e^-$$
$$\overline{Overall: {}^4Ag + O_2 + 2H_2O + 4Cl^- \rightarrow 4AgCl + 4OH^-}$$

It can be seen that the electrolyte, in this case potassium chloride solution, takes part in the reactions, and must be replenished at regular intervals.

9.11.2 Galvanic DO probes

Galvanic probes do not use an external power supply, relying on the natural polarization potential set up between a noble cathode and a basic anode. The most common materials are silver and lead respectively. In this case, the probe reactions are:

$$Cathode: O_2 + 2H_2O + 4e^- \rightarrow 4OH^-$$
$$Anode: Pb \rightarrow Pb^{2+} + 2e^-$$
$$\overline{Overall: O_2 + 2Pb + 2H_2O \rightarrow 2Pb(OH)_2}$$

In galvanic probes, the anode is oxidized and the probe performance depends on the available anode surface area, hence the extensive use of coil-shaped anodes, to maximize the available surface area.

The main difference between polarographic and galvanic probes is price, with the latter being much the cheaper. Polarographic probes may be fractionally faster and may have a longer life, though this depends very much on make of probe and care taken in handling.

Probe responses to step changes in DO level are shown in Fig. 9.9. Response times of 90 s are common and it is doubtful if these probes are sufficiently fast to be used in dynamic $k_L a$ measurements (see Chapter 5) or to study transients. Reducing the response time to below 30 s, which is possible for a limited number of probes, decreases the significance of the probe characteristics and makes the readings more reliable.

Drift and fouling (microorganism sticking to the probe surface) are the two main problems associated with DO probes. Reproducibility is rarely achieved, particularly if a probe is steam sterilized, and repeated calibration is essential.

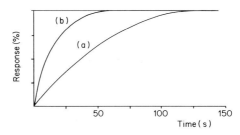

Fig. 9.9 Typical dissolved oxygen probe responses to a step change in dissolved oxygen level. (a) Slow response, and (b) fast response.

Electrolytes: In polarographic probes, KCl is the most common electrolyte with lead acetate used for the galvanic probes. For both, a number of other solutions have been used (see Table 9.2). The choice of electrolyte

Table 9.2   Typical electrolytes for DO probes

KCl
KCl + NaHCO$_3$
KCl + KH$_2$PO$_4$ + AgCl
KOH
KHCO$_3$
NaHCO$_3$
Acetate buffer
KI
K$_2$CO$_3$ + KHCO$_3$

Often chemicals will be added to protect the electrolyte during autoclaving.

may not be critical for probe performance, and it is advisable to experiment with suitable solutions rather than buying expensively from manufacturers.

The loss of solvent through evaporation is a common problem, often leading to premature deterioration of probe performance. It should be remembered that loss of electrolyte occurs during storage as well as during operation.

Membrane properties are vital for good probe response. Membranes with high O$_2$ diffusivity and low CO$_2$ permeability are preferred. Teflon, polyethylene and polypropylene are the most common membrane materials.

A simple analysis of the behaviour of DO probes indicates that their response is related to the probe constant, k:

$$k = \frac{\pi^2 D}{d^2}$$

where D = oxygen diffusivity in the membrane, and d = membrane thickness.

Fast response probes will have a large value for k, implying a thin membrane and/or high oxygen diffusivity. However, this is a gross simplification of the workings of DO probes; the construction of the electrodes (in particular their relative surface area) will greatly influence probe behaviour.

It should be noted that DO probes do not *measure* dissolved oxygen concentration, but the activity or partial pressure of oxygen. For this reason, DO probes are often calibrated to read percentage saturation, using air and oxygen-free nitrogen as the 100 and 0% points for calibration. The actual amount of oxygen dissolved in the liquid can be determined by chemical methods. It is possible, however, to relate the partial pressure of oxygen (or oxygen tension) to dissolved oxygen concentration by using Henry's law, as the oxygen solubility in fermentation broths is very low (see Chapters 4 and 5).

9.11.3 The tubing method

A different approach to dissolved oxygen measurements is provided by the tubing method, which is outlined in Fig. 9.10.

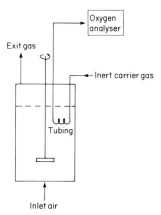

Fig. 9.10   Measuring dissolved oxygen using the tubing method.

The tube, often coiled to increase gas residence time and surface area, is made of a permeable material such as teflon or polypropylene. Inert gas, e.g. N_2 or He passes through the coil at a fixed flow rate, picking up the oxygen which has diffused through the tube wall from the fermenter broth. The oxygen content of the gas is determined by an oxygen analyser, see Section 9.13.

The tubing method is not suited to the control of DO or dynamic measurements, having a response time of the order of 2–3 min, but the simplicity of assembly and operation enhances stability, allowing operation for weeks between calibration. It is necessary to use an analyser which will detect the very low concentrations of O_2 in the carrier gas and erroneous readings can result from microbial fouling and air bubbles contacting the tubing.

9.11.4   Dissolved CO_2 probes

We have probably been very slow to realize the importance of dissolved CO_2 as an affector and indicator in fermentation processes. For this reason, dissolved CO_2 probe technology lags far behind DO probes, and has only recently emerged on the market.

Dissolved CO_2 probes contain a bicarbonate solution encapsulated on to a pH probe by a CO_2-permeable membrane. Present probes are not steam sterilizable, and response times are quoted as about 1 min.

It is also possible to use the tubing method (Section 9.11.3) for dissolved CO_2 analysis. In this case, a high CO_2-permeable tubing is used together with a sensitive CO_2 analyser, see Section 9.13.2.

9.12   Enzyme probes

A major difficulty for the application of computers to fermentation processes is the lack of sensors providing direct information on the state of the culture. In recent years, considerable progress has been made in the design of ion-selective electrodes, but, so far, this has been limited to inorganic ions and is of little direct use in fermentation processes. However, the rapid response and reliability of ion-selective electrodes has been combined with the sensitivity and selectivity of enzyme reactions in the production of enzyme probes.

These are manufactured by immobilizing an enzyme on to the surface of a probe such as DO or pH electrode. The principle involved is illustrated in Fig. 9.11. The substrate to be assayed diffuses into the enzyme layer and the electrode is chosen so that it will respond to reactants or products in the ensuing enzymic reaction.

Normally, estimation of fermentation products requires sampling followed by subsequent sample preparation and analysis. The combination of a fast response probe and the selectivity of enzyme reactions means that, in principle, sample preparation is not necessary with enzyme probes, allowing on-line application. However, most enzyme probes are not steam sterilizable.

9.12.1 Enzyme immobilization methods

The enzyme chosen must react specifically or highly selectively with the substance being assayed. Stability is important; this depends on the enzyme system as well as on the method of immobilization. Most solubilized enzymes are unstable and if used in this form, large amounts may be required. Immobilizing the enzyme reduces both the amount required, and the loss of activity thus cutting down the cost. A number of immobilization techniques are employed.

Chemical immobilization usually involves covalent bond formation. Covalent cross-linking by bifunctional reagents, such as glutaraldehyde, is simple and it is possible to control the physical properties and particle size of the final product, but many enzymes are sensitive to the coupling process and lose activity. Binding to water-insoluble materials places the enzyme in a more natural environment with favourable effects on enzyme efficiency and stability. The carriers used include inorganic materials such as porous glass and natural and synthetic polymers such as cellulose and polyacrylamide.

Physical immobilization is generally quicker and easier than chemical methods. Microencapsulation holds much promise but has not yet found widespread application. In this method the enzyme is inside semi-permeable particles, the particle wall allowing transport of reactant and products through the wall but retaining the enzyme.

Adsorption onto insoluble supports, such as glass or silica gels, is very simple but the enzymes are often easily desorbed by pH, temperature or substrates. This method has been used in the construction of glucose oxidase probes.

Inclusion in gel-lattices is achieved by carrying out a polymerization, e.g. of acrylamide, in an aqueous solution of the enzyme. Very mild reaction conditions favour retention of enzyme activity.

The immobilized enzyme is normally placed on a permeable membrane and secured to the electrode as shown in Fig. 9.11.

9.12.2 The electrode

The electrode must be chosen so that it responds to changes in A, B, C or D in the enzymic reaction:

$$A + B \rightarrow C + D$$

Provided the stoichiometry is known, the reading of the probe can be translated into the concentration of any of the reactants or products. The probe

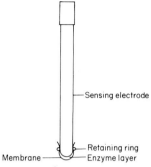

Fig. 9.11   Construction of an enzyme probe.

reading will also be calibrated against a standard concentration of the desired substance.

The electrodes can be divided into two groups according to their principle of operation.

Amperometric electrode measures flux of electroactive species. The response is a linear function of concentration. Amperometric electrodes are used in the detection of substances such as O_2 and H_2O_2 using a polarographic electrode.

Potentiometric electrodes measure concentration of the reactants or products. For these electrodes the response is alogarithmic according to the Nernst equation:

$$E = E^0 + \frac{RT}{ZF}\ln[H^+]$$

where E = electrode potential, E^0 = standard electrode potential, F = Faraday constant, R = gas constant, T = absolute temperature, Z = charge, and $[H^+]$ = hydrogen ion concentration.

This principle is used in ion-sensitive electrodes for hydrogen ion $[H^+]$ and ammonium ion $[NH_4^+]$. It is also used in gas electrodes for CO_2 and NH_3.

Gas electrodes normally have a higher resolution than ion-sensitive electrodes and will, therefore, detect substrates at lower concentrations.

In most cases the limiting factor in designing an enzyme electrode will be the availability of a sensor that can monitor the reaction. Here, other possibilities exist. For example, a thermistor covered with an immobilized enzyme could measure the temperature change resulting from the enzymic reaction, proportional to the substrate concentration.

Generally, potentiometric electrodes are currently the most common and most successfully used electrochemical sensors.

9.12.3 Probe response

The response characteristics of enzyme probes are determined by a number of parameters. The useful lifetime of a probe will vary from one probe to another. In most cases it will depend on probe stability, which can be divided into:

(a) electrode stability
(b) enzyme stability
(c) probe storage

The stability of the electrode is normally not a factor, being more stable than the enzyme. However, electrochemical electrodes may exhibit drift and regular checking of the response is recommended.

The stability of the enzyme depends largely on the method of immobilization. It has been suggested that chemical methods will give the better stability compared with physical techniques. Enzyme probes have been known to be stable for several months. The amount of enzyme present and the operating conditions will also affect probe stability.

In general, stability of enzyme electrodes is difficult to define; both time and number of assays have been used as parameters. It may also be difficult to determine when an electrode is no longer analytically useful. As the enzyme loses activity, daily recalibration may prove enough to extend its lifetime.

Response time. The following scenario can be visualized for enzyme probes. Substrate diffuses through the retaining membrane and into the enzyme matrix. Products or reactants in the enzymic reaction diffuse out of the enzyme matrix and on to the surface of the measuring element in the electrode, producing a probe output which is translated into concentration of the relevant substance.

It will be appreciated that this cannot happen instantaneously. Reported probe response times vary from 30 s to more than 24 h. The response time will depend on many factors, as outlined below.

Rate of diffusion. If the probe is designed to monitor changes which occur in a fermentation broth, it is important that this is well mixed. The substrate for the enzymic reaction will then reach the probe surface quickly. It is also important to use retaining membranes which will not represent a diffusional barrier, the right material normally being found by trial and error.

The parameters which influence response time will be substrate concentration, pH, temperature, amount of enzyme present, etc. If the probe is used on-line it must be designed to operate at the conditions which have been optimized for product concentration.

Some examples of enzyme probes having proceeded beyond the research level are given in Table 9.3. This table is by no means exhaustive, and will rapidly become out of date, but it illustrates the versatility and potential of these instruments.

Table 9.3   Immobilized enzyme probes

Enzyme	Compound	Sensor
Glucose oxidase	Glucose	Polarographic CO
	Lactose	pH
Glucose oxidase/		
β-galactosidase		Polarographic DO
L-amino acid oxidase	L-amino acids	Ammonia
Urease	Urea	pH
Glutamate dehydrogenase	Pyruvate	Ammonia
Uricase	Uric acid	Polarographic DO
β-lactamase	Penicillin	pH
Alcohol oxidase	Alcohol	Polarographic DO

9.13   Gas analysis

9.13.1 O_2 analysers

The importance of oxygen in aerobic cultivation of organisms has been discussed in previous chapters. All students of biotechnology should be familiar with the basic rudiments of O_2 transfer (see Chapter 5).

Oxygen is supplied by passing air (O_2) through the system and there may also be some dissolved oxygen in the various inlet liquid streams. Normally, the latter can be ignored and the difference between the oxygen content of the inlet and outlet gas streams is, therefore, equal to the amount of O_2 transferred into the system. The oxygen content of the gas is normally determined by a gas analyser.

Paramagnetic oxygen analysers are used extensively. Paramagnetism (attraction to the strongest part of the field) is a result of molecules having unpaired electrons. Oxygen has two, and is thus strongly paramagnetic. Compared with oxygen, none of the other gases normally associated with fermentation processes are paramagnetic, as shown in Table 9.4. Hence, this property can be used for selective oxygen analysis.

The most popular analysers are the magnetic wind (thermomagnetic) and the deflection (magnetodynamic) analysers. In the former, the gas enters a cell inside which there is a heated element. The cell is situated in a magnetic field, thus attracting oxygen. As the gas approaches the

Table 9.4 Paramagnetism of gases (in relation to a scale of $N_2 = 0\%$ and $O_2 = 100\%$)

$O_2 = 100\%$
$N_2 = 0\%$
$H_2 = 0.24$
$CO = 0.01$
$CO_2 = -0.27$
$CH_4 = -0.2$
$He = 0.3$
$SO_2 = 0.22$
$NH_3 = -0.26$
$C_3H_8 = -0.86$

filament, the oxygen heats up and suffers a loss in paramagnetism. Cooler oxygen will be attracted by the magnetic field, resulting in a magnetic wind through the cell. The filament is part of a Wheatstone bridge which will become unbalanced when the resistivity of the filament changes due to heat loss to the gas. The imbalance will be proportional to the amount of oxygen in the gas.

These analysers are very susceptible to changes in gas composition and barometric pressure. The main problem concerning the former will be water content, as the gas composition under normal fermenter conditions is relatively constant. The water vapour will affect the analysis in such a way that the measured concentration will be lower than the actual concentration (gas analysers normally estimate vol % in dry gas). This can easily be overcome by drying the gas. Changes in barometric pressure of 1% are said to produce an identical change in O_2 reading. In small-scale operation, where the amount of O_2 consumed is very low, O_2 analysers are normally operated with suppressed zero, being calibrated between 19 or 20 and 21% O_2. Under these circumstances it is very important to take account of pressure changes and water vapour so as not to accentuate errors in the data produced.

Magnetodynamic analysers are less affected by changes in gas composition. In these analysers, the cell contains a body, normally in the shape of a dumb-bell, suspended in a non-uniform magnetic field. The gas is passed through the field; if the paramagnetism of the gas is the same as the dumb-bell, this will have no force acting on it and it will maintain its position in the magnetic field. The paramagnetic property of O_2 will act on the dumb-bell, however, causing a deflection which can be calibrated to measure the amount of O_2 in the gas. By incorporating two cells in the analyser, with one reading the inlet, the other the exit fermenter gas, a direct measurement of the amount of O_2 taken up by the fermenter will be provided. In this case barometric pressure changes are compensated for.

Present and potential users of O_2 analysers should be warned that although many models and makes available are reliable and accurate, albeit relatively slow to respond (response times vary from about 1 to 4 min, more related to purge time than to the analysis itself), they are often misused by insufficient drying and infrequent checking of barometric pressure (if required). Also, the gas may be transported to the analyser using oxygen-permeable tubing. With careful application O_2 analysers can provide meaningful and valuable information on a fermentation process. It must be remembered, however, that pressure and temperature compensation on the gas flow may be required. This is often overlooked, producing results which can lead to failure in scale-up attempts. Particularly when the difference in oxygen content of inlet and outlet gas is small, as in small-scale fermenters, great care must be taken to reduce the potential errors involved in oxygen analysis.

9.13.2 CO_2 analysers

Most gases, including CO_2, absorb infrared radiation (IR) energy at some characteristic wavelength. However, simple gases such as hydrogen, nitrogen and oxygen, will not absorb IR energy, and this principle is used for detection of CO_2 in fermentation processes.

An infrared analyser consists of a light source, optical section and sensor. The most common light sources are the Nichrome coil, the Globar (silicon carbide), and the Nernst glower (a filament of oxides), emitting light at the required wavelength(s) for absorption of the gas. When the gas to be measured enters the analyser, it passes through a cell in the path between the radiation source and the detector. It will absorb some radiation, thus reducing the energy level which reaches the detector. This change in energy is detected and amplified to give an analyser signal related to the partial pressure of CO_2 in the gas, which is translated into CO_2 concentration.

There is a logarithmic relationship between IR absorption and CO_2 concentration, which implies a lack of sensitivity at high CO_2 concentration. For this reason, most analysers are supplied with a range selection switch allowing operation in a relatively narrow range in which the response can be assumed linear.

The production of CO_2 can be used as an indicator of the state of a growing culture. For example, when CO_2 is produced by the microorganisms metabolizing carbohydrates according to the equation

$$C_6H_{12}O_6 + 6O_2 \rightarrow 6CO_2 + 6H_2O + \text{energy}$$

one mole of CO_2 is produced for every mole of O_2 consumed. The respiratory quotient (RQ) is defined as

$$RQ = \frac{\text{rate of } CO_2 \text{ production}}{\text{rate of } O_2 \text{ consumption}}$$

In the high energy yielding equation above, RQ has a value of unity. In fermentation processes, deviations from this value will always occur, for example when some of the glucose is to be converted into biomass and metabolic products. Some examples of the stoichiometry and RQ values obtained for a number of fermentation processes with a range of micoorganisms and substrates are given in Table 9.5 (see also Chapter 3).

Table 9.5 Examples of stoichiometric equations and respiratory quotient for growth of microorganisms

---

Klebsiella aerogenes
$C_3H_8O_3 + 1.87O_2 + 0.32NH_3 \rightarrow 1.47CH_{1.74}N_{0.22}O_{0.43} + 1.43CO_2 + 3.24H_2O$
glycerol   biomass
                                    RQ = 0.76

Candida utilis
$C_6H_{12}O_6 + 3.53O_2 + 0.71NH_3 \rightarrow 3.53CH_{1.84}N_{0.2}O_{0.56} + 2.33CO_2 + 3.67H_2O$
glucose   biomass
                                    RQ = 1.16

Saccharomyces cerevisiae
$C_6H_{12}O_6 + 3.96O_2 + 0.33NH_3 \rightarrow 1.91CH_{1.70}O_{0.50}N_{0.17} + 4.09CO_2 + 4.48H_2O$
                                    RQ = 1.03

Penicillium chrysogenum
$C_6H_{12}O_6 + 3.3O_2 + NH_3 \rightarrow 2.98CH_{1.86}O_{0.62}N_{0.14} + 3.0CO_2 + 4.8H_2O$
                                    RQ = 0.91

Citric acid production by *A. niger*
$C_6H_{12}O_6 + 1.83O_2 \rightarrow 0.91C_6O_8H_7 + 0.5CO_2$
                                    RQ = 0.27

In each case, RQ can be used to monitor and possibly control the fermentation. It must be noted, however, that the composition of biomass is unlikely to remain constant throughout a fermentation and this may affect RQ. In the citric acid material balance, the production of biomass is not included. Most of the biomass is formed in the early part of this fermentation, and during the production phase the amount of sugar converted to cell matter is not very high. It is possible, therefore, to approach the theoretical RQ value indicated, but only for high yielding processes in which few secondary acids are being produced.

The production of CO_2 is also a valuable general process indicator in continuous culture. A steady CO_2 plot is normally the best indicator of a steady state.

In fermentation media with high solids content it can be difficult to determine the cell concentration and, hence, the specific growth rate of the culture. In this case, the data from the CO_2 and/or O_2 analysis can be used to calculate this important process variable. A fixed relationship between CO_2 production, or O_2 consumption, and cell growth rate is usually assumed, implying that it is possible to separate the amount of gas associated with formation of cell material from the gas production/consumption due to energy metabolism and production of metabolites.

9.13.3   Mass spectrometers

Mass spectrometry is based on the separation of ionized molecules under vacuum. The separation, based on the mass to charge ratio, is achieved by magnetic or quadropole instruments. Presently mass spectrometers (MS) are finding increasing application in monitoring fermentations.

Potentially, MS can be used for both continuous on-line gas and liquid analysis. For liquid analysis a probe supporting a strong permeable membrane is inserted into the fermentation broth and dissolved substances such as O_2 and CO_2 and any liquids of sufficient volatility are drawn out of the solution by applying a vacuum. The principle is similar to the tubing method described in Section 9.11.3. A problem to be solved for this method is the choice of membrane which must be structurally sound to withstand the vacuum yet sufficiently thin to allow rapid diffusion of the required substance(s). There are also problems with finding proper conditions for analysis of individual components of a mixture.

These problems are not encountered with direct gas analysis. Potentially, MS can be used to analyse for any vapour phase component simultaneously; in fermentation it is traditionally restricted to O_2 and CO_2 (plus sometimes CH_4, N_2 and H_2), as a sophisticated gas analyser. The advantage over conventional analysers are speed of response (of the order of seconds compared to minutes), greater accuracy, and the increased number of channels, allowing many fermenters to be connected to one instrument. A distinct disadvantage, however, is cost; typically some ten times more expensive than an O_2 analyser.

9.13.4   Gas chromatography

Gas chromatography (GC) has been used successfully in offline analysis of biochemical processes. By proper sample preparation, most components in the process liquid can be determined. This application of GC is beyond the scope of this chapter, however, and the reader is recommended to search in the abundance of relevant specialist literature.

Recently, GC technology has progressed sufficiently to make on-line gas analysis feasible. However, it is not necessary to restrict this to CO_2 and O_2. GC can be used in fermenter head-space analysis, monitoring gas composition and concentration of volatile compounds such as acetaldehyde, ethanol, acetone, etc.

A disadvantage with a GC is its price and the discontinuous nature of the signal produced. (The retention time for the slowest compound can be several minutes.) Gas chromatographs are very sensitive, however, and will detect the presence of intermediates even at very low concentrations. For reliable operations, extensive calibration will be required.

9.14 Determination of cell concentration

It may seem strange that measurements of cells and cell growth have been postponed to the end of a chapter on instrumentation for measurement and control of fermentation processes. The reason is that the technology is still in its infancy; considering the importance of cell growth, this is very unfortunate.

Dry weight measurements are still the most common way of determining the concentration and, hence, cell growth. This involves sampling, separation of the cell from the process liquor and subsequent drying. With the aid of microwave ovens, it is possible to reduce the time lag between sampling and dry weight determination to 20–30 min, providing the broth filters easily. For fermentations lasting 100 h or more, this lag may not seem significant and, indeed, there are a number of fermentation processes which are controlled manually on the basis of dry weight measurements. However, the introduction of on-line computer facilities has put this method out of favour (and in a few cases it has proved to be inadequate).

Dry weight determinations are only possible when the solids content in the medium is low; unfortunately, this is not the case for most industrial fermentation media.

Turbidity measurements have been used for cell determination on a limited scale. The principle involved is to measure the turbidity of a fermentation broth by the amount of light transmitted. This method is limited to systems in which the turbidity of the broth is only due to the cells, i.e. yeast or bacterial single cell fermentation. For on-line process control this method is only applicable to fermentation processes with low cell concentration as the relationship between turbidity and cell concentration rapidly deviates from linearity.

Turbidity readings must be translated into dry weight (or cell number) and the calibration should always be checked. Naturally, it is difficult to read the turbidity of complex fermentation media and relate it to cell concentration. However, the method is more useful in continuous culture where complex media are less commonly used.

Cell numbers require specialist counting methods and are only used to a very limited extent, suffering to a larger degree from the same problems as measurements of turbidity.

Impedance. It is known that microorganisms alter the impedance of their growth medium when actively growing. In suitable media, it is possible to obtain growth curves of a number of yeast and bacteria in a relatively short time. The method, or a related one based on resistance measurements, can be used for detection of contaminants in the food industry but its application in the fermentation industry appears limited; in complex media, changes in capacitance are difficult to relate to cell growth.

Metabolic heat can be related to growth since this is an exothermic process (see Chapter 3). The relationship will be empirical and the errors involved are considerable even on small-scale fermenters. Very accurate temperature control is essential, but given that temperature sensor technology is very advanced, there is some justification for considering this method, as metabolic heat is not directly affected by scale or medium composition, etc.

CO_2 production (O_2 consumption) can be empirically related to cell growth. However, this approach requires intimate knowledge of the fermentation process.

DNA content of cells varies very little with growth rate, environment, metabolism or nutrient uptake. DNA analysis is more complex than dry weight determination, but the methods are well documented. In media with high solids content DNA analysis may be one of the most reliable methods for estimation of cell concentration.

Other methods, such as measurement of ATP content, have been examined. Most have not proceeded beyond the stage of determining single cells growing at low concentrations in synthetic media. However, the importance of developing on-line or automatic sampling with fast, automatic off-line methods for cell growth determination has been realized, and rapid development should be expected in this field. At the moment, gas analysis appears to be the most promising method.

Reading list

Atkinson, B. and Marituna, F. (1983). 'Biochemical Bioengineering and Biotechnology Handbook'. Nature Press:

Brown, D. E. (1977). The management of fermenter power input. *Chem. Ind.* **684.**

Fleischaker, R. J., Weaver, J. C. and Sinskey, A. J. (1981). Instrumentation for process control in cell culture'. *Adv. Appl. Microbiol.* **27,** 137.

Flynn, D. S. (1983). Instrumentation and control of fermenters. *In* 'Filamentous Fungi', Vol. 4 (Eds J. E. Smith, D. R. Berry and B. Kristiansen) pp. 77–100. Edward Arnold: London.

Huskins, D. J. (1981). 'Quality Measuring Instruments in On-Line Process Analysis'. Ellis Horwood:

Johnson, C. D. (1977). 'Process Control Instrumentation Technology'. John Wiley: New York.

Tannen, L. P. and Nyiri, L. K. (1979). Instrumentation of fermentation systems. *In* 'Microbial Technology', Vol. 2 (Eds J. H. Peppler and D. Perlman). Academic Press: New York and London.

Part II | Practical Applications

10 | Microbial Biomass as a Protein Source

J. OLSEN and K. ALLERMANN

10.1 Introduction

10.1.1 Historical background

Single cell protein (SCP) is the accepted term for microbial cell material intended for use as food or feed. The term is somewhat misleading, as what is produced is normally not a pure protein, but variously treated cells from a variety of microorganisms, both mono- and multicellular, bacteria, yeasts, filamentous fungi or algae (Table 10.1).

Table 10.1 Composition of some microorganisms used as single-cell protein

Analysis	*Paecilomyces varioti*	*Fusarium graminearum*	*Candida utilis*
Dry matter (%)	96	94.2	91
Crude protein (% N × 6.25)	55	54.1	48
Fat (%)	1	1.0	1.35
Ash (%)	5	6.1	11.2
Lysine (g per 16 g N)	6.5	3.5	7.2
Methionine (g per 16 g N)	1.9	1.23	1.0
Cystine and cysteine (g per 16 g N)	1.0	0.75	1.0

From Forage and Righelato (1979). *In* 'Economic Microbiology', Vol. 4 (Ed. A. H. Rose). Academic Press: London and New York.

Since ancient times, in both Western and Oriental cultures, microorganisms have been used to transform or produce food, and so have always been part of the diet for man and animals. However, the first *industrial* production of a microorganism for nutritional purposes took place in Germany during World War I, when 'Torula Yeast' was produced. After the war German interest waned, but was revived in the mid-1930s, and during World War II about 15 000 tonnes per year of yeasts were produced and incorporated into army and civilian diets, mainly in soup and sausages. Interest in producing fodder yeast also developed in the United States and in Great Britain, and continued after the end of World War II in a rather desultory way, until a world-wide interest quickened in the mid-1950s.

The first international conference on SCP was held in 1967 at the Massachusetts Institute of Technology, while most of the projects were still at the experimental level; only British Petroleum (BP) presented results from SCP fermentation on an industrial basis. At the second MIT conference in 1973 many companies in different countries had started large-scale production and were able to demonstrate their technical capability.

10.1.2   Why produce SCP?

Interest in producing SCP was stimulated by many publications from the international agencies concerned with health, food and agriculture about a world-wide protein shortage. In the developed countries the rising standard of living has resulted in an increasing demand for high-quality proteins for compounded feeds, which are fundamental in modern techniques for the production of eggs, poultry, veal and pork. These compounded feeds, designed to satisfy the entire nutritional requirements of the animal, usually contain between 10 and 30% protein by weight. This is normally provided by the incorporation of oil-seed meals such as soya or fish meal, and SCP could be a valuable alternative to some of these traditional sources. By reducing the flow of soya, fish proteins and also cereals into animal feed, use of SCP might make more of these products directly available for human consumption. Furthermore, in Europe, the Soviet Union, Japan and other areas where soya crops cannot be grown, large-scale production of SCP would make animal production in these areas less dependent on imported proteins.

As compared with the traditional methods for producing proteins for food or feed, large-scale industrial production of microbial biomass for the same use has some characteristic advantages: microorganisms in general have a high rate of multiplication, a high protein content (in terms of dry weight microbial biomass may contain 30–80% protein), they can utilize a large number of different carbon sources (some of which are traditionally regarded as wastes), strains with high yield and good composition can be selected or produced relatively easily, production installations occupy limited areas and give high yields, and (except for algae), microbial production is independent of seasonal and climatic variations and therefore more easy to plan.

10.2   The SCP process

Regardless of type of substrate or organism employed the production of SCP always involves certain basic steps (Fig. 10.1):

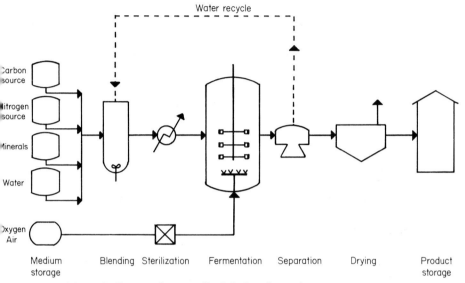

Fig. 10.1   Schematic diagram of a generalized single-cell protein process.

(1) provision of a carbon source, usually requiring some combination of physical or chemical treatments of the raw materials;
(2) preparation of a suitable medium containing the carbon source and sources of nitrogen, phosphorus, and other essential nutrients;
(3) prevention of contamination of medium or plant;
(4) cultivation of the desired microorganism;
(5) separation of the microbial biomass from the spent medium;
(6) after-treatment of the biomass with or without specific purification operations.

To carry through these basic steps a number of auxiliary operations are also required, and some of these are mentioned in the following discussion.

10.2.1   Cultivation

As the fermentation is normally run either sterile or under clean conditions, precautions must be taken to avoid contamination. These may include heating or filtration of the components of the medium and sterilization of the fermentation equipment (Chapter 7). SCP processes (other than with algae) are highly aerobic, and aeration is therefore an important operation. Heat is generated (Chapter 3) and has to be removed by a cooling system. The choice of fermenter is not easy (Chapter 5); on the laboratory scale any fermenter that can give sufficient oxygen transfer and mixing is satisfactory, but for the industrial scale, capital and running costs must also be considered.

Biomass fermentations are always run continuously, for maximum economy, and usually at dilution rates approaching the maximum (μ_{max}) so as to give near maximal productivity (biomass concentration × dilution rate) (Chapter 4). This gives maximum cell protein, but also (see later) maximum RNA content. If the dilution rate is too close to μ_{max} ('washout'), unused substrate limits the process economics.

10.2.2 Biomass recovery

Single-cell organisms like yeasts and bacteria are normally recovered by centrifugation. As explained in Chapter 6, bacteria require higher centrifugal energy for separation, and agglomeration, by flotation or settling in presence of suitable flocculating agents, may be essential. Filamentous organisms may be recovered on rotary filters, which are cheaper. It is important to remove as much water as possible prior to final drying, as this is normally a costly procedure both in capital and in energy requirements, except where solar drying and low-cost labour can be used (usually giving a lower-grade product). Equipment for harvesting microorganisms on a large scale cannot be operated sterile, but is normally designed for hygienic operation and requires regular cleaning. Handling of the broth that leaves the fermenter is crucial for bacteriological quality, since the O_2 tension drops and the pH rises, providing good growth conditions for very undesirable bacteria. However, this danger is minimized by hygienic handling and by various heat treatments during the final stages of harvesting, which ensure inactivation of heat-sensitive organisms (including the produced microorganism); to increase the nutritional value the recovery procedure will often include a step to ensure breakage of the cell walls. The final dried products are normally bacteriologically stable if handled properly. Depending on the type of substrate and the type of SCP product, it may be necessary to include after-treatments, to remove substrate components or more generally to reduce the content of unwanted compounds in the product (e.g. nucleic acids), or even to isolate the protein. The RNA content of rapidly-growing cells is high (Chapter 2, Fig. 2.20) and practical methods for the reduction of nucleic acids in SCP include: alkaline hydrolysis, chemical extraction, manipulation of growth and cell physiology, and activation of endogenous RNases (usually by brief heat treatment).

Finally, measures must be taken to prevent the release into the environment of large quantities of microorganisms, living or dead. Where the spent medium is still high in BOD (Biochemical Oxygen Demand) this has to be treated in order to avoid pollution of the environment. One obvious way to do this is by recycling the spent growth medium, a procedure that will simultaneously help to minimize fresh water requirements and costs. Estimated water requirements between 18 and 45×10^6 litres (dependent upon substrate and organism) have been given for a 100 000 tonnes per year SCP plant.

10.3 Selection of microorganisms

A microorganism that is to be grown as a protein source for animal or human food must have certain basic properties. The more important of these are: non-pathogenic to plants, animals or humans; good nutritional value; acceptability as food or feed; absence of toxic compounds; and low production costs. Production costs in turn depend on such features as growth rate, yield, protein content, requirement for supplementary nutrients, selective advantage on medium, and separating and drying properties. Some data on protein quality are given in Table 10.2.

10.3.1 Algae

Algae used for biomass production mostly belong to the genera *Chlorella*, *Scenedesmus*, or *Spirulina*. They can be grown either photosynthetically and autotrophically (light and inorganic carbon sources) or heterotrophically (with organic compounds as energy and carbon sources). The most rational method for mass cultivation of algae must be the former, and the limiting factor on a large scale is then illumination; the most practical way to produce algal biomass for protein is in open ponds in sunlight. In large-scale systems of this kind aseptic conditions cannot be maintained at reasonable costs and the risk of contamination is serious. Another problem is the low cell density (in large-scale systems 1–2 g of dry weight per litre) which is necessary for optimal light conditions, but which with unicellular algae necessitates costly harvesting procedures. Filamentous algae (e.g. *Spirulina*) can be recovered more economically on filters or even skimmed from the surface of open ponds.

The macromolecular composition of algae is very dependent on the growth conditions. The content of crude protein (total $N \times 6.25$) can be

Table 10.2 Amino acid composition of yeasts and other selected foodstuffs

Amino acid	Content (g per 16 g nitrogen) of amino acid in:							
	SCP				Traditional feed proteins		Food proteins	
	Toprina	Soviet Union	Pekilo	Saccharomyces cerevisiae	Fish meal	Extracted soya bean	Egg	FAO reference
Isoleucine	5.1	5.9	4.8	4.6	4.6	5.4	6.7	4.2
Leucine	7.4	7.2	7.4	7.0	7.3	7.7	8.9	4.8
Phenylalanine	4.3	4.4	4.0	4.1	4.0	5.1	5.8	2.8
Tyrosine	3.6	4.0	3.5	—	2.9	2.7	4.2	2.8
Threonine	4.9	5.2	4.1	4.8	4.2	4.0	5.1	2.8
Tryptophan	1.4	1.1	1.1	1.0	1.2	1.5	1.6	1.4
Valine	5.9	6.3	5.1	5.3	5.2	5.0	7.3	4.2
Arginine	5.1	5.0	5.3	—	5.0	7.7	—	—
Histidine	2.1	2.4	1.9	—	2.3	2.4	—	—
Lysine	7.4	7.6	6.5	7.7	7.0	6.5	6.5	4.2
Cystine	1.1	0.9	1.0	—	1.0	1.4	2.4	2.0
Methionine	1.8	1.6	1.9	1.7	2.6	1.4	5.1	2.8
Total sulphur-containing acids	2.9	2.5	2.9	—	3.6	2.8	7.5	4.8

From Levi *et al.* (1979). *In* 'Economic Microbiology' Vol. 4 (Ed. A. H. Rose). Academic Press: London and New York.

up to about 60%, and the amino acid profile of algal SCP is generally good though somewhat low in sulphur-containing amino acids. Algae will have a high content of photosynthetic pigments, which may be of interest for some compounded feeds, but is undesirable for human consumption. Feeding studies with different animals have generally shown that micro-algae are suitable as a protein additive if not used in very high proportions, and given adequate quality control. Experiments with the incorporation of *Chlorella* and *Scenedesmus* in human diets have shown serious problems, but *Spirulina* seems to be more suited for human consumption (see Section 10.4.2).

10.3.2 Bacteria

The higher growth rates that bacteria in general offer as compared with other microorganisms may often be of interest in SCP production. Given the wide range of possible substrates, the number of bacterial species that has been considered for SCP production is rather large. As the pH of most bacterial SCP fermentations is controlled between 5 and 7 it is important to obtain and maintain sterility during the process, otherwise contamination with pathogen bacteria is always a risk. The recovery of bacteria by centrifugation is not without problems, and improved methods of recovery are an essential part of practical SCP processes.

The crude protein content of bacteria can be very high (above 80%), but the normal content of nucleic acids, especially RNA, is also high (up to about 20%) and must be reduced. The amino acid profile of bacterial SCP is generally good although sometimes with a small deficit of the sulphur-containing amino acids. The possibility of endotoxin production by many Gram-negative bacteria has to be considered if they are chosen for SCP production.

10.3.3 Yeasts

The technology of large-scale yeast production has been developed over a period of more than a century and in particular species from the genera *Saccharomyces*, *Torulopsis*, and *Candida* have been studied thoroughly. In general the growth rates of yeasts are quite high, though slower than the most rapidly-growing bacteria. The pH in yeast fermentations is usually maintained at values from 3.5 to 5.0, and this is useful in reducing the risk of bacterial contamination. Yeast cells can be recovered readily from the spent medium by continuous centrifugation.

The crude protein content of yeast can be 55–60% and the nucleic acid content up to 15% on a dry weight basis, so nucleic acid reduction may again be necessary. The amino acid profiles of yeast SCP, shown in Table 10.2, are generally good, although the deficit in sulphur-containing amino acids is normally more pronounced than in bacteria; supplementation with methionine is easily done and may become common practice with most SCP products for feed. Yeast SCP normally has a usefully high content of B-group vitamins.

10.3.4 Filamentous fungi

Although the growth rates of higher fungi are generally lower than those of bacteria and yeasts, it is possible to isolate many microfungi with growth rates approaching those of yeasts. In general, the growth rate will vary considerably depending on the substrate. Many filamentous fungi grow well at a pH range of 3–8, and again it is possible to operate at a pH below 5 to minimize bacterial contamination; contamination with yeasts is always a risk unless sterile operations are used. The growth form of fungi in submerged culture can be either yeast-like, filamentous, or in 'pellets', and each form may have advantages and disadvantages in a SCP process. Filamentous or pellet forms are more easily recovered from the growth medium by simple filtration.

The crude protein content of fungi varies widely but can be up to 50–55%. Fast-growing filamentous fungi will also have a high content of nucleic acids (RNA up to 15%). Chitin, which is part of the cell wall, may account for a substantial proportion of the total nitrogen. The amino acid profile of micro-fungal SCP is generally good but low in sulphur-containing amino acids.

Toxic compounds (mycotoxins) are produced by many fungi, and any strain which is proposed as a food or feed must be examined very thoroughly for possible mycotoxin production.

10.4 Substrates and processes

The sources of reduced carbon compounds that can be used for SCP production can be divided into two groups, namely fossil and renewable, though certain carbon compounds, e.g. methanol, ethanol, can be produced from either. The largest investments in research and development have been with utilization of fossil carbon sources, while the use of renewable carbon sources as substrates has two main aspects: (1) use of high grade carbohydrates for the production of food or feed, (2) use of waste materials, when the production of SCP becomes a tool in pollution control.

10.4.1 Fossil carbon sources

Liquid hydrocarbons
The liquid hydrocarbons which are considered as substrates for production of SCP are the *n*-alkanes (saturated, straight-chain hydrocarbons). The lowest homologues of the *n*-alkanes that are liquid at room temperature (*n*-pentane to *n*-octane) are generally believed to be toxic, through their solvent action on cell membranes or membrane proteins. Thus, the liquid *n*-alkanes normally considered as substrates are those having from 9 to about 18 carbon atoms. The proportion of *n*-alkanes in crude oils varies from 0–30%. If required, they can be selectively extracted from kerosene or from gas oil. The direct use of gas oil as substrate has also been investigated, but its use is problematic; only about 10% of the gas oil is fermented, and the rest must be recovered, separated from the SCP, and returned to the refinery for further processing. Possible contamination with carcinogenic polycyclic aromatics makes this route unattractive.

The C_{10}–C_{18} *n*-alkanes are almost insoluble in water and the transfer of the substrate into the cell is of fundamental interest. Hydrocarbons in the fermentation broth are dispersed by the shearing action of agitation and aeration to a macro-emulsion, which by biochemical activity of the cells is transformed into a micro-emulsion. The production of emulsifying surfactants by hydrocarbon-utilizing organisms has been demonstrated, and these are thought to assist dispersion of the hydrocarbon molecules and their passage across the cell membrane.

Microscopic evidence suggests that direct contact between cells and hydrocarbon droplets is necessary for growth. Many different microorganisms, including bacteria, actinomycetes, yeasts and moulds, are able to use liquid hydrocarbons as carbon and energy sources for growth; all are capable of using other substrates as carbon- and energy-sources. The ability to oxidize hydrocarbons is inducible, and the process is necessarily highly aerobic. A detailed list of species which have been reported to utilize hydrocarbons is given by A. H. Rose (see Reading list).

Factors that affect growth rate, yield and productivity on liquid hydrocarbons include oxygen transfer, mass transfer of the substrate to the cell, and heat production. Although the production of a given amount of cells from a hydrocarbon fermentation requires smaller quantitites of the carbon substrates than from a carbohydrate, much more oxygen is required. This stresses the need for fermentation systems in which very high oxygen-transfer rates are obtainable (transfer coefficient up to 10–$15\,\mathrm{kg\,m^{-3}\,h^{-1}}$) (Chapter 5), while the need to disperse the immiscible substrate has already been mentioned. Heat production with hydrocarbons is significantly greater than with carbohydrates and this too is a factor in the fermenter design.

British Petroleum (BP) were involved in much of the pioneering work on the production of SCP from liquid hydrocarbons, and developed two processes, one using gas oil and another one using purified *n*-alkanes as the substrate.

A gas oil based pilot-plant (industrial scale) was established at Cap Lavera near Marseille, France, with a continuous production capacity of 50 tonnes of *Candida* yeast per day. Air-lift fermenters were used with continuous feeding of gas oil, ammonia and mineral nutrients, and the yeast produced was harvested continuously. The process was operated under clean but non-sterile conditions and contamination was controlled by maintaining low pH. The primary yeast product contained unmetabolized gas oil, which was removed by counter-current solvent leaching and by washing with surfactants before the final product was recovered. The need for this somewhat complicated recovery process, which moveover lowers the metabolic energy of the product when used for feed, is probably one of the reasons why BP preferred the alternative *n*-alkane process, represented schematically in Fig. 10.2, which was developed at Grangemouth, Scotland.

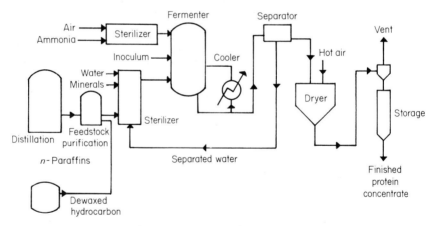

Fig. 10.2   Schematic diagram of the Toprina process using *n*-alkanes. Redrawn from company information (BP Proteins Ltd).

The substrate was pure (>97.5%) *n*-alkanes, C_{10} to C_{23}, produced by a molecular sieve separation process. This also was a yeast process, using *Saccharomycopsis (Candida) lipolytica* operated continuously in a mechanically agitated, fully baffled air-sparged fermenter with turbine mixers. The scale was up to $300\,m^3$ and the production capacity 4000 tonnes per year. The fermentation was run under aseptic conditions; the *n*-alkane, minerals and water were heat-sterilized and air and gaseous ammonia were sterilized by filtration. Yeast from the fermenter was separated by continuous centrifugation to a cream containing about 15% solids, which was spray-dried to give a stable product with a moisture content of about 5%. Since the *n*-alkanes were almost completely utilized by the yeast the recovery of the SCP product was easier and cheaper than from the gas oil process. The product was marketed by BP under the brand name Toprina; its nutritional value compared favourably with soya meal and fishmeal and its crude protein content was about 60%.

The product was subjected to a long series of nutritional and toxicological tests at independent institutes. The results were so satisfactory that BP and ENI, the Italian state-owned hydrocarbon company, in a joint venture (Ital-proteine) constructed a 100 000 tonnes per year SCP plant with three 1000 m^3 fermenters at Sarroch, Sardinia, based on the Grangemouth process. Completed in 1976, the plant was never given an operating permit, due to a dispute with the Italian government on health issues, which BP contested in vain. The plant was closed down in 1978 and BP has now withdrawn from the SCP business. Toprina was approved for use as an animal feed supplement in many other countries, and another SCP plant built simultaneously by Liquichimica Biosintesi at Saline de Montebello, Italy, intended to produce 100 000 tonnes per year of *n*-alkane grown *Candida* yeast using a Japanese process (Kanegafuchi), was closed down due to the same dispute. Plans for industrial-scale SCP production in Japan were suspended from 1973 because of public protests. Several other yeast-based processes for producing SCP from liquid hydrocarbons have been run at pilot or production scale, and bacterial processes have also been investigated, but today (1985) no industrial-scale processes using these substrates are in operation in Western Countries. Nevertheless, several OPEC countries are monitoring developments, while in the Soviet Union several plants are in operation (see below).

Gaseous hydrocarbons
Of gaseous hydrocarbons methane is the most studied substrate for SCP production. It is a major constituent of natural gas, and would be of interest as an SCP substrate because natural gas in some parts of the world is flared at the wells. Its advantages as a SCP substrate are:

(1) it is available in high purity;
(2) it leaves no residues, being easily removed from the fermentation medium; and
(3) high productivities and yield coefficients can be achieved in continuous cultivation.

However, problems with SCP production from methane derive especially from the necessity of transfer of the two gases, methane and oxygen, and it is possible to have both single and double gaseous substrate limitation. The need for high oxygen transfer accompanies a high heat production and necessitates cooling. Other problems include possible formation of inhibitory products in the medium and potential explosive hazards in operating with over 12% (v/v) of oxygen. The special problems result in significantly higher capital requirements, and may be one explanation why so many projects designed to use methane as an SCP substrate have never passed pilot plant scale.

Only a limited number of microorganisms, probably only bacteria, are able to utilize methane as a carbon and energy source (see Chapter 2); they include *Pseudomonas methanica*, *Methanomonas methanica*, the thermophilic *Methylococcus capsulatus*, and *Pseudomonas methanitrificans* which is able both to utilize methane and to fix atmospheric nitrogen.

An illustrative example of suspended methane SCP projects is the former project of Shell Research Ltd in the UK. They undertook research to pilot plant-scale using *M. capsulatus* and also a mixed culture of *Pseudomonas*, *Hyphomicrobium*, *Acinetobacter* and *Flavobacterium* spp., but the programme, like others in Europe and the USA, was cancelled prior to process development. In Japan, Kyowa Hakko Kogyo Ltd developed processes with a *Brevibacterium* sp. utilizing gaseous hydrocarbons (C_1 to C_4) and also mixtures.

Methanol

The production of biomass from methane can alternatively be pursued by an indirect route via methanol, using bacteria or yeasts. Methanol can be made by very efficient chemical conversion from methane, thus saving surplus natural gas, and alternatively from a range of other sources, such as coal, gas oil, wood and naphtha. Its use has some advantages: it is fully water soluble, there are few explosive hazards, and the range of possible microorganisms is greater. Not only the bacteria that utilize methane, but several others that will not do so are able to use methanol. Examples include: (bacteria) *Methylomonas methanolica*, *Pseudomonas utilis*, *P. extorquens*, *Hyphomicrobium* sp.; (actinomycetes) *Streptomyces* sp.; (yeasts) *Hansenula polymorpha*, *Candida boidinii*, *Torulopsis glabatra* (see Chapter 2).

The most advanced process for SCP production from methanol, and the first to be commercialized, has been developed by Imperial Chemical Industries Ltd (ICI). They use a bacterium, *Methylophilus methylotrophus*, which is able to grow in the temperature range of 35–40°C. The process requires fully aseptic conditions, and a novel 'pressure cycle' fermenter has been developed (Chapter 5). The production plant installed at Bill-ingham (UK) has a capacity of about 50 000 tonnes per year (Fig. 10.3).

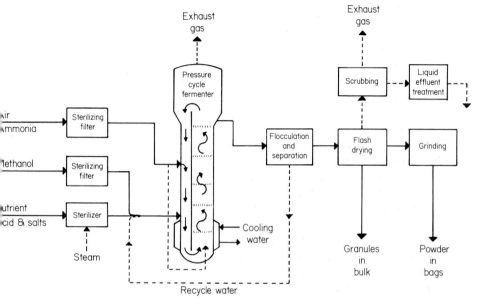

Fig. 10.3.   Schematic diagram of the ICI protein process using methanol. Redrawn from company information (ICI Denmark A/S).

In continuous biomass production from methanol it is essential to keep the steady-state methanol concentration minimal; otherwise the conversion efficiency (yield constant) is dramatically reduced. In a small fermenter the local high concentration near the point of substrate addition is very rapidly dispersed but on scale-up this mixing effect soon becomes limiting. To overcome this, the ICI fermenter, whose total working volume is about $1000\,m^3$, is fitted with several thousand substrate feed inlets throughout its working volume. To promote both oxygen transfer and bulk mixing, the fermenter is also fitted with numerous sieve-plates to enhance turbu-lence.

ICI also developed a new separation technique, in which the bacteria are first recovered by flocculation and froth flotation, to give about 10% solids, followed by centrifugal dewatering and air-drying. Process water is recycled. The dried biomass contains 71% protein and is marketed under the name 'Pruteen', particularly for use as a milk replacer in calf feeding.

By genetic manipulation, the original NH_3-assimilating pathway of the bacterium, which is driven by ATP breakdown, can be replaced by one which is less efficient in terms of NH_3 uptake but which does not consume

ATP; the biomass yield (on carbon substrate) is thereby significantly increased.

Other important methanol-based projects using bacteria have been developed by the partnership of Hoechst AG/Uhde GmBH (West Germany) and by the Norprotein Group (Norway/Sweden). L'Institut Français du Petrole and Mitsubishi Gas Chemical Co. Inc. (Japan) have projected methanol processes with yeasts.

Ethanol
Ethanol can be obtained from petroleum by catalytic water addition to ethylene. It has advantages as a carbon substrate for SCP production similar to those mentioned for methanol, and it may even have a greater acceptability as a raw material for producing a food grade product.

Ethanol can be used by certain bacteria, yeasts, and mycelial fungi. Only one project is known to be working at production scale and this is run by Amoco Foods Co. with a 5000 tonnes per year plant in Minnesota, USA. The organism used is *Candida utilis*, and the product is used for food purposes.

10.4.2   Renewable carbon sources

Carbon dioxide
Unlike the reduced carbon substrates, carbon dioxide can only be used by organisms with an independent energy-yielding mechanism, as in photosynthesis. Carbon dioxide is present in normal air at too low a concentration (0.03%) for optimal photosynthesis by algae, but it is possible to supply additional CO_2 in an inexpensive form like combustion gas. Another approach is to exploit alkaline lakes with a high concentration of sodium carbonate. Algal pond systems incorporate some of the advantages of SCP production as compared with conventional agriculture. The utilization of the sunlight is more efficient, as it is possible to have year round cultivation and optimal utilization of mineral nutrients. It is also possible to use arid and desert areas and to cultivate algae in seawater. However, pond systems also have many of the disadvantages of conventional agriculture, such as dependence on climate (including sunlight), the use of land, invasion by herbivores, weeds and pathogens.

Some of the most promising algal projects have been inspired by traditional uses of algae in man's diet. Since ancient times the local population at Lake Chad in Africa and also the Aztecs at Lake Texcoco in Mexico, 10 000 km apart, have harvested the same kind of micro-algae, namely the blue–green alga *Spirulina*, which is easily harvested because of its filamentous habit, can be dried in the sun, and is used for food. *Spirulina* grows well in alkaline lakes (up to pH 11) and at this pH it is possible to keep it almost as a monoculture. Today the Sosa Texcoco Co. in Mexico is running a production plant and the daily output should approach 5 tonnes dry weight. The cultivation is in ponds up to 1 m deep, with harvesting by two-stage filtration, first on screens and then in a rotating strainer. The harvested biomass is further concentrated by vacuum filtration and drum dried. The yield of a *Spirulina* pond is 50 tonnes ha^{-1} $year^{-1}$ compared to soya-bean with a yield of 6 tonnes ha^{-1} $year^{-1}$. Attempts to improve the yield further have been made, using closed systems of clear plastic. Such a closed system keeps out contaminants and can also function as a solar collector, but other practical issues may arise such as wall growth and pumping problems.

Other algal projects have been described from countries all over the world, and today about 1500 tonnes per year are produced from many small plants in Japan and Taiwan, mostly with *Chlorella* sp. Research has been carried out for many years in West Germany and India, but many serious problems have still to be solved.

Molasses
Molasses is the main substrate for the traditional production of yeasts such as *Saccharomyces cerevisiae* (baker's yeast) and *Candida utilis* (Torula yeast). Molasses is a by-product of cane- and beet-sugar production, and is widely used as the substrate for ethanol production, when dried yeasts are often a by-product. Baker's yeast can be regarded as a specialized form of single-cell protein, and the world production is considerable (more than 200 000 tonnes per year, dry weight), but the main consideration in its production is its properties in the dough fermentation and not its use as a single-cell protein. Molasses is an excellent substrate for SCP production, but because of the high demand for this substrate in agriculture and in other fermentation industries, it is normally considered too expensive to use in this manner.

Whey
Whey is the liquid by-product in cheese manufacture, after removal of fat and casein from whole milk. If dried, the main composition of whey is 70% lactose, 9–14% protein, and 9% ash. The world production of liquid whey was in the order of 74 million tonnes per year in 1973 and at least half of it was wasted, turning a high quality substrate into an effluent problem. More economically, the protein can be separated and sold as a food supplement; this also makes the residual lactose solution more easily fermentable. Protein separation can be done by coagulation (heating at 95°C or acidification), by ion exchange, or by ultrafiltration. Already during

World War I amounts of mycelium of *Geotrichum candidum* (formerly *Oidium lactis*) were produced, but today the dominating organisms are yeasts. Some of the major projects today are: (1) Fromageries Le Bel process (France) using *Kluyveromyces fragilis*; (2) Kiel Process (West Germany) using a mixed culture of *Lactobacillus bulgaricus* and *Candida krusei*; (3) the Vienna Process (Austria) using a fast-growing acid-resistant strain of *Candida intermedia*. Fromageries Le Bel produces about 2300 tonnes per year of SCP and the product has been used for more than 10 years for human consumption for dietetic purposes. The origin of the raw material makes microbial biomass produced from whey more generally accepted for human consumption.

Polysaccharide hydrolysates
Both cellulosic and starchy materials can be hydrolysed into monosaccharides, mainly glucose. It then becomes possible to use a large range of microorganisms for microbial biomass production, including the well-known fodder yeast *Candida utilis*. Hydrolysis of the polysaccharide can be by enzymes or by acid treatment. In the case of cellulose, cellulase from fungi such as *Trichoderma viride* has been much studied and is often advocated, but in practice complete hydrolysis of cellulosic materials is an expensive process because of the refractory nature of the polymer, and perhaps the most promising projects today are those which use solid and semisolid fermentation systems (see below).

The problem with cellulosic materials is the amount of pretreatment that is needed before the native plant products (lignocellulose) can be effectively hydrolysed by enzymes, or alternatively the extent of sugar destruction and unwanted by-product formation during acid hydrolysis. The problem has been very extensively researched, because of the very large amounts of potentially available substrates, and the reader is referred to the Reading List for some recent review analyses.

Food grade starch would normally be regarded as too expensive a substrate for microbial biomass production. Despite this, Rank Hovis McDougall (RHM) have developed a process based on starch hydrolysate. The microorganism selected for this process, after extensive screening, is the mould *Fusarium graminearum*. The main reason for choosing a mould is the fibrous texture of the product, which assists its recovery and makes it easier to manufacture final products with a meat-like structure. It is even possible to control the mean fibre length by the dilution rate in the fermentation, and thereby manipulate the eventual texture of the product. This organism has been produced on a pilot scale in three continuous fermenters at High Wycombe, UK for some years (Fig. 10.4), and after extensive trials the product 'Mycoprotein' has been approved for human

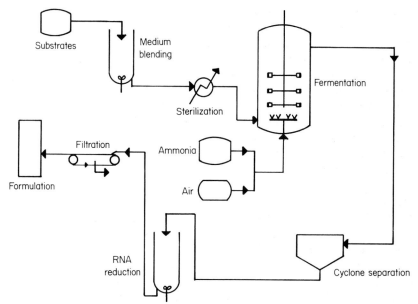

Fig. 10.4. Schematic diagram of the RHM Mycoprotein process using starch hydrolysate. Redrawn from company information (RHM Ltd).

consumption by the UK Government and food products containing Mycoprotein are now marketed in the UK. The mould will grow up to $\mu = 0.28\,h^{-1}$ at 30°C on a variety of mono- and oligo-saccharides, and the pilot plant running on starch hydrolysate is operated at a biomass density of 15–20 g 1^{-1} and a specific growth rate of $0.17\,h^{-1}$.

After fermentation the RNA content is reduced, from about 10% of the dry weight to about 1%, by holding at 64°C for 20 min (endogenous RNase activation); the mycelium is then harvested on a vacuum filter. Recently an agreement between RHM and ICI has been made to produce Mycoprotein in the pressure cycle fermenter at Billingham, UK.

Combined effluent treatment and biomass production
In many food- and drink-processing industries it is possible to isolate effluents suitable as SCP substrates, e.g. in breweries, distilleries, confectionery industries, potato, and canning industries (for whey, see above). Another important effluent is sulphite liquor from wood pulp mills. Overall, very large amounts of carbohydrates and other fermentable organic compounds that could be useful in SCP production are wasted and become pollution problems. If waste streams from manufacturing processes are

isolated at an early stage, without mixing or dilution with unwanted streams, they are often useful substrates for SCP.

A project with the utilization of a confectionery effluent producing *Candida utilis* has been commercialized in UK at Bassett, Ltd. The process treats 140 m³ effluent per day by continuous fermentation, reducing the Biochemical Oxygen Demand (BOD) by 81% and producing 1.5 tonnes of dry yeast daily. The Symba process was developed in Sweden to treat starch effluents from potato processing, and an interesting aspect was the use of a combination of two microorganisms, *Endomycopsis fibuliger* and *Candida utilis*. The first organism degrades the starch and the second assimilates the sugar more rapidly. The process was started up in 1973 but has now been abandoned, mainly because changing methods of potato processing give a lower starch wastage.

Sulphite-waste liquor (Table 10.3) has been widely used as substrate

Table 10.3.   Organic composition of spent spruce sulphite liquor

		Content (% w/v)
Lignosulphonic acids		43
Hemilignin compounds		12
Incompletely hydrolysed hemicellulose and uronic acids		7
Monosaccharides		
D-glucose	2.6	
D-xylose	4.6	
D-mannose	11.0	
D-galactose	2.6	
D-arabinose	0.9	22
Acetic acid		6
Aldonic acids and other substrates		10

From Forage and Righelato (1979). *In* 'Economic Microbiology', Vol. 4 (Ed. A. H. Rose). Academic Press: London and New York.

for SCP production in North America, Europe, and the Soviet Union, mainly with *Candida utilis*. The most advanced process based on this effluent is the Pekilo process using the mould *Paecilomyces varioti* (Fig. 10.5), for which a production plant treating 100 m³ h⁻¹ of sulphite liquor and producing 10 000 tonnes per year of dried mycelium was built at Jämsänkoski in Finland.

Inhibitory sulphur dioxide is steam-stripped from the sulphite liquor before fermentation, which is operated at pH 4.5. The Pekilo process claims several advantages. The BOD of the sulphite effluent is reduced by over 80%, because in addition to both pentose and hexose sugars *Paecilomyces* also uses the acetic acid content of the sulphite waste (see Table 10.3).

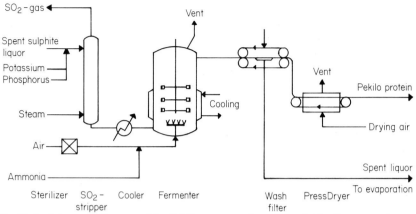

Fig. 10.5   Schematic diagram of the Pekilo process using spent sulphite liquor as substrate. From company information (Tampella Process Engineering).

The mycelial mat is easily harvested by filtration, and mechanical dewatering (to 35% dry solids) reduces drying costs. Pekilo has a crude protein content of 52–57% and is cleared for animal feed uses; it is being tested for human consumption. A change in wood pulping method, from the sulphite process to the sulphate (Kraft) process, closed down the original plant at Jämsänkoski, but a new plant has been constructed in 1982 at Mänttä (Finland) with an annual production capacity of 7000 tonnes. The investment has been about £5.5 million; the plant uses both sulphite-waste liquor and stillage from an alcohol production plant, which is another waste of potential use for SCP production.

Solid substrates

The fermentation of solid substrates such as starchy or cellulosic materials can be approached in two different ways. The solid material can be suspended in an aqueous medium. A general problem with such preparations is that gelatinization of the starches, etc. can generate highly viscous broths causing poor mixing and O_2 transfer at dry weight concentrations higher than 2% w/v. There is also the obvious diseconomy of adding so much water to the substrate only to remove it from the product. Alternatively, then, the problem is approached through 'solid state fermentation', the growth of microorganisms on solid materials without the presence of free liquid. In practice the boundary between the two ways of fermenting solid substrates is not sharp and the term 'semi-solid' fermentation is often seen. Such fermentations usually require quite different bioreactor designs; for example trays, towers and bulk-flow tunnels or drums. The fermentation of solid substrates for food products has in fact been known for centuries, as in the European mould-ripened cheeses, the Oriental *Tempe* (soya beans and/or groundnuts fermented with *Rhizopus* and other moulds), or the African *Gari* (cassava starch fermented by a mixture of organisms, mostly *Corynebacterium manihot* and the mould *Geotrichum candidum*). The industrialized manufacture of such traditional products should be practicable, especially where the products are already locally acceptable. One serious problem in poorly-controlled solid state fermentations is infection with toxin-producing moulds, usually avoided by using a large and active inoculum of the 'correct' species.

A world-wide traditional solid fermentation is the production of mushrooms, the fruit-bodies of higher fungi, which are cultivated on different substrates ranging from straw (Chinese or Straw mushroom, *Volvariella volvaceae*), and composted straw (Common edible mushroom *Agaricus bisporus*) to pieces of wood (Shiitake, *Lentinus edodes*). World production of the common species exceeds 1×10^6 tonnes per year (fresh weight). An example of a new project is that developed at the University of Waterloo, Canada. This is designed for conversion of agricultural and forestry wastes (straw, bagasse, sawdust) into microbial biomass, based on cultivation of the fungus *Chaetomium cellulolyticum*, which degrades cellulose rather rapidly. The cellulosic material is thermally and chemically pretreated, and semi-solid fermentations using different reactor-types have been tested, including low-technology systems. Important progress in the utilization of cellulosic materials is marked by the fact that residence times as short as four hours are claimed.

10.5   SCP production in the Soviet Union

Compared to the voluminous literature on all aspects of SCP projects in the Western countries, information on actual processes in the Soviet Union is very scarce, even though the Soviet Union is the largest producer of SCP in the world; production of feed yeast in 1979 was estimated as 1.1×10^6 tonnes. The Soviet literature is difficult to follow, especially the figures on production size, location, and any exact information on actual production plants. Nevertheless it is known that at least 86 SCP plants are in operation in the Soviet Union, and for 56 of these the type of substrate has been identified. New production capacities for an extra 1.2×10^6 tonnes was planned for 1981–1985. According to the Soviet Minister for Microbiological Industry, several plants are operating on purified *n*-alkanes on scales from 70 000–200 000 tonnes per year. Soviet dependency on imported fodder protein may explain the Russian interest in SCP production; it is claimed that the demand for SCP, in order to be independent of imported protein products for feed, is 10–12 × 10^6 tonnes per year.

10.6 Nutrition and safety evaluations

The most important problem for the future of microbial biomass production is its acceptability by the community as a safe and valuable protein product when produced by an economic route; this necessitates high standards in testing and production control.

10.6.1 General characterization

The first requirement is for a detailed chemical, physical and microbiological characterization. Chemical composition must be described by the content of protein, amino acids, nucleic acids, lipid, trace elements, vitamins and by analysis for substrate residues and toxic substances (e.g. heavy metals, mycotoxins, polycyclic hydrocarbons). Protein content stated as 'crude protein' determined as N × 6.25, can be misleading because of a high non-protein nitrogen content in many microorganisms. Relevant physical properties include density, particle size, texture, colour, storage and functional properties. The microbiological data must define the species and strain used, and also give data from tests for contaminants. A careful declaration provides valuable information as a basis for planning biological tests, and makes it possible to ensure that the product standard is kept constant.

10.6.2 Biological evaluation

The actual availability of nutrients and the toxicological safety can only be tested in a biological way. The safety question can be divided into safety for the animals and the safety of the animal products for the human consumers. The nutritional value is tested most meaningfully on the target species, which means that products intended for human consumption will have to be tested also on humans, but all tests must include more than one species. The rat is always used at an early stage because of the great experience with this as a standard laboratory animal and because multigeneration tests are easier to arrange. Biological tests must also be made with target animals of different ages, and possible teratogenic effects investigated with pregnant animals. Microbial biomass products for human consumption must obviously be tested through a long-term multistaged process. It is important to detect possible toxic or carcinogenic compounds, whether synthesized by the microorganism, remaining as residues from the substrate, or due to processing of the product.

A special problem with products for human consumption is the nucleic acid content. Because of the lack of uricase in man there is a danger of a damaging accumulation of uric acid, and so steps to reduce nucleic acid content in SCP products for human food use must be taken. Safety problems in production must also be covered, for example to avoid dust generation leading to allergic reactions in plant operatives. Guidelines for evaluation of SCP have been developed by the International Union of Pure and Applied Chemistry (IUPAC) and also by the United Nations Protein Advisory Group (PAG). Of special interest in this context are PAG No. 6, on the requirement for preclinical testing, PAG No. 7, on food specifications and human testing, and PAG No. 15 on animal feeds. Most SCP producers try to fulfil these standards, even though it is argued that no traditionally used bulkfeed ingredient could pass them!

10.7 Economics and future prospects

In free market economics an SCP product competes with traditional protein sources; for example for incorporation in compounded animal feeds the product will have to compete primarily with soya bean meal and fish meal. The market price will be set on a performance basis relative to that of the alternatives in a least-cost formulation.

Microbial protein for food use will compete primarily with soya bean concentrates (70–72%) and soya bean protein isolates (90–95% protein), which are sold to the food industry for their functional value in food rather than for their nutritional value; SCP for food will probably have to compete on a similar functional basis as well as on an equivalent cost basis. The texture of some SCP products from fungi seems promising in that respect, but the key factor for any future success will be acceptance both by the food industry and by the consumers.

The scale of the utilization of microbial biomass as a protein source will be determined by the economic parameters of its production, and it is therefore important to understand the technical factors that will most significantly affect the overall manufacturing economics. Examples of such an analysis are given in Table 10.4.

Two major future trends for SCP production are:

(1) Efforts to produce high-quality protein products for which a much higher selling price is justified; for example RHM's Mycoprotein, intended for human consumption; ICI's Pruteen and the yeast product from Fromageries Le Bel, which can replace milk protein in calf feeding.

Table 10.4   Approximate relative costs[a] for SCP production from molasses and sulphite waste liquor

Process	Yeast/molasses	Mould/sulphite waste liquor
Variable costs (% of total)		
Carbohydrate source	35	15
Other materials	15	30
Aeration	⎫	15
Cooling	⎪	2
Recovery and drying	⎬ 16	4
Miscellaneous	⎭	2
Labour	9	9
Fixed costs	23	23
Organism	*Candida utilis*	*Paecilomyces varioti*
Y-carbohydrate (kg kg^{-1})[b]	0.5	0.47
Y-oxygen (kg kg^{-1})	1.4	—
Productivity (kg m^{-3} h^{-1})	1	4

[a] Data based on Moo-Young (1977) and Forage and Righelato (1978).
[b] Y-values indicate the efficiency of conversion of substrate into product.
From Forage and Righelato (1979). *In* 'Economic Microbiology' Vol. 4 (Ed. A. H. Rose). Academic Press: London and New York.

(2) Attempts to reduce running and capital costs considerably, using low-cost carbon sources (including wastes and effluents) and low-technology fermentation systems. An example of this approach is the Waterloo process already described. If SCP technology also serves the purpose of pollution control, the economic viability of the process is effectively reduced by the costs of alternative effluent treatment schemes; this was the basis of the investment in the new Pekilo plant at Mänttä Pulp Mills in Finland. On the longer-term basis, it must not be overlooked that by the year 2000 a deficiency of soya beans of the order of 70×10^6 tonnes per year is still forecast. Even with increased acreage devoted to protein crops there will be a very serious protein deficit unless either new protein sources such as SCP are massively developed or there are major changes in human dietary trends.

Reading list

Birch, G. G., Parker, K. J. and Worgan, J. T. (Eds) (1976). 'Food from Waste'. Applied Science Publishers Ltd: London.
Carter, G. B. (1981). Is biotechnology feeding the Russians? *New Scientist* **90** (1250), 216–218.

Davis, P. (Ed) (1974). 'Single Cell Protein'. Proceedings of the International Symposium, Rome, Italy, 7–9 Nov., 1973. Academic Press: London and New York.

Kharatyan, S. G. (1978). Microbes as food for humans. *Ann. Rev. Microbiol.* **32,** 301–307.

Kihlberg, R. (1972). The microbe as a source of food. *Ann. Rev. Microbiol.* **26,** 427–466.

Litchfield, J. H. (1977). Comparative technical and economic aspects of single-cell protein processes. *Adv. Appl. Microbiol.* **22,** 267–305.

Litchfield, J. H. (1979). Production of single-cell protein for use in food or feed. *In* 'Microbial Technology', 2nd Edn, Vol. 1. (Eds H. J. Peppler and D. Perlman), pp. 93–155. Academic Press: London and New York.

Moo-Young, M., Daugulis, A. J., Chahal, D. S. and Macdonald, D. G. (1979). The Waterloo process for SCP production from waste biomass. *Process Biochem.* **14** (10), 38–40.

Rose, A. H. (Ed.) (1979). 'Economic Microbiology', Vol. 4: Microbial Biomass. Academic Press: London and New York.

Solomons, G. L. (1983). Single cell protein. *In* 'Critical Reviews in Biotechnology', Vol. 1. (Eds G. C. Stewart and I. Russel), pp. 21–58. CRC Press: Boca Raton.

Tannenbaum, S. R. and Wang, D. I. C. (Eds) (1975). 'Single-Cell Protein II'. MIT Press: Cambridge, MA.

Zanetti, R. J. (1984). Breathing new life into single-cell protein. *Chem. Engng* **91** (3), 18–21.

11 | Industrial Alcohol

E. A. JACKMAN

11.1 Introduction

Few economists would dispute that the effect of increased energy prices has been a major cause of world-wide recession and that it must in the end be balanced by new technology in the fields of energy saving and substitute fuels. Inevitably, the idea of 'biomass' as a replacement for fossil fuels has been popularized beyond the bounds of feasibility. Biomass can never provide more than a partial, though none the less invaluable, solution to the energy problem. Within that partial solution, fermentation of the actual or potential sugar content of the biomass is a main route either in producing alcohols and other volatile compounds to replace conventional automotive fuels, or at some future date replacing oil feedstocks for the manufacture of olefins and derived products. This chapter deals, therefore, with industrial ethanol fermentation as it operates today, and in necessarily practical terms. The resulting emphases are not necessarily the same as those in more idealized accounts.

Cellulosic waste is by far the most abundantly available form of biomass throughout the world, but at present it is more economic to reclaim its energy content by direct combustion. The problem of developing commercially attractive systems for releasing utilizable sugars from native cellulose is receiving much attention worldwide, and its economic solution should transform the whole field of biomass utilization, but until this objective is realized, fermentation fuel production will be at a modest level in the temperate industrialized countries where the potential for large-scale agricultural crops, other than food, is limited.

The only realistic substrates for fermentation industry at present are energy crops yielding sugars or readily hydrolysed polysaccharide, either grown specifically for biomass conversion, or diverted from food production where surpluses and spoilt crops products can be made available. Such raw materials will mainly originate where the sun shines best, and particularly in developing countries where labour costs are relatively low.

It is sometimes argued that where many people live at a subsistence level, all crops should be retained as food supplies. Yet, despite the 'green revolution' in food production, want has not been eradicated. It is at least as arguable that the added value from industrialization produces prosperity which is the precursor to stimulating agriculture. Setting up a stable industrial base is an important step if the world-wide exodus from rural areas is to be halted. On this view, agro-industry as opposed to pure agrarian reform offers a better prospect for developing countries. In terms of infrastructure, only the large undertakings, such as sugar from sugar-cane, into which sufficient financial, technical, agricultural and organizational inputs can be injected, have been successful. Schemes involving low energy usage are not necessarily associated with small-scale undertakings. In fact, the reverse is true in the agro-industries where scales of operation can benefit the net energy ratio.

Ethanol has now become synonymous with energy. As such, microbiological considerations are not the only criteria, and traditional fermentation industries which were superseded or reduced in magnitude by the emergence of synthetic petrochemical products will not necessarily be revitalized simply by a return to old established processing methods.

Three major problems confront the modern distiller:

(1) energy consumption.
(2) efficiency in conversion.
(3) effluent pollution.

All are interlinked, starting from the nature of the raw material used and terminating at a benign environment. Yet within all the activities involved, the fermentation process itself is still the crucial step which sets the design parameters and operating requirements for the overall distillery.

11.2   Fermentation feedstocks

The more readily utilized energy crops can be grouped into five basic categories:

(a) By-product sugars from crop processing (molasses, sweet sorghum syrups, spent sulphite liquor, etc.).
(b) Sugar crops (sugarcane, sugar beet, sorghum, etc.).
(c) Cereals (maize, wheat, rice, etc.).
(d) Tubers (cassava, yams, potatoes, etc.).

(e) Miscellaneous sources (sago, polysaccharide residues from nut-oil extraction, etc.).

Less readily utilized at present are the cellulosic materials which can be classified as follows:

(a) Direct forestry products (eucalyptus trees, pine, etc.).
(b) Cellulosic wastes (sawdust, chips, bark, straw, stover, waste-paper, etc.).

Out of all the above possible sources, problems of their present availability, price structures and technological developments in processing reduce the selection in terms of economic feasibility to a very small number. Although crops, such as sugar-cane, are presently being converted into ethanol, such ventures all depend upon strategic intervention by governments. Within a free market economy, substrates can be curtailed to three basic raw materials, namely molasses, maize and cassava.

Under special circumstances, other raw materials will demonstrate immediate feasibility and when the unit value of energy increases at a faster rate than most other commodities, other energy crops will soon become attractive propositions. However at present, only the three above-mentioned feedstocks warrant detailed consideration.

The nature of the particular raw material used in fermentation has considerable bearing upon the processing methods to be employed. All the constituents of the feedstock are important, and there is a wide disparity between industrial substrates and those usually used in research. The ideal substrate, a single hexose sugar such as glucose, at the correct strength in a sterile water solution containing the necessary proportions of nutrients, does not exist naturally. Moreover, the influence of the potable spirits trade has imposed its tradition on industrial alcohol manufacture even though flavour character of the distillate is no longer a consideration. Too often industrial alcohol projects based on molasses or maize are seen as simple extensions of the manufacture of rum or whisky.

This is understandable since the potable industries have developed their technologies over a large number of years into sound, practical and profitable systems. However, they serve a high-priced luxury market and operate on a lower scale of production to that expected for the energy-orientated alcohol projects of the future. Most of their techniques were established when the costs of energy and raw materials were low and when pollution abatement was not considered a social necessity.

All practical raw materials require pretreatment to make them suitable for fermentation and in the past insufficient attention was given to producing a substrate as near ideal as possible within the physical limitations of its original composition. To take advantage of fermentation techniques offering efficient conversion with low energy product recovery and where an effective system can be employed in stillage effluent treatment, the liquid substrate should meet the following conditions:

(1) Fermentable sugars concentration should be correctly adjusted to suit the particular fermentation method and to ensure that residual sugars after fermentation are kept to a minimum level.

(2) The substrate should be clarified, at an acceptable temperature and optimum pH, and contain adequate nutrients for yeast.

(3) Microorganisms other than the main inoculum should be eliminated by pasteurization, antibiotics, antiseptics or sterilization; the degree and method of elimination depending upon the fermentation system employed.

(4) Substances with a toxicity to the yeast should be removed or reduced to an acceptable level.

(5) Adverse effects of osmotic pressure should be kept within acceptable limits.

11.3   Specific substrates

11.3.1   Sugar juice

Sugar juices, whether from cane (which ranks particularly high in efficiency of photosynthesis per hectare) beet or sweet sorghum stalks, are interesting feedstocks since they immediately supply a readily fermentable substrate, but this can also be a drawback.

After harvest, the sugars are subject to natural deterioration from enzymes and bacteria. Only after the juice has been evaporated to a syrup is it possible to hold raw material for processing over an extended period. More conventionally, a sugar juice distillery only operates during the harvest season, with consequent inefficient use of plant and manpower. Moreover, in the preparation of sugar juice as a feedstock for alcohol, the technology of conventional sugar manufacturing methods has usually predominated, whether appropriate or not.

With conventional sugar-cane milling or diffusion methods for sugar beet, the addition of 'imbibition water' is necessary for efficient extraction. Generally, cane will then yield its own weight of mixed juice, typically containing only about 12.5% w/v of fermentables, expressed as hexose sugars, so that without evaporation, the maximum ethanol level after fermentation is only some 6% w/v.

Fermentation conditions set the design and operating parameters for the overall distillery and there is little flexibility. A conventional sugar-cane distillery producing 60–70 litres of ethanol per tonne of cane is, necessarily, a simple installation based upon traditional methods using high steam loads for ethanol recovery, large numbers of fermenter vessels and discharging a large volume of pollutive effluent. The available energy from the moist bagasse (the cane residues after sugar extraction) makes such thermally inefficient installations economic, and, indeed, to prevent a build-up of bagasse, it actually becomes essential to operate the steam generators inefficiently.

In sugar manufacture, the mixed raw juice is clarified prior to evaporation to a concentrated syrup for subsequent sugar crystallization and recovery. Lime is normally added, raising the pH of the juice from 5.0–5.7 to 7.0 or above. Heat treatment with coagulation removes fibre and suspended colloids. By raising the pH, inversion of the sucrose during further processing is halted. Impurities in the sugar juice passing clarification, together with calcium salts, will remain as residue in the final molasses.

For the alcohol fermentation, however, the liming process is a retrograde step, since:

(a) Optimum pH conditions for the fermentation are at 4.5–5.0, closer to the original juice pH.
(b) Calcium salts can cause severe scaling in heat exchangers and distillation equipment.
(c) Liming removes phosphates and nitrogenous compounds from the juice, which are yeast nutrients.
(d) Inversion of sucrose is a benefit.

Consequently for alcohol production in Brazil it has become standard practice to retain the normal cleaning system of screens and sedimentation, but using cold raw juice and without liming. The cleaned juice is still relatively turbid, containing colloids and small fibre particles which are extremely retentive to bacteria, and tend to cause blockages in the nozzles of yeast centrifuges where used.

Improvements in clarification methods for sugar juice are difficult to justify on economic grounds where the concentration of fermentable sugars limits the fermentations, and the bagasse is only useful as fuel. For example, steam injection into the raw juice could be used to raise the temperature for pasteurization and at the same time to bring more colloidal materials out of solution, enhancing the clarification by entrapment of the finer fibre particles. However, heat recovery from the pasteurized juice then requires capital investment in effective and reliable heat exchangers. Thus, although various improvements are under trial, no standard system which is acceptable to the majority of distillers has yet been put into commercial operation.

11.3.2 'A' molasses

Sugar manufacture proceeds in three stages where 'A', 'B', and 'C' sugars are crystallized and recovered leaving a final 'C' molasses from which maximum sucrose has been crystallized.

Many sugar technologists have put forward the concept of dual 'A' sugar and ethanol manufacture to combat the volatility of sugar prices since the price of ethanol is relatively stable in the market place and can only improve. 'A' molasses is an ideal substrate for fermentation when conventional sugar juice clarification methods are employed. It has a high sucrose purity and at about 85% solids content should be capable of producing 385 litres ethanol per tonne (allowing a reasonable level of losses). Its stability in storage can reduce year-round distillery capital costs, while more efficient processing methods can be applied in fermentation, distillation and effluent treatment.

11.3.3 'C' molases

Despite final molasses production being geared to crystal sugar manufacture whereby its supply is limited, this substrate is extensively used for both industrial and potable alcohol, particularly where its use for animal feedstuff compounding is restricted by transport costs. If 'C' molasses is used for alcohol production, the overall sugar:alcohol ratio is about 12:1; 'C' molasses is the basic feedstock for those developing countries with a sizeable sugar industry, and no organized animal feed market.

Such molasses normally contains a high bacterial count due to non-hygienic handling and storage. A typical 'C' molasses analysis at 80 wt% solids is given in Table 11.1. Fermentable constituents, expressed as invert sugar,

Table 11.1   Composition of 'C' molasses

Component	Wt%
Sucrose	35.0
Other reducing substances	19.0
Other organic matter	14.0
Inorganic ash	12.0
Total solids	80.0
Suspended solids	8.0% v/v

are usually in the vicinity of 50–55% w/w. Non-fermentable residues are concentrated in the molasses and cause problems in distillery operations; in particular, suspended calcium and soluble inorganic salts, colloids, fibres, gums, organic acids and earth residues. Therefore, on dilution to achieve a substrate suitable for fermentation, suspended solids should be removed to a reasonable level and bacterial growth curtailed.

Most common methods of pretreatment use hot acid clarification, dosing at about 0.5% by weight of concentrated sulphuric acid, where pasteurization is accompanied by precipitation and followed by hydrocyclone or centrifugal separation. Since the diluted molasses substrate will need to be acidified for fermentation, the acid addition in pretreatment does not substantially influence the overall cost.

Solids separation is carried out at 70–95°C where the majority of the sludge is the calcium salt, $CaSO_4.2H_2O$, the solubility of which decreases with temperature above 38°C. During thermal treatment, a mild hydrolysis takes place; fermentability of some of the polysaccharides is improved, and certain volatile organic acids, sulphur dioxide and other harmful components can be removed in the vented steam. Sugar losses are kept to a minimum by decanter sludge washing incorporating the primary dilution water. Basically, the pretreatment of molasses is a precursor to efficient fermentation by a particular technique and cannot be economically justifiable in itself.

11.3.4 Cereals

On a dry basis, corn (maize), wheat, sorghums (milo) and other grains contain around 60–75% w/w of starches hydrolysable to hexose with a significant weight increase (stoichiometrically, the starch to hexose ratio is 9:10) and offer a high yielding ethanol resource.

Most cereal starches contain a mixture of α-amylose (20–30%) and amylopectin (70–80%). The former is a water soluble linear polymer while the latter is a water insoluble branched polymer. The saccharification of amylose is much faster than that of amylopectin, but since amylopectins predominate, the overall conversion to fermentable sugars is governed by their breakdown.

Originally, cereals were hydrolysed by acid catalysis. Conversions were incomplete and the requirement for high temperatures over an extended time period led to undesirable by-products. Viscosity problems hindered the process; often, the cooked mash would also contain 'limit dextrins', requiring a lengthy fermentation time for yeast dissimilation. Gelatinized starches were prone to 'retrogradation' on cooling, where the starch molecules re-aggregate, forming insoluble crystallites accompanied by increases in mash viscosity.

Today, most conversion processes use a cooking phase, helped by the addition of enzymes for complete conversion. Sometimes acid hydrolysis followed by enzymic conversion is practised, but the present trend is to rely upon direct enzymic hydrolysis carried out in two stages.

In the first stage of liquefaction, α-amylase is used in dextrinization of the starch occurring over a relatively short time period at high temperatures.

Dried grains are milled to a fine grist and mashed with warm water. The slurry, normally at about 2.5–3.0 kg water per kg milled grist and adjusted to pH 6.0–6.5, is dosed with a portion of the enzyme, heated by steam jet injection to c. 110°C and held for about 20 min under pressure, normally in continuous holding tubes at fixed flow rates to ensure that the critical factors of time and temperature are maintained.

In this stage, the α-1,4 links of the starch molecules are broken down at random; shorter chains of glucose units (dextrins) are formed, resulting in a rapid viscosity reduction of the gelatinized starch. Conversion is relatively low and normally results in a dextrose equivalent (DE) of 10–15. (This measure of conversion is defined as the weight percentage of reducing sugars calculated as dextrose on the dry available starch.)

The mash is then cooled to about 85–90°C. Further α-amylase is added and the mash is held in agitated vessels for about 90 min to complete post-liquefaction. Total enzyme dosing rates are normally set at about 1.5 kg α-amylase per tonne of starch. The dosing rate varies amongst distilleries, but economics govern the quantities that are used; it is extremely difficult to exceed a DE of 20.

Completion of saccharification using glucoamylase is carried out at a lower temperature, normally in the region of 55–60°C, with longer residence times to maximize conversion. Both the α-1,4 links from the non-reducing end of the molecular and the α-1,6 links are hydrolysed passing through the stages of dextrins, maltose and glucose. DE levels of 99 can be achieved, which is usually 10–15% greater than by acid hydrolysis.

The mash is acidified below pH 5 and glucoamylase dosing rates depend upon temperature, residence time in the reaction vessels, the DE level after post-liquefaction, and the duration of fermentation. For a two-day residence time at 55–60°C a dosing rate of 1.5 kg glucoamylase per tonne of original starch is typical. Again, the economics of enzyme cost dictate the optimum usage. Dilution to an appropriate strength for fermentation is normally carried out after this stage. However, the extent to which saccharification is completed before fermentation is a matter of judgement.

Changes in temperature influence the enzymic reaction rate in accordance with the Arrhenius equation; as a rough rule-of-thumb enzymic activity will halve for every ten degree Celsius reduction in temperature. This means that the continuing saccharification activity of the glucoamylase during the fermentation stage, at about 30–35°C, is considerably diminished. However, during the initial stages of fermentation there is also a rapid removal of the available sugars. This actually promotes the glucoamylase action and tends to counterbalance the retardation due to a reduction in temperature. As fermentation slows down when the ethanol concentration increases, the glucoamylase reaction may become rate-limiting overall.

Where saccharification is not complete before fermentation, the residence time during fermentation will be extended to achieve reasonable final ethanol concentrations; fermentations where 'limit dextrins' are present may require several days to complete. At dilute conditions to suit fermentation levels, vessel volumes become an important cost factor.

In this respect, some of the new fermentation techniques show interesting possibilities. Where ethanol can be removed as it is formed, rapid fermentation rates can be continuously maintained, with a corresponding improvement in the kinetics of enzymic saccharification. If ethanol removal and recycling of the exhausted substrate can be effected without thermal destruction of the glucoamylase, a high enzyme concentration can be built up for the same total enzyme usage.

11.3.5 Mash retention

In conventional cereals processing, the mashed fermented grist is retained in suspension and is pumped direct to the primary distillation column where, in the stripping section, the suspended solids are effectively washed, and final traces of ethanol are removed. Spent grist is recovered as an animal feed by screen and decanter separation. The system is simple and cheap and is the traditional method; however mash retention has disadvantages:

(a) The presence of fibres and other suspended solids hinders fermentation.
(b) Fermentation methods are restricted to traditional batch operations in which 'once-through' yeast pitching is necessary.

Compared with clarified conditions in fermentation, where it is possible to recover and recycle yeast, mash fermentations are less efficient in ethanol conversion, and require longer residence times in correspondingly larger vessels.

11.3.6 Tubers

Temperate zone root crops have little or no application in industrial alcohol manufacture because of their greater value in well-established food uses.

However, this is not the case with the root crop cassava (manioc) which is widely cultivated in most tropical countries. Since it is easy to grow, resistant to drought and pests, and can acclimatize to soils low in fertilizer nutrients, it provides a staple food, but one with a very low protein content. Attempts to upgrade its nutritional value have had little success and it remains a problem crop, particularly with bitter varieties which contain cyanogens, requiring primary water leaching before being edible.

Yet, compared with sugar-cane, cassava has the potential to produce up to $2\frac{1}{2}$ times the quantity of ethanol per tonne of harvested crop, and it is much cheaper to grow. Against this must be balanced two main adverse factors. Traditional crop yields are low per harvested area and external energy sources are needed in processing.

However, agricultural improvements, raising cassava from a village husbandry level, have been rapidly developed and varieties cropping at 30–40 tonnes per hectare with adequate above-ground woody portion to supply energy needs, are now available.

A typical tuber composition is given in Table 11.2 for a bitter variety (C.M.C.–84).

Table 11.2   Typical tuber composition

Tuber components	Wt%
Moisture	61.3
Starch	30.5
Total sugars	2.6
Ether extract	0.13
Others	1.7
Fibre	3.0
Ash	0.77
Total	100.00

Available reducing sugars = 0.7%
Crude Protein (N × 6.25) = 1.06%
HCN   = 575 p.p.m.
                         (dry matter basis)

Cassava starch differs from cereal starch. Although the amylose content is similar, between 17 and 30%, the amylopectins have a much shorter chain length than those found in cereals, which simplifies its mash preparation and enzymic breakdown. Pretreatment involves prewashing to remove dirt, stones and surface peel from the roots which then undergo prechopping and final comminution in disintegrators containing rasps on a rotating cylinder. Once mashed to a pulp, the methods for liquefaction and saccharification of tubers are similar to conventional cereal treatment, but the demand for water is less. Cassava tubers are capable of being mashed without the addition of water, since the starch granules are already dispersed in the water phase within the root. Hydrocyanic acid liberation has no influence in processing, and cyanides are normally flashed off with the steam during cooking.

The spent mash is normally less than the residue from grain hydrolysis and its reduced protein content gives it less value as an animal feedstuff. In large-scale ethanol ventures these wastes will be of most use when squeezed to a reasonable moisture content and used as a solid fuel, similar to bagasse.

To increase the radius of operation and to overcome seasonal variation in the supply, tropical air-drying the tuber roots for two days will reduce the moisture to 15% with around 70% fermentables. Chips or meals can be stored with a stable shelf life, require less storage volume with ease of handling, and at about 55% moisture loss, have reduced transportation costs.

11.3.7 Cellulosic substrates

The most abundant potential source of utilizable sugars is cellulose, obtained either directly as forestry products or indirectly in wastes such as straw, corn stover, bagasse, or paper-waste. Conceivably the oceans could yield vast quantities of cellulose as seaweed. However, cellulose is difficult to hydrolyse and is always accompanied by hemicellulose and lignin. These make the recovery of hydrolysable cellulose more difficult and pose additional problems in devising their own economic exploitation.

The acid hydrolysis of cellulose to fermentable sugars is technically possible and was widely used in state-controlled war-time economies, but the free-economy practicality of the primitive technology used is doubtful. However, research is widespread and the successful development of one or more processes combining economic pretreatment with rapid hydrolysis and efficient recovery of utilizable sugars seems probable. The successful development of purely enzymic routes for cellulose hydrolysis seems more problematic, but considerable interest attaches to the possibility of direct conversion of suitably pretreated cellulose into ethanol or other fermentation volatiles using selected mixed cultures of cellulolytic and fermenting bacteria, such as *Clostridium* species, some of which are usefully thermophilic. None of these developments, however, has yet impinged upon practical alcohol production.

11.4 Fermentation

11.4.1 Conversion mechanism

The fermentation mechanism was first quantified by Gay-Lussac, based on the stoichiometric conversion of a hexose sugar into ethanol and carbon dioxide:

$$C_6H_{12}O_6 \rightarrow 2C_2H_5OH + 2CO_2$$

hexose sugars	ethanol	carbon dioxide
180	92	88

Therefore: 100 kg hexose sugar = 51.1 kg ethanol + 48.9 kg carbon dioxide.

This theoretical yield at 51.1% by weight is termed the Gay–Lussac coefficient, and represents the basic datum in conversion efficiency.

The next step in understanding the mechanism was made by Louis Pasteur, and his experiments of over a century ago represent a watershed in the microbiological sciences. Yields obtained from his ideal fermentation of glucose are given in Table 11.3.

Table 11.3   Yields in the ideal Pasteur alcohol fermentation

	Wt%
Ethanol	48.4
Carbon dioxide	46.6
Glycerol	3.3
Succinic acid	0.6
Cell matter	1.2
Total	100.1

A larger appearance of products compared with the original sugar is due to oxygen (air) and nutrients from an external source being necessary for cellular growth.

Therefore, the Pasteur coefficient at about 94.7% of the theoretical GL (Gay–Lussac) yield is considered to be the maximum possible ethanol yield that can be achieved in fermentation.

However, the Pasteur coefficient may be exceeded by the re-use of yeast, or where yeast growth can be achieved from carbohydrates not naturally fermentable into ethanol. In commercial practice, where non-ideal substrates are used, sustained conversion efficiencies for a reasonable length of time are normally in the region of 90% GL. Typical product concentration ranges per 100 g of fermented glucose in yeast based ethanol processes are given in Table 11.4.

Specific yeast properties, such as high alcohol and CO_2 tolerance, rapid growth and fermentation capacity, reduce the number of microorganisms suitable for industrial-scale operation to a small number, of which by far the most important are selected strains of *Saccharomyces cerevisiae* and *Saccharomyces carlsbergenesis* (uvarum).

Table 11.4 Expected product concentrations range per 100 g fermented glucose in a yeast-based ethanol process

Product	Concentration range (g per 100 g fermented glucose)
Ethanol	45–49
Carbon dioxide	43–47
Glycerol	2–5
Succinate	0.5–1.5
Fusel oil	0.2–0.6
Acetate	0–1.4
Butylene glycol	0.2–0.6
Cell material	0.7–1.7

11.4.2 Fermentation processes

Both batch and continuous processes are employed, and many of these incorporate some form of cell recycle. As well as reducing the fermentation time, cell recycle reduces the amount of substrate converted to cell matter. In the Melle–Boinot batch process, which has found widespread application, the yeast cells recovered after the fermentation are recycled to the fermentation vessel to which fresh substrate has been added. Operations at yeast levels well above 15 kg dry matter per m^3 are common, with fermentation times typically between 8 and 18 h, depending on yeast strain, nature of the substrate and operating temperature.

Most continuous, rapid fermentation techniques available today as commercial units, comprise 3–5 closed vessels, arranged in cascade. Each vessel will have its own cooling arrangement and ethanol concentrations build up in each stage, usually commencing at 4% v/v in the first vessel and reaching 10% v/v at final exit. Productivities in the region of 10–20 kg ethanol $m^{-3} h^{-1}$ (based on effective operating volume) are normal; much higher productivities are often quoted for laboratory systems but these always relate to ideal substrates and leave such essential factors as efficient sugar conversion and cheap reliable installation out of their account. For comparison, conventional batch methods normally achieve productivities of only 1.5–3 kg ethanol $m^{-3} h^{-1}$. However, a high productivity rate unrelated to yield efficiency expressed as the Gay–Lussac coefficient is meaningless. In economic terms, a reduction in waste can be far more beneficial than an increase in haste.

For large-scale ethanol production and where flavour characteristics are not the criterion in the method of fermentation, it is widely accepted that a distillery should be totally continuous in all its unit operations which, apart from fermentation, can be readily achieved.

Advantages of continuous fermentation are:

(1) Fermenter vessels are reduced in size (or number) with intermediate holding capacity no longer being required.
(2) The distillery can be run at steady state conditions and at set-points in control within finer range limits, with greater ease in operation.
(3) Without fluctuations during operations, utility requirements are constant, giving a greater economy in usage. Without peak demands, related equipment costs can be reduced.
(4) High productivity (kg ethanol $m^3 h^{-1}$) as compared with batch.
(5) Reductions in manpower.

Against this, batch operations have the following benefits:

(1) Higher final product concentration.
(2) In many countries, manpower intensive conditions may be of greater importance.
(3) Less reliance upon automatic controls and less demand on technical competence.
(4) Overall cheapness and simplicity in plant construction and maintenance.

Continuous fermentation requires a high degree of asepsis to prevent ingress and build-up of infection. Batch systems have the flexibility for rigorous cleansing during the turnaround cycle.

It must always be borne in mind that by far the most costly element in alcohol manufacture is the cost of the raw materials—some 2/3 to 3/4 of the total. Even within the remaining cost, the installation and operating costs in the fermentation stage are relatively minor elements. Consequently, any effect of improving the production efficiency of the fermentation can easily be offset by an increase in the capital cost, and even more easily nullified if the efficiency of raw material conversion is not fully maintained. This is the main reason why continuous fermentation processes have not been introduced with anything like the alacrity which their inventors have expected.

11.4.3   Other nutrients

Compositions of yeasts vary, but a typical *S. cerevisiae* will have an average dry matter composition of 47% C, 6% H, 32.5% N, 1.1% P, 2.1% K, plus other minor constituents. The composition of the carbohydrate fraction conforms closely to $(C_6H_{10}O_5)n$ combined with complex amino acid units, lipids and inorganic constituents.

Therefore, in yeast propagation, apart from the C, H and O components in the original substrate and from the oxygen provided during aerobic growth, additional nutrients are added to provide constituents such as N, P, K, etc.

Generally, potassium and other minor inorganic constituents will be available in excess within the original substrate. It is only in the provision of nitrogen and phosphorous where nutrient addition becomes necessary.

Some of the nitrogen may be available as crude proteins in the substrate. However, when proteins are broken down in yeast formation there is a tendency for volatile impurities to be formed, such as esters, mercaptans, etc. Therefore, it is common practice to add nutrients in slight excess of their stoichiometric demand. Solutions of ammonium sulphate and/or urea are used to provide the nitrogenous constituents, whereas acids or acid salts of phosphorus are added to provide the P_2O_5 content.

11.4.4 Operating temperature

As with most biological phenomena, temperature has two effects on yeast growth; with increasing temperature all yeast functions are accelerated, including their breakdown. Consequently there is an optimum temperature for yeast growth. The picture is more complicated if ethanol production rates are measured, since the temperature optimum for ethanol production is generally some 2–4°C higher than that for growth. In addition, temperature affects the sensitivity of both growth and ethanol production to ethanol inhibition (see Section 11.4.5). Consequently, the ideal temperature for a batch fermentation varies through the batch duration, while that for a continuous fermentation will depend on the balance between growth and ethanol production at the steady state. In practical operations, however, other considerations also intervene, including the not inconsiderable cost of providing accurate temperature control and the difficulty of effective cooling in hot climates.

With most commercial yeast strains temperatures between 32 and 38°C are maintained, but when there is a danger of bacterial infection, as in tropical countries, lower temperatures are recommended. Under these conditions operation at pH 4.5–5.0 is also helpful in restricting bacterial growth.

Proper temperature control must take into account not only the relatively small heat evolution associated with the anaerobic process, but also, where appropriate, the much larger evolution of heat associated with aerobic yeast propagation.

With extended residence times, as in traditional cereals mash fermentations, heat removal can be accommodated naturally. The substrate itself can take up some of the heat, and where the fermenter vessel is in an open-sided structure with a reasonable degree of external air movement, practical heat transfer coefficients for convection and radiation at the vessel walls can be in the region of $80 \, kJ \, m^{-2} \, h^{-1 \circ} C^{-1}$. Fermenter size, and the ratio of surface area to volume now become important, as do the ambient conditions which dictate the rate of heat dissipation.

In rapid fermentation processes with large fermenter vessels having a low area/volume ratio, and also where high ambient temperatures prevail, some form of assisted cooling becomes necessary. In older installations this was normally achieved by cooling coils or external spray rings. Both provide little flexibility in control, and also complicate cleaning regimes. Most modern systems use externally located water-cooled heat exchangers, particularly of the plate type, with a facility for ease of dismantling, cleaning and altering surface area, coupled to a pump and piping to the fermenter; pumped circulation of the fermenting substrate has the added advantage of providing bulk mixing in the fermenter.

Table 11.5 shows the various methods used for temperature control on industrial rapid fermentation units.

11.4.5 Ethanol inhibition effects

The most important inhibitory effects on the yeast fermentation of sugars to ethanol are those due to ethanol itself. Their discussion is not simple and it is important to understand exactly what processes the inhibition affects.

The conversion of glucose into ethanol and CO_2 also produces ATP, in the proportion 2 moles ATP per mole glucose transformed. At low ethanol levels this ATP is used for the assimilation of further glucose into new cells (compare the Pasteur stoichiometry, Table 11.3), and the growth rate is now the highest possible fraction of the rate of substrate utilization. At higher ethanol levels—for example, as a batch fermentation proceeds, two things happen. First, the rate of substrate utilization falls (and consequently, so does the rate of ethanol formation, since this is stoichiometrically linked to substrate use). This is because ethanol exhibits, by mechanisms which are still uncertain, the overall pathway of sugar uptake and catabolism. Second, however, the extent to which sugar catabolism provides energy for yeast growth is also reduced, apparently because more ATP is diverted to repair processes needed to maintain cell function. Consequently, as the ethanol level rises, the effect on yeast growth is even more marked than the effect on ethanol production rate.

If the initial sugar level is high enough, the alcohol level eventually becomes so high that the cells no longer grow at all; indeed, they begin to die, and their viability depends on the length and severity of their exposure to higher ethanol levels. However, ethanol production by the non-growing cells continues, and, in fact, does not stop altogether until a considerably higher level is reached.

Thus, in a batch fermentation provided with excess substrate, the successive events are: yeast growth initially in parallel with ethanol production; ethanol production decelerating but yeast growth slowing down even more; yeast growth ceasing and viability beginning to fall, while slower ethanol production continues; ethanol production eventually ceasing. The ethanol level at which the yeast ceases to grow measures one sort of limit, its growth tolerance to ethanol; the level at which it ceases to produce ethanol measures a different limit, its production tolerance. In general, growth tolerances are in the range 6–9% w/v but production tolerances with some yeasts may be 15% w/v or even higher.

The underlying kinetic effects are best brought out by plotting specific rates against ethanol level, as in Fig. 11.1, in which the specific growth rate, μ (g cells (g cells)$^{-1}$ h^{-1}), and the specific production rate, ν (g ethanol (g cells)$^{-1}$ h^{-1}) are both plotted against the ethanol level P. The maximum values μ_m and ν_m are both observed at zero ethanol level, where the ratio μ/ν—that is, the ratio of growth to ethanol formation or, what is equivalent, to substrate consumption, has its highest value. P_1 is the growth tolerance, at which μ and μ/ν both $\rightarrow 0$; P_2 is the ethanol production tolerance at which $\nu \rightarrow 0$. The particular form of the curves in Fig. 11.1 is somewhat arbitrary and here, the linear forms:

$$\mu = \mu_m(1 - P/P_1)$$
$$\nu = \nu_m(1 - P/P_2)$$

have been chosen for simplicity; though many observations are consistent with this, other equations are often proposed—often without sufficiently accurate data to validate them.

An important consequence is that any single-stage continuous fermentation, which depends upon there being some yeast growth to sustain its operation, can only be run at ethanol levels less than the growth tolerance limit P_1, whereas a corresponding batch fermentation can give ethanol levels up to P_2. Where the substrate itself is available in sufficient concentration to afford the higher level, this is obviously a major consideration.

11.4.6 Yeast separation

Most rapid fermentation techniques are based upon the Melle–Boinot principle in which the yeast cells at the end of fermentation are harvested

Table 11.5 Temperature control in industrial alcohol fermentations (Alfa-Laval Publications, 1980)

Condition	Natural convection and radiation in still air	Spray-ring water cooling	Internal coils and other extended surface media	External plate heat exchanger and pump system
Starting wash temperature 28°C, cooling water temperature at inlet 27°C				
Fermenter size for Melle–Boinot technique	Laboratory size up to 1000 litre capacity	Fermenter size up to 20 000 litre capacity A/V ratio > 2.0	All sizes providing coil surface is adequate	All sizes providing plate heat exchangers and pump sizes are adequate
Increased air movement	Improvement in heat removal	Slight improvement in heat removal	Assisted improvement in heat removal	Assisted improvement in heat removal
Rise in ambient conditions	Reduction in heat removal	Slight reduction in heat removal	Negligible effect	Negligible effect
Heat transfer coefficient kJ h^{-1} m^{-2} °C^{-1}	20–40	100–200	800–1600	~12 500
Inclusion of agitator in fermenter	Negligible effect	Considerable improvement during end fermentation stage	Considerable improvement during end fermentation stage	Not required as forced recirculation gives constant heat removal throughout fermentation
Design flexibility	None	None, improvement only by using refrigerated water	None, improvement only by using refrigerated water	Flexible. The plates can be increased (or decreased) should design conditions be altered to change the heat load

Protective lining to internal fermenter walls to prevent vessel corrosion	Act as thermal insulation with reduction in heat removal	Acts as thermal insulation with reduction in heat removal	Negligible reduction in heat removal	Negligible reduction in heat removal
Materials of construction and corrosion	If in carbon steel, vessel has limited life due to corrosion, if lined, see above	If in carbon steel, vessel has limited life due to corrosion, if lined, see above	Vessel walls can be internally lined. If coils in carbon steel they have limited life due to corrosion because of low pH values	Vessel walls can be internally lined. Plate heat exchanger in stainless steel, bronze pump, external piping and fittings can be in inert materials such as rigid plastic. System has minimal corrosion problems and extended life
Cleanliness and control of infection	No fermenter internals and easy to clean	No fermenter internals and easy to clean. External sprays cause algae and other growths in fermenter room. Hardness in cooling water causes incrustation at ext. walls and spray jets	Supports for coils and frames provide traps for contaminating microorganisms. Coils and surfaces require labour intensive cleaning between batches. The internal surfaces to pipes and plates cannot be cleaned should hardness in cooling water cause scaling	No fermenter internals and easy to clean. The external plate heat exchanger can be quickly cleaned by pumping through or dismantled and reassembled should thorough cleaning be occasionally required

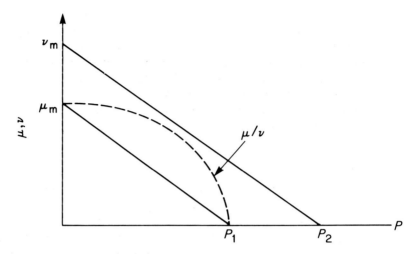

Fig. 11.1   The specific growth rate μ and the specific ethanol production rate v plotted against ethanol concentration P. The dotted line is the ratio μ/v; P_1 is the tolerance level for growth, and P_2 the tolerance level for ethanol production.

and returned to a fresh substrate. Theoretically, if all the yeast is recovered and retains its original level of activity, high yeast concentrations can be maintained, giving rapid conversion rates, and no sugar is used for yeast propagation, giving an increased conversion efficiency. Yeast recovery is equally essential for rapid conversions in continuous fermentation processes.

In practice, recovery is never total; a proportion of the yeast dies from natural causes and part of the separated yeast must always be discarded to prevent a build-up of suspended matter.

Most yeast cells used industrially have a size range between 5 and 20 μm, and their density is not much greater than that of water. Such cells will not settle out from the suspension at any useful rate. Only by flocculation where the cells form flocs with a combined particle size suitable for a reasonable settling rate, can the yeast be withdrawn from the fermenter heel by natural methods.

The phenomenon of yeast flocculation is not fully understood but the pH and the presence of calcium ions are important factors. The ability to flocculate naturally is determined genetically. A disadvantage is that flocculating yeasts tend to separate before fermentation is complete; conversely some yeasts with low flocculating properties remain suspended until very late in a batch fermentation but can only be separated by mechanical centrifugation.

Nozzle type centrifuges operating at up to $5000 \times g$ are usually used where the presence of other solids in the mash allows. These are capable of recovering up to 99.9% of the viable yeast cells in one stage, with yeast slurries for recycling at concentrations of 40–60% pressed yeast. Most of the Brazilian distilleries operating the Melle–Boinot technique employ centrifuges for yeast recovery. Savings in sugar losses as available ethanol pay back the capital and operating costs in a very short time, measured in months rather than years.

11.4.7   Ethanol recovery

Attaining a high ethanol concentration at the end of the fermentation is an essential requirement in keeping operating costs low. As the ethanol concentration increases, less steam is needed for primary distillation (the beer column) and the volume of the outgoing stillage is reduced. Where new rapid fermentation techniques are being developed, suitable for continuous operation, levels in ethanol concentration must be retained or improved upon as these energy costs can far outweigh the capital-related cost savings from a shorter residence time (cf. Section 11.4.2 above).

11.4.8   Immobilized cells

Considerable research and development has been carried out on the use of immobilized cells held in suitable carrier media whereby the substrate is passed through highly reactive conversion zones. The basic philosophy is to control the fermentation by fluid flow conditions through a catalyst, which is the normal procedure for most chemical conversions. However promising, such systems have not yet found practical applications. So far, the engineering problems of securing proper flow distribution in the reactors, particularly those which are caused by the need to provide for disengagement of large volumes of CO_2, have not been solved. The need to provide for an additional facility to prepare and replace or regenerate the catalyst is also seen as a major drawback.

11.4.9   Continuous ethanol removal

If ethanol can be removed continuously from the fermentation concurrently with its formation, fermentation can proceed at low ethanol concentrations, i.e. without significant inhibition. Not only will this give faster conversions, but a higher quality of ethanol can be expected since the majority of volatile impurities such as fusel oils, esters, etc. are produced at terminal fermentation conditions. One proposal to achieve this result was to reduce the operating pressure, thereby lowering the alcohol boiling point and the steam required for distillation. However, the energy needed to maintain the low pressure while venting the evolved CO_2 to atmosphere is prohibitive.

By maintaining the fermenter as an atmospheric pressure vessel with normal carbon dioxide venting, and applying reduced pressure to a circulating liquid slipstream, overall reductions in energy and equipment costs should give improved process economies. Yet even with these improvements the yeast is in contact with the liquid undergoing evaporation at reduced pressures, where temperatures therefore should not exceed 47°C. Also, the system requires a high degree of asepsis since the evaporation is carried out at optimum temperatures for mesophilic bacteria multiplication.

This final restraint is removed by the Biostil technique in which the circulating slipstream from which the ethanol is removed is first freed from yeast. Ethanol evaporation is no longer restricted by the need to maintain yeast viability and, indeed, evaporation is accompanied by pasteurization. The technique has been developed since 1977 in a semi-scale pilot plant at 800–900 litres ethanol day^{-1}, and the first commercial unit was commissioned in Australia using molasses in April 1981. Productivity rates of over 20 kg ethanol m^{-3} h^{-1} are possible depending upon substrate strength, cellular levels and ethanol concentrations set points. Yield efficiencies in the region of 90% Gay–Lussac can be maintained in continuous operation.

Most cane final molasses can be operated at 43% w/v substrate strengths, rising to 55% for 'A' molasses or cane juice syrups. Starch mashes can also be processed at high concentrations. Similarly, stillage effluents from the primary stripper are reduced in volume and at high solids concentrations. Apart from molasses, most stillages can be in the region of 60% dry solids content, suitable for direct use as an animal feed or as a fuel.

11.5   Distillation methods

In assessments of net energy ratios for ethanol manufacture, processing cost is the largest component compared with agriculture and transportation. A high proportion of this processing energy is needed for ethanol recovery by distillation.

Stages in conventional distillation consists of:

(1) Primary distillation of the dilute fermented substrate in a beer column to concentrate the strength to 85% v/v or greater, with the removal of unwanted volatiles, such as aldehydes.

(2) Rectification of the beer column distillate to achieve close to the constant boiling point ethanol–water azeotrope (96.5% v/v). At the

same time, fusel oils can be removed by water decantation above the feed plate.

(3) In the production of ethanol as a gasoline blend component, dehydration to 99.4% v/v, or greater, is an additional stage using a ternary water entrainer such as benzene or cyclohexane. In this case, the azeotropic distillate is decanted with a further stage of rectification for recovery of the entrainer allowing its return to the main azeotropic column.

A conventional four column Coulter distillation unit system is shown in Fig. 11.2.

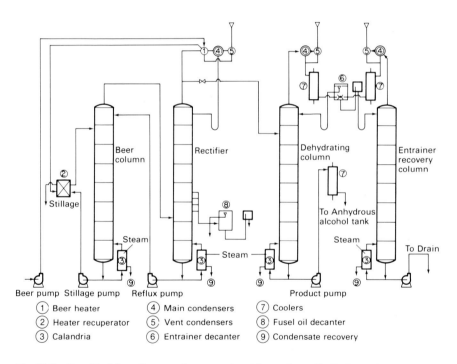

① Beer heater	④ Main condensers	⑦ Coolers
② Heater recuperator	⑤ Vent condensers	⑧ Fusel oil decanter
③ Calandria	⑥ Entrainer decanter	⑨ Condensate recovery

Fig. 11.2 Simplified flow diagram of a conventional four column distillation unit.

The conventional system as an extension of traditional potable spirit production is an intensive thermal energy user. Steam demand is normally in the region of 4 kg per litre of ethanol produced. A reduction of 1 kg per litre ethanol can result by increasing the alcohol in beer concentration from 6% to 10% v/v. Thus, savings in steam costs have been a stimulus in developing yeasts which can ferment to high ethanol concentrations.

However, the physical nature of the vapour–liquid equilibrium relationship for ethanol and water enforces a restriction in the ability to concentrate above 95% v/v by fractional distillation. Where the beer column is combined with rectification to achieve a 96.5% v/v distillate overhead, little advantage is achieved in fermenting to a high feed composition since the rectifying operating line begins to approach the equilibrium line. This demands a high reflux ratio and a large number of theoretical plates.

Other methods for final dehydration are by chemical desiccants and adsorption. The use of lime was practised over a century ago, but ethanol losses at 5% rendered the process uneconomic. Adsorbants, such as starch, sucrose, cellulose, NaOH, molecular sieves, etc., have been proposed as alternatives to final azeotropic distillation or even in replacing the distillation function altogether. Their commercial applications do not yet appear to be feasible in terms of capital cost investment, physical layout arrangements and media recovery.

Cars with alcohol engines are now produced having higher compression ratios and advanced ignition timing closely matching gasoline powered cars in performance and fuel consumption. Alcohol strength need not be greater than 95% v/v. This development in using ethanol solely for automative purposes will reduce the steam consumption in distillation down to about 2.2 kg per litre of ethanol produced, and improve the net energy ratio.

However, even to produce anhydrous alcohol, distillation plants can be designed for total steam consumptions in the region of 2.5 kg per litre ethanol. With vapour recompression, this ratio can be reduced still further and ultimately it is expected that steam consumptions of 2.2 kg per litre anhydrous ethanol will be commonplace. Thermocompression applies the principle that vapours leaving a boiling surface are only slightly less in total heat than the head load used to supply the driving force. If only small temperature differences exist, recompression by a centrifugal compressor can be used to re-use vapours. Naturally, electrical or other power sources are needed for the work done. If high pressure boiler steam requires an initial pressure reduction to operate the system, steam-jet ejectors can be used for this recompression stage. Basically, the principle of multiple stage evaporation applies where the vapours and heat sources from one distillation column are used to reboil another. A large number of distillery equipment suppliers now have these systems available with high thermal economies.

Distillation technologies are highly developed, particularly in tray and packing design. It is expected that future distillation equipment will resemble petroleum and petrochemical units, with a lowering of capital costs by using cheaper materials for shells and internals. After the beer column, the use of stainless steel construction is no longer essential when the product is of a non-potable nature.

11.6 Effluent processing

Even an ideal fermentation of a hexose sugar will result in a population equivalent (PE) pollution load in the waste-water effluent of 1.5 per litre of ethanol produced. With impure substrates, such as molasses, this pollution load can rise to a PE of 15. Hence, a molasses distillery of reasonable size operating at 60 000 litres of ethanol day^{-1}, could carry a pollution load equal to a city of up to one million inhabitants.

It is, therefore, implicit that traditional municipal methods of stillage treatment by aeration are totally uneconomic, even when lagooning, activated sludge, deep shaft aeration, or high rate biological trickling filter methods are employed.

Other aerobic systems, such as fungal biomass production, also demonstrate uneconomic feasibility.

In the treatment of distillery effluents there are only three practical systems that can be considered:

(1) Direct use on the land.
(2) Anaerobic digestion for methane-rich gas generation.
(3) Evaporation to a syrup for use as an animal feedstuff or as a fuel.

In all cases the economic effectiveness of the system is highly sensitive to the liquid bulk being discharged from the beer column. The quantity and quality of this highly polluting stillage is dependent upon the raw material being processed, the fermentation technique and particular distillation method.

11.6.1 Direct use on the land

Only where the quality and quantity of the components in the stillage are of benefit to a specific soil can disposal on the land be of advantage; otherwise the application of stillage becomes an expensive method of soil irrigation, particularly where road tanker transport is involved.

In circumstances where the N/P/K ratios are suitable and where soils are low in humus, spraying over a period of time, previous to or during growth, can improve crop yields. However, the radius of operations between the distillery and the fields being treated is critical to the economics.

Blending of stillage into the soil, its pH, temperature and depth of penetration are also important factors. Aerobic microorganisms are mainly concentrated at the surface, whereas at a short depth down the micro-organisms rapidly become anaerobic in function. Soil saturation should be avoided, which often limits the periods when spraying can be applied, particularly during and after heavy rainfalls when run-off to rivers and other areas can occur.

In many countries initial benefits have been found by stillage spraying, but crop yields can fall off after a number of years without proper agronomic control, enforcing an extended radius of application or alternative methods of disposal.

11.6.2   Anaerobic digestion

This system has become increasingly popular since around 95% of the Biochemical Oxygen Demand (BOD) load can be converted into gaseous fuel and sludge. However, the methane in the gas only represents about 60% of the original organic matter in terms of available calorific value. Nevertheless, the system is relatively cheap to install and simple to operate when the stillage is of an acceptable quality (see Chapter 12).

The main drawback is that the BOD in the treated effluent may not be sufficiently degraded for legislative limits and a final aerobic finisher is often required.

11.6.3   Evaporation

Production of animal feed syrups from stillage has been the most common method of treatment in industrialized countries, particularly where cereals constitute the raw material for fermentation.

Molasses stillage is in a different category. Dissolved salts being concentrated to a high level have a laxative effect upon animals, and a stillage syrup must be admixed in small quantities with other rations.

However, evaporation is expensive, both in capital cost and thermal energy, and the extent of the solids content in the original stillage dictates the economics of application. The main advantage is that the condensate requires only minimal aeration for discharge to waterways.

Figure 11.3 illustrates the steam demand in a 5-stage effect evaporator related to the solids content in molasses stillage to produce a syrup at 60 wt% DS. At up to about 6 wt%, the steam demand for evaporation is greater than the available steam that could be generated by syrup combustion.

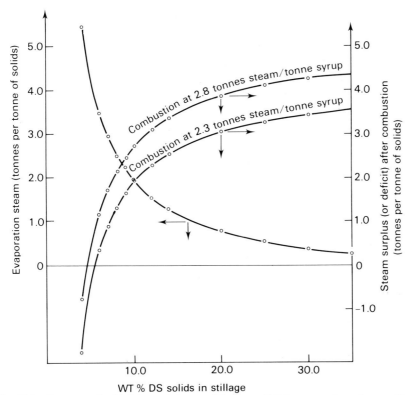

Fig. 11.3 Steam surplus/deficit conditions at differing wt% DS contents in molasses stillage where evaporation is carried out in a 5-stage effect evaporator at 0.233 tonnes of steam per tonne of water evaporated (syrup concentration ~60 wt% DS).

When the solids concentration in stillage approaches about 12 wt%, the net energy, expressed as steam, exceeds the equivalent energy release from methane production by anaerobic digestion. At this solids level it is expected that the BOD_5 will be greater than 35 000 p.p.m., depending upon the substrate quality.

11.6.4 Stillage syrup combustion

Modern stillage syrup boilers must be capable of having a high thermal efficiency without excessive fouling from inorganic ash. In the past when thermal energy recovery was not the criterion, dry distillation using a 'Por-ion' furnace has been used to recover carbon and a potash-rich ash from molasses stillage syrup. However, high temperature combustion units in which an ash can be recovered in a non-fused form are now available.

By the use of new fermentation techniques, it will be possible to produce stillages at high dissolved solid (DS) contents whereby evaporation to a suitable strength for combustion can be carried out by scrubbing the waste hot combustion gases being discharged to the stack. In this case, efficient thermal and equipment economy is achieved.

Depending upon the substrate used, a distillery should be capable of operating without an external fuel source and at the same time totally destroying its own pollutive waste.

11.6.5   Selection of stillage treatment system

An evaluation of the various stillage disposal systems is given in Table 11.6 based upon probable economic feasibility.

Table 11.6   Stillage treatment systems

Disposal method	Stillage type[a]	Wt% DS in stillage		
		Up to 9% Conventional[b]	9–16% Stillage recycling[b]	Over 16% Fermentation developments[b]
Direct use	M	1	1	1
on the	G	3	2	2
land	C	3	2	2
Anaerobic	M	1	2	3
digestion	G	1	1	2
	C	1	1	2
Evaporation	M	3	2	1
to a feed	G	2	1	1
syrup	C	3	2	2
Evaporation	M	3	2	1
and syrup	G	3	3	2
combustion	C	3	2	1

[a]M = Molasses, G = Grains, C = Cassava.
[b]1. Considered economically and technically suitable.
2. Economic and technical feasibility depends upon the particular conditions and end uses.
3. Considered economically and technically unsuitable.

12 | Anaerobic Digestion

F. R. HAWKES and D. L. HAWKES

12.1 Introduction

Anaerobic digestion is the name given to the production of methane plus carbon dioxide from organic matter by a mixed population of microorganisms in the absence of oxygen. The major commercial benefits can be energy production, pollution reduction and odour control. In principle any organic waste may be treated in this way, and the process has been applied to a wide range of industrial, agricultural and domestic wastes. However, despite the large-scale use of anaerobic digestion in sewage treatment works for over 60 years, and the much publicized success of the biogas plants in India and China, development of commercial digesters in the Western world has been retarded by a lack of mechanically reliable digesters and of the knowledge of how to operate them at optimal efficiency. Advances during the past decade in digester design and in our understanding of the digestion process mean that digester technology is now set for expansion. These advances are described in this chapter.

12.2 Biochemistry and microbiology of anaerobic digestion

12.2.1 Interspecies interactions

Successful anaerobic digestion of organic wastes always involves a mixed culture of bacteria with a complex interdependency, terminating in the production of methane by methanogenic bacteria. These bacteria utilize a very limited number of substrates for methane formation, chiefly acetate and H_2 plus CO_2, so that hydrolytic and fermentative bacteria must first act on the organic waste, producing as their end-products the substrates for methanogenesis.

A further level of interdependence is involved, since the scavenging of H_2 by methanogens to convert CO_2 to CH_4 allows fermentative bacteria the option of producing H_2 rather than reduced fermentation end-products from their reduced coenzymes such as NADH (see Chapter 2, Section 2.2.6) by the reaction

$$NADH + H^+ \rightarrow NAD^+ + H_2$$

This reaction is only feasible at low partial pressures of H_2; methane bacteria have a low saturation coefficient, K_s for H_2 (around $5\,\mu M$) keeping the H_2 partial pressure around 10^{-4} atmospheres in a well-balanced digester. Methanogens thus actually bias the fermentation pathways of other bacteria away from reduced end-products and towards the methanogens' substrate, acetate, from which most ($\sim70\%$) of the methane produced in digestion is derived.

There are two advantages the fermentative bacteria may derive from this arrangement of interspecies hydrogen transfer. Firstly, an organism which at higher partial pressure of H_2 reoxidizes its reduced coenzymes by the production of reduced end-products from pyruvate (e.g. ethanol, lactate, butyrate and propionate) may instead be able in a digester to use protons as electron acceptors and generate H_2. This allows the pyruvate to be metabolized via acetyl CoA to produce ATP by a substrate level phosphorylation step commonly found in anaerobes (see Chapter 2, Table 2.2). Thus utilization of H_2 by methanogens enables fermentative bacteria with the right complement of H_2-evolving enzymes to improve their yield of ATP per gram of substrate.

Secondly, the low hydrogen partial pressure permits growth on reduced substrates of bacteria which have no other means of oxidizing their reduced coenzymes but by H_2 evolution. This type of metabolism was discovered by Bryant and coworkers in the late 1960s while investigating what had been thought to be a pure culture of a methanogen '*Methanobacillus omelianskii*' cultured on ethanol. It was shown that the true methanogen in this culture, *Methanobacterium* strain MoH (*M. omelianskii*, hydrogen-utilizing) was, like all other methanogens so far discovered, unable to grow on ethanol, but that an ethanol-fermenting bacterium, S organism (syntrophic organism), was coculturing with MoH. The S organism remained undetected for so long because it was not on its own capable of growth by oxidizing ethanol to acetate, since it needed a low partial pressure of H_2 to allow reoxidation of NADH.

12.2.2 Microbial populations in digesters

To identify the species of microorganisms occurring in digesters is an enormously difficult task, since it is never certain that the culture conditions used allow growth of all the types of organism present. To count the numbers of each species is even more difficult, especially since most wastes are particulate and bacteria occur attached to the particles. Presently, therefore, our understanding of digester microbiology is based on a conception of broad trophic groups apparently common to all digestion processes operating on complex wastes. These trophic groups with their substrates and products, as shown in Fig. 12.1, and are the hydrolytic and fermentative

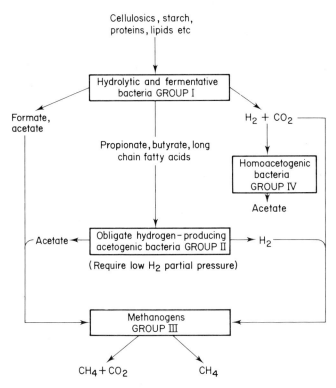

Fig. 12.1 Microbial populations in digesters.

bacteria (Group I), the obligate hydrogen-producing acetogenic bacteria (Group II) and the methanogens (Group III). A fourth group able to synthesize acetate from H_2 plus CO_2, the homoacetogenic bacteria, is of uncertain significance.

Hydrolytic and fermentative bacteria
This group contains both obligate and facultative anaerobes and is responsible for removing small amounts of oxygen, introduced for example by feeding the digester. Populations of 10^8–10^9 hydrolytic bacteria per cm^3 of sewage sludge digester are quoted by Zeikus (in Stafford *et al.*, 1980b). Hydrolysis of proteins, cellulose, hemicelluloses and pectins occurs in digesters and it is commonly found that hydrolysis, particularly of complex polysaccharides, is the rate-limiting step. Cellulosic biomass frequently contains lignin which is practically non-biodegradable in anaerobic systems and may protect the polysaccharide fraction of the biomass from enzymic attack so that only perhaps half of the available polysaccharides may be digested.

Obligate H_2-producing acetogenic bacteria
These bacteria are also referred to in the literature as obligate proton-reducing bacteria because their sole means of oxidizing NADH is by the H_2-evolving reaction given above. Populations of approx 4×10^6 per cm^3 of sewage sludge digester have been reported but they have proved extremely difficult to investigate because they are unable to grow freely except in conditions where H_2 is removed as it is formed. Several species likely to be important in digestion have been isolated in Bryant's laboratory from sewage sludge digesters and the rumen, for example *Syntrophomonas wolfei* which oxidizes butyrate to acetate when grown together with a methanogen, and *Syntrophobacter wolinii* which degrades propionate to acetate and H_2 only in coculture with a sulphate-reducing, H_2-removing species. For thermodynamic reasons the degradation of propionate requires a lower partial pressure of H_2 than the degradation of butyrate, which may explain the requirement for a sulphate-reducer rather than a methanogen as a partner, since the K_s for H_2 recently determined for one sulphate-reducing species is lower than that for methanogens.

   The β-oxidation of longer chain fatty acids to acetate, known to occur in digesters, and the fermentation of aromatic compounds are also due to this type of metabolism, and enrichment cultures carrying out these reactions have been characterized.

Methanogens
Populations of 10^6–10^8 methanogenic bacteria per cm^3 have been detected in digesters. The methanogens are among the strictest anaerobes known, and their culture has required the development of techniques able to maintain this strictly oxygen-free environment. These techniques, initiated by Hungate in the 1950s and developed in the last decade chiefly in the laboratories of M. P. Bryant and R. S. Wolfe at the University of Illinois, have

opened up a wholly new field, as the methanogens appear to belong to a novel group of bacteria, the Archaebacteria, with many unique features.

After the resolving of the *M. omelianskii* culture into two components in the late 1960s it became apparent that pure cultures of methanogens could only use a very limited range of compounds as an energy source. McInerney *et al.* (in Stafford *et al.*, 1980b) summarize the taxonomic scheme proposed for the methanogens isolated up to that time, showing that all species could obtain energy by the following reaction:

$$4H_2 + CO_2 \rightarrow CH_4 + 2H_2O \qquad \Delta G^{0'} = -139 \, \text{kJ mol}^{-1}$$

while nine species could grow by converting formate to methane plus CO_2 and only three (all of the species *Methanosarcina barkeri*) could grow on acetate, methanol and methylamine. Since several studies of carbon flow in digesters have shown that the bulk (~70%) of the methane arises from acetate, it is surprising that so few of the methanogens so far isolated can actually utilize it for growth. However, considering the reaction involved:

$$CH_3COO^- + H_2O \rightarrow HCO_3^- + CH_4 \qquad \Delta G^{0'} = -30 \, \text{kJ mol}^{-1}$$

it is perhaps surprising that there is enough energy available for growth at all—see Chapters 2 and 3 for a consideration of the energy requirements for ATP synthesis. Certainly the doubling time of *M. barkeri* grown on acetate of about 50 h is slower than for growth on H_2 plus CO_2. Indeed, during digestion of a simple substrate such as sucrose in sugar beet processing waste-waters, it is likely that growth of methanogens will be the rate-limiting step.

Methanogens contain several novel coenzymes including at least two (F_{420} and F_{430}) which are fluorescent, their assay forming the basis for methods designed to indicate the size of the methanogenic population in digesters. As it appears that the content of these compounds varies from species to species and with growth conditions, the assay may only be useful to detect changes in previously stable populations. For an account of these novel coenzymes and energy coupling in methanogens, the reader is referred to the section 'Metabolism and Growth Physiology of Methanogens' in Dalton (1981). A general account of the biochemistry and microbiology of anaerobic digestion may be found in Buvet *et al.* (1984).

12.3 Types of digester

Digesters may be designed for use in low technology rural situations or for sophisticated industrial applications. The digester may be operated on a batch or a continuous basis; in most circumstances continuous processes are favoured, since the waste is usually produced continuously and there is a steady demand for the gas. The classic design for an industrial (including agricultural) digester is a variant of the continuously stirred tank reactor (CSTR). Because the contents of this type of digester are completely mixed, the effluent will contain some fraction of freshly added, undigested feed material, and will also include some of the still active microbial population.

The major alternatives to the CSTR seek to overcome these problems. These are (a) 'plug-flow' type or tubular digesters where it is hoped each increment of feed added through the inlet will pass through the digester in a sequential manner, emerging fully treated at the outlet; and (b) digesters involving microorganism retention, such that a large active population is retained in the digester and not washed out in the effluent. The most successful forms of microorganism retention are found in the upflow anaerobic sludge blanket (UASB) digesters in which microorganisms spontaneously form rapidly settling granules, and the various forms of fixed-film reactors in which microbial populations are selected which will adhere to a solid support medium. Examples of these digester types are discussed below.

12.3.1   Low technology digesters

It has been estimated that the waste produced in an average Western home might yield sufficient biogas to bring to the boil three litres of water per person per day. This energy resource, negligible in Western terms, might give, for example, a significant reduction in water-borne diseases in a tropical country. Robust designs on a batch or semicontinuous (once-daily feeding) basis, built of local materials, have developed and have been particularly successful when backed by some form of instructions to users, (Bulmer *et al.*, 1984). Government-backed digester building schemes in Nepal, Korea, China and India have led to the construction of many thousands of simple rural digesters in Asia. These digesters are normally unheated and mixed manually, insulation with local materials being a necessary feature of those in colder climates.

12.3.2   Batch digesters

It is energetically wasteful to treat organic matter arising in a solid form (20–40% dry matter), e.g. plant material, poultry manure, domestic refuse, by slurrying with liquid and digesting in a CSTR. The less the liquid that is added, the less the energy required to raise it to the operating temperature. A further advantage is that producers of dry wastes already have the handling facilities necessary, so that loading an on-site digester requires little extra equipment.

Rijkens (in Palz *et al.*, 1981) proposes a solid-bed reactor where hydrolysis occurs, the leachate of which is recirculated through a methanogenic digester and back over the bed. A series of these reactors at different stages in digestion could be used in large-scale applications. Land-fill sites for domestic refuse are also potential batch digesters producing methane, and gas has been extracted from one such site in the Los Angeles area and used for district heating.

12.3.3   Continuously stirred tank reactors

These completely mixed continuously fed digesters are commonly used to treat waste of low solids content (2–10% dry matter). Mixing serves (a) to distribute the incoming feed, (b) to prevent build-up of non-digested solids, (c) to distribute the applied heat, and (d) to prevent scum accumulation. Mixing may be mechanical or by gas recirculation, where digester gas is bubbled back through the digester contents. A quiescent area or tank may be incorporated to allow collection of settled microorganisms for return to the digester.

Since utilization of digester gas is not normally constant throughout the day, a gas holder of sufficient capacity to store some of the gas produced in, say, 12 hours is advisable—as a rule of thumb this may be half the digester volume. Instead of separate holders, some digesters have floating roofs or butyl bag collectors integral with the digester itself.

CSTRs are the most common type of reactor applied to sewage and manure treatment. Digesters currently operating on sewage vary widely in size, for example 12 000 m^3 in Dusseldorf Sud, West Germany and 30 m^3 on a Wessex Water Authority site in the UK. Farm digesters have been constructed in the USA to treat the waste arising from 100 000 head of cattle, while in the UK digesters have been developed for 50–200 cow herds.

12.3.4   Plug flow type digesters

It is impossible to achieve plug flow in a digester where solids are settling out and gas evolution gives lateral mixing. Indeed if true plug flow were to occur digestion would be impossible, since each increment of feed must be inoculated with bacteria from the digester population for the process to succeed. At present the name plug flow is given to designs which are unstirred and tubular; while the solid material may tend to move through the digester sequentially, it appears that the liquid fraction mixes rapidly, hence the name plug flow is inappropriate and the designation 'tubular' is preferable.

Full size tubular digesters have been operated on farms by Fry in South Africa and Jewell in the USA. Though they are simple in construction, the absence of mixing can give rise to scum problems, particularly where a digester tank with a large surface area is involved. Inclined tubular digesters have been operated in the authors' laboratory and at pilot scale, with a minimum surface area for scum formation, and full-scale developments are under way. A number of other research groups are also now working on inclined tubular digesters.

12.3.5   Upflow anaerobic sludge blanket reactors (UASB)

This type of digester was developed by G. Lettinga in the Netherlands (see Lettinga *et al.* in Stafford *et al.* 1980b), and is particularly suited to treating low strength (~1% dry matter) soluble wastes. It has been applied on a commercial scale by the CSM sugar beet company, some 15 plants with average volumes of over 100 m^3 now being operational.

The digester design provides for an even distribution of the entering waste-water over the base of the digester; the liquid flows upwards through a sludge of settled bacteria and a region in the digester where flocculating bacteria sediment in granules about 4 mm in diameter. The treated waste emerging from the sludge blanket passes into a quiescent zone free of gas bubbles where settling of bacteria detached from the blanket occurs.

There are certain limitations inherent in the UASB design. The bed can be disrupted if the influent flow rate is too fast, or if gas production is too vigorous. The digester may not treat particulate wastes effectively since particles appear to interfere with flocculation and may also accumulate in the bed, thus reducing its effectiveness per unit volume. Inefficient operation may also occur if the influent forces channels through the sludge blanket instead of passing uniformly through the whole volume of the bed.

12.3.6   Film reactors

These can be subdivided into those with a stationary support medium for microorganism attachment, anaerobic filters, and those where the support itself is in motion in the liquid stream.

Anaerobic filters
In these digesters waste is passed over a population of bacteria attached to an inert solid support medium such as gravel, commercially available PVC supports, glass beads, baked clay or needle-punched polyester. L. van den Berg in Canada has been active in investigating these support

media and this type of digester has been used to treat vegetable processing wastes of between 1 and 10% dry matter, and animal wastes.

Fixed film reactors have the advantage over UASB reactors in that they are not susceptible to washout by high hydraulic shock loads and can work with particulate wastes. However they can be subject to clogging and channelling, and microbial attachment during start-up may be slow. Hence the nature and arrangement of the support material very much affect the process and an optimal support medium has not yet been found; this topic is discussed by van den Berg and Kennedy (1981).

Fluidized or expanded bed reactors
Waste water flowing up through these reactors fluidizes, or at least expands, the bed of particles to which the bacteria are attached. The suspended particles are in constant motion, channelling or clogging should be prevented and very efficient substrate transfer achieved. The support granules should be of low density to minimize energy requirements for fluidization. Again an optimum support medium has not been found although sand, PVC particles and carbon granules have been used.

The concept of an expanded bed reactor has been developed by W. J. Jewell in Cornell University and used to treat primary settled sewage (see Jewell in Stafford *et al.* eds, 1980). Since unlike the anaerobic filter or UASB processes clogging should be eliminated, fluidized or expanded bed reactors could be used to treat particulate waste.

12.4 Types of waste for digestion

12.4.1 Solids content

The most practical classification of wastes for anaerobic digestion purposes is into low, medium and high strength wastes and solid wastes. To subdivide arbitrarily on the basis of dry matter or total solids (TS) content, these categories correspond roughly to 0.2–1%, 1–5%, 5–12% and 20–40% TS respectively. When deciding which of the various digester designs is appropriate it is also necessary to know whether the waste is totally soluble or contains particulate matter.

Not all of the dry matter in the waste is actually biodegradable—for example inorganic compounds, grit, etc. To distinguish the organic content two methods are routinely used: determination of volatile solids (VS) involving heating dry matter at 500°C to remove organic material as CO_2, and determination of COD (Chemical Oxygen Demand), usually applied to low strength wastes, involving a quantitative chemical oxidation of organic material. Both methods are based on assumptions and subject to interference errors and should be interpreted with caution. Other methods of determining the organic content of wastes, e.g. BOD_5 (Biochemical Oxygen Demand measured over a 5-day period) or TOC (Total Organic Carbon, only applicable to solubilized samples), are less frequently used. For cellulosic wastes the organic content may not give a valid estimate of digestibility, since lignin is organic but not degradable to a measurable extent anaerobically.

12.4.2   Chemical composition

The waste should be a suitable growth medium for the digester population and nutrients such as phosphate, trace elements and a nitrogen source must be added if not already present. A carbon to nitrogen ratio of below 40:1 is desirable. The waste may, as in the case of many animal manures, already have a good buffering capacity around neutrality, attributable to phosphate, ammonium ions, bicarbonate and carbonate, volatile fatty acids and other organic acids and amines. In this case pH control during digestion is not necessary.

Concentration of potentially inhibitory compounds should be noted, e.g. ammonia on high protein wastes, sulphate in wastes from paper processing and heavy metals in tannery wastes.

12.4.3   Source of wastes

Industrial and food processing wastes
Laboratory work on brewery and distillery wastes shows the potential for anaerobic digestion applications in this industry. In the sugar processing industry sugar beet washing water is already a classic substrate for anaerobic digestion. Cannery fruit and vegetable wastes have also been treated; like beet these are seasonal in their production. Other wastes known to be successful substrates for digestion include potato processing waste-waters, whey from cheese production and abbatoir wastes.

Animal manures and agricultural biomass
Anaerobic digestion has potential in agriculture for pollution and pathogen reduction, odour control and energy production. Intensive farming methods may give large amounts of manures with no associated land for disposal by traditional means. Land spreading of pig manure in any case gives rise to many complaints as to the odour, and farmers may be forced to install treatment plants or close down. Note that cattle manure contains solids which have undergone microbial attack during an approximately 4-day stay

in the rumen so will contain a higher proportion of recalcitrant lignocellulo-sics and will give less favourable gas yields than non-ruminant manures.

Dry agricultural biomass—straw, corn stover from maize production, bagasse from sugar cane, etc.—has been the subject of laboratory experiments on digestion. The digestibility is poor and the C:N ratio may be high; it appears there are better ways of utilizing these materials. Green biomass, however, has been successfully digested. In the UK such material would probably be ensiled for animal feed, but in countries with more land per head of population, e.g. New Zealand, the growing of energy crops specifically as substrates for digestion has been investigated.

Domestic and municipal wastes
Sewage sludge has been digested anaerobically on municipal sewage works since the turn of the century in many parts of Europe. Large, well-con-structed and well-maintained generators utilize the biogas to provide power for the plant, though pollution reduction is the primary aim, as untreated sludges are greasy, odourous and contain pathogens. The dilute sewage waste-water remaining after sludge settling may also be a suitable waste for digestion, though not in a CSTR. Figure 12.2 shows a digester of 80 m³

Fig. 12.2 Photograph of a small sewage works digester operating at Treborth Sewage Works in Bangor, Wales, courtesy Farm Gas Ltd, Bishop's Castle, Shropshire, UK.

capacity on a small sewage works in Bangor, Wales. This type of digester, originally designed for farm applications, is increasingly being used on small sewage works for treating the sewage sludge.

Several attempts have been made to sort domestic refuse and extract the 10–15% of putrescible material with the objective of treating it anaerobically, for the expense involved in such sorting means that the overall economics of the process are marginal.

12.5 Digester operation

12.5.1 Temperature dependence of digestion

There are two ranges of temperature over which working digesters have been operated, the mesophilic (20–25°C to 40–45°C) and the thermophilic range (50–55°C to 60–65°C), each involving different bacterial species. Gas production falls off rapidly at the limits of these ranges and the effects of a short (usually accidental) period of raised temperature from the mesophilic to the thermophilic range may take many weeks to overcome. A similar period is required for a mesophilic population to adapt to thermophilic conditions.

Thermophilic digesters may be less stable, respond adversely to accidental cooling and are more susceptible to ammonia inhibition. However, the destruction of pathogens and animal parasites is greater in the thermophilic range, an advantage if digester effluent is to be sprayed on the land, and thermophilic digesters may operate on shorter retention times than the corresponding mesophilic process.

12.5.2 Retention time and loading rate

The hydraulic retention time (HRT), i.e. the length of time liquid added remains in the digester, is given by the expression

$$\text{HRT} = \frac{\text{Digester volume}}{\text{Volume of waste fed daily}}$$

The solids retention time will be longer than this if particles are retained in the digester, e.g. in a plug flow type design. Simple wastes may pass through digesters with microorganism retention, e.g. the UASB, with retention times of merely hours, while complex wastes such as animal manures are digested at RTs of 10 days or more. In digesters with no provision for microbial retention care must be taken to prevent washout, especially of slower growing methanogens at retention times below three days.

The loading rate or rate at which solids are added to the digester is a function both of the volume added daily and the solids content of this feed material. The loading rate is expressed as weight of organic matter (usually kg VS or COD) per m³ of digester per day, and the maximum loading rate which can be applied to a digester is a function of the digester design and the type of waste. For a full-scale UASB treating sugar beet waste-water, more than 15 kg COD m⁻³ digester day may be achieved; for a farm digester operating on animal manures at 35°C about 4 kg VS m⁻³ day⁻¹ would be a good performance.

12.5.3 Gas yield

This important parameter is a measure of the degree of digestion of the organic matter, as it is the volume of gas generated per unit weight of organic matter added to the digester. Units are usually m³ gas kg VS⁻¹ or m³ kg COD⁻¹.

If pure organic substrates were added and completely broken down to $CO_2 + CH_4$, then the following gas yields could be calculated per kg of substance added:

starch/cellulose/glucose	0.8 m³ kg⁻¹
fatty acids (based on stearate)	1.5 m³ kg⁻¹
protein (based on $C_5H_7NO_2$)	0.9 m³ kg⁻¹

Gas yields as high as this are rarely reached in practice, since generally the conversion of organic matter to gas is incomplete. Values of the order of 0.6 m³ per kg VS added are obtainable for sewage sludge, and somewhat higher values for some yeast brewery wastes and distillery wastes. Pig manure may give gas yields of 0.4 m³ per kg VS added, whereas cattle manure may give values of around 0.2 m³ per kg.

It is difficult to quote representative gas yield values in isolation, since as is discussed in Section 12.6.1, gas yield varies with waste composition, retention time and inhibitor concentration, among others. However for a waste of fixed composition the gas yield per weight of organic matter *destroyed* should be constant.

12.5.4 Volatile fatty acids as an indication of performance

Based on empirical observations on sewage sludge digestion over many years, it has become the practice to monitor the concentration of volatile fatty acids (chiefly acetate, propionate and butyrate) as an indicator of digester performance. Concentrations of these acids in stable digesters may remain constant at values usually between 100 and 2000 p.p.m unless a fluctuation in digester operating conditions, e.g. in feed composition or temperature, occurs. As the utilization of propionate and butyrate by the obligate H_2-producing acetogens is dependent on H_2 consumption by the methanogens, which also consume acetate, any decrease in activity of the methanogenic population may be followed by an increase in the VFA concentration. Thus raised levels of VFA may be an effect, not a cause of methanogen inhibition.

12.5.5   Two-stage digestion

This process modification is also referred to in the literature as two-step or two-phase digestion. It is argued that the different population groups in digesters have different optimal operating conditions, and that efficiency could be improved by separating the acidification stage involving the hydrolytic and fermentative bacteria from the methanogenic stage involving the obligate hydrogen-producing acetogens and the methanogens. This can be done by keeping the pH of the initial reaction vessel around pH 6 or below, where methane generation is inhibited but hydrolysis and fermentations still proceed. Gas produced in the first stage is thus a mix of H_2 and CO_2.

Laboratory experiments with glucose as substrate suggest that a two-stage system can treat three times the organic loading of a one-stage process and that it is more stable to shock loading. The relative retention times for each stage depend on the type of waste; with simple sugars and starch, methanogenesis may be the slower process and during growth on glucose retention times of 10 and 100 h for the acidification and methanogenic steps respectively are used. During growth on complex cellulosics it is the initial acidification step which will require the longer retention time.

This two-stage operation is frequently used in the digestion of simple wastes, e.g. sugar beet and potato processing wastes, as these, unlike manures, are not highly buffered and so the first stage can easily be maintained at pH 6 or below.

12.6   Energy from digesters

The revival in interest in anaerobic digestion during the 1970s was largely brought about by the sharp increase in oil prices and the desire to find other 'natural' sources of energy. The gas produced by the process is usually about 70% methane and 30% carbon dioxide, with an energy content of about $26 \, MJ \, m^{-3}$, and is therefore combustible. The energy which is of interest in any practical application is not just the gross gas produced by the bacteria but the net energy left over after the digest has been heated, the contents mixed and after the energy losses have been allowed for.

12.6.1 Parameters affecting gross gas production

The volume of gas that can be produced from a given quantity of waste, the gas yield, (measured as m³ per kg VS added or m³ per kg COD added, see Section 12.5.3) is a critical parameter for the design of digesters. The most important factors affecting gas yield will be discussed here and are the type of waste, the digester operating temperature, the retention time, and the presence of inhibitors. The volume of gas produced by a digester is in turn directly proportional to the mass of biodegradable solids fed, within operational limits.

Type of waste
The volume of gas will depend upon the mix of various components in the waste (see Section 12.5.3), some wastes such as those from ruminants giving lower gas yields than others. For complex substrates such as animal manures it is usually not possible to calculate the amount of gas that could be produced and figures have to be determined empirically.

Operating temperature
It is generally found that maximum gas is produced by mesophilic anaerobic bacteria at temperatures around 35°C and by thermophilic bacteria at about 55°C, although these figures do vary somewhat with the waste being digested. Gas production is less at other temperatures but the exact relationship is not yet known. In the range 20–35°C, an important one for practical mesophilic digestion, the relationship for sewage sludge, for example, is thought to be as shown in Fig. 12.3. Compared with a particular gas production at 35°C the volume produced at 25°C is still 89% of this and at 20°C it is about 80%. Clearly this must be taken into account when deciding at what temperature to operate a digester since large savings in the energy required to heat the feed may be possible for only a small reduction in gross gas production. The result would be a higher net energy from the plant. Although Fig. 12.3 applies to the digestion of sewage sludge, a similar relationship exists with other wastes.

Retention time
At a constant solids content an increase in retention time will give a greater gas yield although the relationship is not linear. Since the feed material will often be a complex substrate, as for example in the case of animal manures, individual components will be broken down at different rates. This results in a gas yield against retention time graph of the form shown in Fig. 12.4. In a real situation, increasing the retention time required to treat a fixed volume of waste in order to give a greater gas yield incurs

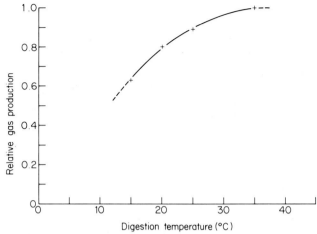

Fig. 12.3   Relative gas production plotted against temperature for sewage sludge digestion.

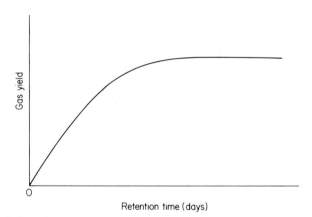

Fig. 12.4   Typical graph of gas yield against retention time.

disadvantages, notably the increase in cost of the larger digester. If there is no limit to the amount of material available for digestion then putting more through the digester each day, thus shortening the retention time, will obviously produce more gas per day per volume of digester, but at a lower gas yield.

Volume of material fed

The higher the biodegradable solids content of the feed the higher will be the gross gas production up to the limit of acceptable loading rate.

The limit on solids content will usually be a practical one, either on what can be achieved by settling in the case of sewage sludge, or what can be handled by the available pumps in the case of certain high-solids manures. Pumps are usually only capable of handling up to 10–12% solids and much higher solids content requires water to be added to facilitate handling or needs specially designed digesters, often of the batch type, (see Section 12.3.2). Where high solids material can be fed, then retention times must be long enough to avoid exceeding the maximum allowable loading rate.

Toxicity

(a) NH_4^+/NH_3. In agricultural wastes high levels of ammonia (e.g. 3000–4000 p.p.m. as ammonia-N) may be found in digester feedstock, particularly pig and poultry manures. van Velsen (see Stafford *et al.*, 1980b) pointed out the difficulty of assigning hard and fast toxic limits to the effect of NH_4^+/NH_3 on the digestion process; adaptation of the bacterial population and natural selection of strains able to operate at higher ammonia levels took place over several months, so that digestion could then occur at NH_4^+-N levels of 5000 p.p.m. Ammonia toxicity is pH-dependent, being more pronounced at alkaline pHs, as it is thought that NH_3 is more toxic than NH_4^+. Note that apart from loss as NH_3 gas, all N is retained during digestion, so the effluent loses little of its fertilizer value.

(b) SO_4^{2-}. Some industrial wastes, e.g. from paper processing, may be high in sulphate. Work on anaerobic muds has demonstrated that SO_4^{2-} and CO_2 compete for H_2, sulphate-reducing bacteria carrying out the reaction:

$$SO_4^{2-} + 4H_2 + H^+ \rightarrow HS^- + 4H_2O \qquad \Delta G^{0\prime} = 152 \, \text{kJ mol}^{-1}$$

Recently the K_s for H_2 for a sulphate-reducing bacterium and a methanogen both isolated from sewage sludge were determined, showing that the former has a lower K_s for H_2 and is likely to out-compete the methanogen.

(c) Industrial and agricultural contaminants. In the USA antibiotic feed additives have occasionally caused failure of farm digesters. In the UK the use of antibiotics is strictly controlled but an allowed ruminant feed additive, monensin, inhibits wasteful side reactions producing CH_4, and so is a potential digestor inhibitor. Chloroform, a methane analogue, and other chlorinated hydrocarbons, can inhibit methanogenesis at extremely low concentrations (3 μM for chloroform). Disinfectants and detergents may enter the waste in washing water but inhibitory levels are not yet established. Heavy metal inhibition in digesters may prove less of a problem than anticipated if the waste contains sufficient sulphide, as most metals form precipitates. In agricultural wastes the addition of copper sulphate to pig feed as a growth stimulant has not reportedly affected digestion.

12.6.2   Parameters affecting net energy production

The net energy produced by a digester is the difference between the gross energy produced and that used to maintain the process. Usually the material must be pumped into the digester, it must be heated to the operating temperature, there are heat losses through the walls, and the digester has to be mixed; each of these operations is energy consuming.

The largest energy input for digesters operating in temperate climates with wastes at ambient temperature is usually in raising the incoming feed to the operating temperature, often a rise of 25°C or more. The expression

$$\text{Mass} \times \text{Specific heat} \times \text{Rise in temperature}$$

gives the quantity of heat required to raise the temperature of the feed.

For the digester with a daily input of $2\,m^3$ of waste at a temperature of 10°C being fed to a digester operating at 35°C, the daily energy required to raise the temperature would be:

$$(2 \times 1020) \times (4094) \times (35 - 10) = 208.8\,\text{MJ}$$

assuming the sludge density is $1020\,\text{kg}\,\text{m}^{-3}$ and the specific heat of the sludge is $4094\,\text{J}\,\text{kg}^{-1}{}^{\circ}\text{C}^{-1}$.

Heat loss through the digester walls for a well lagged tank with a U-value of about $0.8\,\text{W}\,\text{m}^{-2}{}^{\circ}\text{C}^{-1}$ is usually small in comparison, and the heat loss through the roof in contact with the gas is even less. For the case shown above, with a conventional digester operating at a retention time of 15 days and an ambient temperature of 10°C, a typical annual mean for the UK, the heat lost through the digester walls would be in the order of 67 MJ per day and through the roof 17 MJ per day.

The U-value for the insulation can be much higher than that calculated if care is not taken in its installation, especially where flanges or pipes protrude, and whereas the thicker the insulation the less the heat loss, this saving in energy is offset by higher capital cost.

If the digester is operated at a lower temperature then there are a number of consequences: the gross gas production will be lower (see Section 12.6.1), the heat losses through walls and roof will be lower and the energy required to raise the temperature of the incoming feed will be less. The result may well be that more net energy will be produced.

12.7   Economics

The process of anaerobic digestion, whilst sometimes being utilized for pollution control, is often also considered for energy production. It is almost impossible to put an accurate monetary value on pollution control, and even when energy production is the main consideration, a thorough economic evaluation is still not easy, especially if one is not able to forecast accurately the quantity of gas that could be produced.

The difficulties in predicting the net energy yield and the economics of the situation are obvious when one reviews the factors which have to be taken into consideration. The gross gas production varies directly as the volume of the digester whilst the heat losses through the sides vary with the surface area; also the gas yield varies with the temperature of operation and with the retention time in a non-linear way. Increased retention time means an increased digester volume for a constant volume of waste treated, and hence a more costly digester; the capital cost varies with size but the relationship again is not a linear one. In addition, it is believed gas yield varies with the solids content. The heat required to raise the feed to operating temperature varies directly as the daily feed volume. Thus to obtain an accurate value for the net energy produced involves a somewhat complex calculation now usually performed using a computer.

Figure 12.5 is an example of the output from a computer model developed

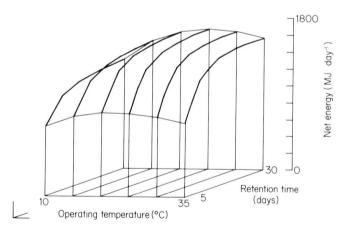

Fig. 12.5 Net energy produced plotted against operating temperature and retention time for the slurry from 1000 pigs at 7% solids. The maximum net energy occurs in this case at 25°C and 30 days retention time.

by the Anaerobic Digester Research and Development Unit of the Polytechnic of Wales to take account of some of the factors just discussed. It shows a graph of net energy produced for various retention times and operating temperatures calculated for a digester with an input of slurry from 1000 pigs. The waste is assumed to be at 7% total solids with a volume fed per day of 4 m³ at a temperature of 10°C. The gas yield varies as the operating temperature and with the retention time. From the graph one can see that in these circumstances the maximum net energy of 1665 MJ day⁻¹ would occur when the operating temperature is 25°C and the retention time is at its maximum of 30 days. Figure 12.6 however is

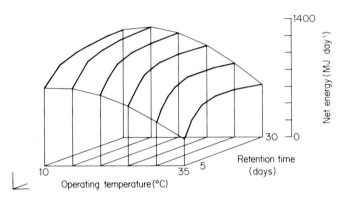

Fig. 12.6 Net energy produced plotted against operating temperature and retention time for the slurry from 1000 pigs at 3.5% solids. The maximum net energy occurs at 15°C and 30 days retention time.

for the same situation except that the waste is now only 3.5% total solids due to the addition of extraneous water, hence the volume is thus increased to 8 m³ day⁻¹. The maximum net energy obtainable is now not only lower, at 1308 MJ day⁻¹, but would occur under a different set of conditions; again at the longest retention time considered but now at an operating temperature of only 15°C. If an operating temperature of 35°C were used in this case then only about 600 MJ day⁻¹ would be produced.

Although in this example the maximum net gas production would occur at the longest retention time considered, this means using the largest and most costly digester and so may not be economic. The cost of various size digesters would have to be taken into account before deciding upon the retention time that is acceptable. Further, the operating costs cannot be ignored and in some instances can be considerable. Pre- and post-digestion processes are also of great importance in assessing the overall economics of a digester installation. Whether the gas is to be used to generate electricity or is to be burnt directly in an efficient heat exchanger will, for example, determine the equipment required and the maintenance costs for this part of the operation.

Despite these many constraints there are numerous examples of digesters which the operators regard as economic. Figure 12.7 shows an example

Fig. 12.7 Digester treating sugar wastes at East Grand Forks, USA. Photograph courtesy AB Sorigona, Staffanstorp, Sweden. The plant produces $17\,600\,\mathrm{m}^3$ of methane per day under design load conditions giving a net gain of the equivalent of 13.2^{-1} tonnes per day of fuel oil. The treatment efficiency is 97.5% BOD_5 reduction.

of a digester used for pollution control and also giving a return on the gas produced. The plant is the ANAMET anaerobic–aerobic waste-water treatment process from Sorigona, and is used for treating the wastes from the American Crystal Sugar Company plant at East Grand Forks, USA.

Acknowledgements

We thank our colleagues of the Anaerobic Digester Research and Development Unit for many helpful discussions, and in particular wish to acknowledge the contribution of Dr B. L. Rosser to Figs 12.5 and 12.6.

Reading list

Bulmer, A., Finlay, T., Fulford, D., Lau-Wong, M. (1984). 'Biogas—Challenges and Experience from Nepal'. United Mission to Nepal, PO Box 126, Kathmandu, Nepal.
Buvet, R., Fox, M. F. and Picken, D. J. (1984). 'Biomethane, Production and Uses'. Turret-Wheatland Ltd: Rickmansworth.

Dalton, H. (Ed.) (1981). 'Microbial Growth on C_1 Compounds'. Heyden and Son Ltd: London.

Hobson, P. N., Bousfield, S. and Summers, R. (1981). 'Methane Production from Agricultural and Domestic Wastes'. Applied Science Publishers: Barking.

Hughes, D. E., Stafford, D. A., Wheatley, B. I., Baader, W., Lettinga, G., Nyns, E.-J., Verstraete, W. and Wentworth, R. L. (Eds) (1982). 'Anaerobic Digestion 1981'. Elsevier: Amsterdam.

Meynell, P. J. (1982). 'Methane: Planning a Digester', 2nd Edn. Prism Press: Dorchester.

Palz, W., Chartier, P. and Hall, D. O. (Eds) (1981). 'Energy from Biomass'. Applied Science Publishers: Barking.

Stafford, D. A., Hawkes, D. L. and Horton, R. (1980a). 'Methane Production from Waste Organic Matter'. C.R.C. Press: Boca Raton.

Stafford, D. A., Wheatley, B. I. and Hughes, D. E. (Eds) (1980b). 'Anaerobic Digestion'. Applied Science Publishers: Barking.

van den Berg, L. and Kennedy, K. J. (1981). Support materials for stationary film reactors for high rate methanogenic fermentations. *Biotechnol. Lett.* **3**, 165–170.

Wentworth, R. L. (Ed.) (1985). 'Proceedings Third International Symposium on Anaerobic Digestion AD83'. 14-19 August 1983, Boston, MA, USA. 99 Eire Street Cambridge, MA, USA.

13 | Organic Acids and Amino Acids

J. L. MEERS and P. E. MILSOM

13.1 Introduction

This chapter describes the production of tonnage quantities of chemicals using moulds, yeasts and bacteria under either aerobic or anaerobic conditions.

The organic acids mentioned include citric, gluconic, itaconic, 2-ketogluconic, erythorbic, tartaric, lactic and acetic and other volatile fatty acids. The amino acids treated are L-glutamic acid, L-lysine, L-aspartic acid, L-tryptophan, and D-arylglycines (which are used in the production of semisynthetic penicillins). In terms of world production volume, the most important of these are citric acid, glutamic acid, and lysine (Table 13.1). In

Table 13.1 World production of major fermentation acids (approximate 1980 data).

Product	Unit value ($ kg^{-1})	Total production (tonnes year^{-1})
Glutamic acid	3.5–4.0	300 000
Citric acid	1.5	300 000
Lysine	5.0–5.5	40 000
Lactic acid[a]	1.0–1.5	40 000

[a] US production mostly synthetic.

each case the type of organism used, mechanism of product accumulation, important fermentation factors, and method of recovery of product are

359

outlined. Throughout, however, the general observations which should be borne in mind are:

(1) For molecules of this kind, chemical routes are always likely to compete effectively with biological routes unless the latter are highly efficient and economically conducted.

(2) For an efficient fermentation it is necessary to select the correct organism, strain, or mutant and to grow it under the correct nutritional conditions; the economy of the process is then mainly governed by substrate costs.

(3) Most of these fermentations require rather careful control of all conditions relevant to both fermentation and recovery stages.

(4) In all cases, a successful fermentation is one which yields the product in such concentration, and purity, that its efficient recovery is most readily carried out.

13.2   Citric acid

13.2.1   General

$$\begin{array}{c} CH_2COOH \\ | \\ HO-C-COOH \\ | \\ CH_2COOH \end{array}$$

Though citric acid was formerly produced from lemon or lime juice, successful fermentation processes were developed from the 1920s onwards, and today upwards of 300 000 tonnes are produced annually. The product is mainly used as an acidulant in soft drinks and confectionary, but to an increasing extent also as a metal complexing agent, to reduce oxidative deterioration and for metal cleaning. As the fermentation organism, strains of the mould *Aspergillus niger* have nearly ways been used, though more recently, certain yeast strains have also been employed. Strains may be freshly isolated from soils or plant materials, or obtained from culture collections, and the selected strains are further improved by mutation and selection.

Beet molasses was for many years the preferred substrate. Sucrose, cane molasses or purified glucose syrup from maize are sometimes used according to availability and price. Processes employing *n*-paraffins as starting material, and with strains of *Candida* or other yeasts as the fermentation organism, have been fully developed, but became uneconomic with the rise in the prices of petroleum products, and have never been run on a large scale.

13.2.2   Biochemical mechanisms

The process based on carbohydrate using *A. niger* has received the most attention, and mechanisms are outlined in Fig. 13.1. The exact reasons

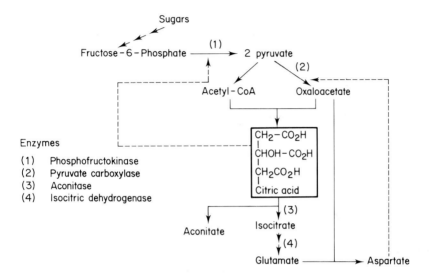

Fig. 13.1   Citric acid formation. Dashed lines indicate feedback controls.

for citric acid accumulation in industrial strains of *A. niger* are still being discussed, but it involves an incomplete version of the normal tricarboxylic acid cycle.

Citric acid is formed by condensation of oxaloacetic acid with acetyl coenzyme A, both derived from pyruvate. High yields of citric acid from sugar necessitate the carboxylation of one molecule of pyruvate by the CO_2 released in the decarboxylation of another molecule of pyruvate to give acetyl CoA. The blockage of the citric acid cycle at citric acid results in a depletion of aspartate which would otherwise inhibit the enzyme causing pyruvate carboxylation. The inhibition by citrate of phosphofructokinase, an essential enzyme in the conversion of glucose or fructose to pyruvate, is nullified by an accumulation of NH_4^+ ions in the cells. This state of affairs is brought about by a manganese deficiency (see below), which also adversely affects the synthesis or activity of aconitase and isocitric dehydrogenase, enzymes which normally catalyse the catabolism of citric acid. Citric acid also inhibits mitochondrial isocitric dehydrogenase.

Because citric acid only accumulates when the concentrations of manganese and iron are low, it is necessary to pretreat the media unless very pure raw materials are used. Glucose syrups or 'Hitest' molasses are treated by ion exchange, while beet or cane molasses are treated with ferrocyanide to precipitate the heavy metals. Additions of ammonia, phosphate, and other inorganic nutrients are made where necessary, and the medium is sterilized by heat.

Inoculation is carried out using spores of *A. niger* obtained from a sporulation culture grown on a solid medium, or a pregrown vegetative inoculum. For yeast fermentations, an actively growing culture is obtained on a smaller scale.

The original citric acid fermentation was based on surface cultures of *A. niger* and about one-fifth of world production still uses this method, with molasses substrate in aluminium or stainless steel trays. These trays are stacked in fermentation rooms supplied with filtered air which serves both to supply oxygen and to remove reaction heat. In newer plants, whether *A. niger* or a yeast is used, the fermentation is carried out in aerated stainless steel tanks, either agitated reactors or sparged towers. The use of oxygen instead of air to improve oxygen transfer has been reported in some cases where unagitated vessels are employed. Foaming of the broth is counteracted by mechanical or chemical foam breakers.

13.2.3   Fermentation

The fermentation is carried out at a temperature of about 30°C. Heat is produced and some provision for cooling is required.

When *A. niger* is used, the starting pH is dependent on the medium employed. With molasses media, the initial pH must be neutral or slightly acid for germination and growth to occur. Where the medium is based on relatively pure glucose or sucrose and inorganic salts, the starting pH may be in the range 2.5–3.5. The pH falls during the fermentation, with final pH often about 2.0; when yeasts are used, the pH is often controlled near neutrality by addition of lime, calcium carbonate, or sodium hydroxide. Reported yields of citric acid (g citric acid monohydrate per 100 g sugar supplied) are in the range 70–90%.

13.2.4   Product recovery

At the end of the fermentation, lasting 6–15 days, the organism is separated from the liquor either by filtration or centrifugation. Yeast-based fermentation liquors kept neutral with $CaCO_3$ or lime need to be acidified with a mineral acid before this step. The citric acid is then precipitated from the filtrate as insoluble calcium citrate tetrahydrate by addition of lime. The filtered washed calcium salt is decomposed with sulphuric acid and the calcium sulphate filtered off. The citric acid solution so produced is concentrated by evaporation and crops of crystals obtained in a conventional manner. Alternatively, citric acid may be extracted from the filtered broth using either tributyl phosphate or long chain secondary or tertiary amines. The acid is extracted into the solvent at a low temperature and re-extracted into water at a higher temperature. The purified aqueous solution is then concentrated and crystallized in the usual way.

The molasses process is illustrated by Fig. 13.2.

13.3   Gluconic acid

13.3.1   General

$$
\begin{array}{c}
\mathrm{COOH} \\
| \\
\mathrm{H-C-OH} \\
| \\
\mathrm{HO-C-H} \\
| \\
\mathrm{H-C-OH} \\
| \\
\mathrm{H-C-OH} \\
| \\
\mathrm{CH_2OH}
\end{array}
$$

Gluconic acid is used as a complexing agent, particularly for calcium in detergent formulations. Calcium gluconate has some veterinary applications, and the δ-lactone is employed as a slow release acid in dry baking mixtures.

13.3.2   Biochemical mechanisms

Gluconic acid can be made by electrolytic oxidation of glucose or by catalytic oxidation using air or oxygen and a platinum catalyst. However, at present it is made using strains of *A. niger* in a fermentation process, but a catalytic oxidation of glucose using immobilized glucose oxidase may well be used in the future.

Redissolved glucose monohydrate crystals or a glucose syrup is employed as substrate, and the essential reaction is the removal by glucose oxidase of two atoms of hydrogen from glucose to produce glucono-δ-lactone. This is hydrolysed to gluconic acid either spontaneously or by the enzyme gluconolactonase.

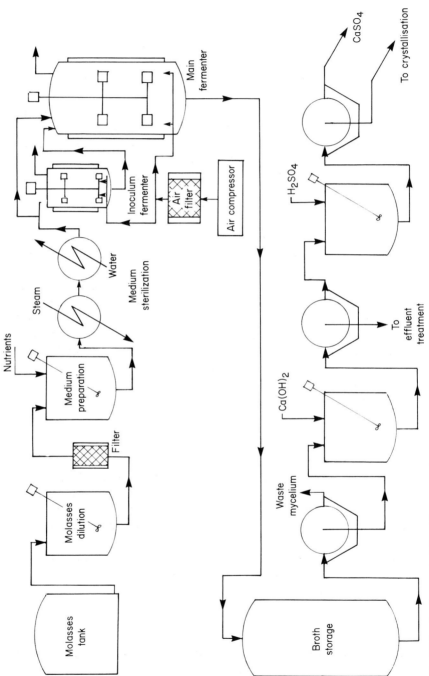

Fig. 13.2 A submerged citric acid process using molasses.

13.3.3 Fermentation

For the fermentation process, the medium consists of a solution of up to 40% w/v glucose monohydrate (or equivalent starch hydrolysate) containing sources of nitrogen, phosphate, potassium, magnesium, and trace elements. Corn steep liquor or yeast extract are sometimes used as additional nutrients. The fermentation is inoculated either with a suspension of spores of *A. niger* or with about 10% of a pregrown vegetative inoculum obtained by germinating such a spore suspension in a medium similar to the fermentation medium. Fermentation is carried out in stirred aerated baffled tanks of stainless steel, with provision for pH and temperature control.

The fermentation process is carried out at about 30°C. If the broth gets too acid the formation of gluconic acid is inhibited. It is therefore necessary to add alkali such as calcium carbonate, in sufficient quantity at the start of the fermentation, or sodium hydroxide, added automatically as 50% w/v solution during the fermentation according to signals from an *in situ* pH electrode. With high initial glucose concentrations, the fermentation products (calcium or sodium gluconate) tend to crystallize and cause problems. One solution to this problem (Pabst, British Patent 1 249 347) is to add only part of the glucose at the beginning of the fermentation and to control the pH at about 6.5 by sodium hydroxide addition. Further additions of glucose are then made without pH control and the pH falls to about 3.8. A non-crystallizing mixture of gluconic acid and sodium gluconate is made by this method, giving yields in excess of 95% in 72 h.

13.3.4 Product recovery

Gluconic acid recovery
Where calcium carbonate is used as a neutralizing agent and the concentrations are not too high for the calcium gluconate to remain in solution, the mycelium is removed by filtration and sulphuric acid added. The precipitated calcium sulphate is filtered off and the gluconic acid evaporated to a 50% w/v solution.

Where, as in the process of the Pabst patent, sodium hydroxide is used for pH control in the fermentation, the sodium ions may be removed by means of cation exchange.

Sodium gluconate recovery
Where sodium hydroxide is used in the fermentation, the mycelium is first filtered off, the pH is adjusted (if necessary) to neutrality, and the sodium gluconate crystallized by evaporation.

Glucono-δ-lactone recovery
The lactone crystallizes from gluconic acid solutions after evaporation under controlled temperature conditions.

13.4   Itaconic acid

$$CH_2{=}C{-}{-}{-}CH_2$$
$$\underset{COOH}{|}\ \underset{COOH}{|}$$

13.4.1   General

Itaconic acid is used, either as free acid or as an ester, to improve the properties of vinyl polymers, for example those used in fibre production to change dyeing characteristics, or in manufacture of polymers used in emulsion paints.

Although itaconic acid accumulation was originally observed in *Aspergillus itaconicus*, commercial production uses selected strains or mutants of *Aspergillus terreus*.

Carbohydrates are used as the carbon source. Pure sucrose or glucose may be used together with inorganic salts, or purified 'Hitest' molasses, or media containing a proportion of beet molasses. Additions of calcium, zinc and copper are important.

13.4.2   Biochemical mechanisms

The work of Bentley and coworkers using radiotracer and other techniques showed that the carbohydrate was metabolized by Embden–Meyerhof glycolysis to pyruvate, which was further converted through citric acid (see Fig. 13.1) to aconitic acid. The latter was converted to itaconic acid by the enzyme aconitic acid decarboxylase, an enzyme which was isolated and found to be very sensitive to lack of oxygen. Thus, itaconic acid is produced by a diversion reaction from a tricarboxylic acid cycle which, like that in *A. niger* (Fig. 13.1), is incomplete. Under unfavourable conditions the itaconic acid may be irreversibly oxidized by the enzyme itaconic oxidase to give itatartaric acid or its lactone β-hydroxyparaconic acid.

13.4.3   Fermentation

The medium is made up and sterilized by heat. A spore inoculum is prepared on a solid medium and the harvested spores grown in an inoculum medium to produce a vegetative inoculum which is transferred to the main fermentation medium. Although surface fermentation in trays has been used, the submerged method is preferred. Aerated and agitated stainless steel tanks are employed, and provision for cooling is necessary.

The fermentation is carried out at relatively high temperatures, i.e. up to 40°C. Before itaconic acid accumulation will begin, a pH of about 2 must be attained, but once itaconic acid accumulation is well under way, higher yields are obtained if the medium is partially neutralized, for example with calcium hydroxide. The fermentation is highly aerobic, and even short interruptions of aeration lead to severe losses. Yields of the order of 60% based on carbohydrate supplied are obtained in about 72 h.

13.4.4   Product recovery

The mycelium is separated from the broth by filtration and the resultant liquor clarified. The itaconic acid is then recovered by evaporation and crystallization, or alternatively by ion exchange or solvent extraction.

13.5   2-Ketogluconic acid

13.5.1   General

$$
\begin{array}{c}
\text{COOH} \\
|\\
\text{C}=\text{O} \\
|\\
\text{HOC}-\text{H} \\
|\\
\text{H}-\text{C}-\text{OH} \\
|\\
\text{H}-\text{C}-\text{OH} \\
|\\
\text{CH}_2\text{OH}
\end{array}
$$

2-Ketogluconic acid is used as starting material in the chemical synthesis of erythorbic acid (Section 13.6). Strains of *Acetobacter* or *Pseudomonas* have been used for 2-ketogluconic acid production, but *Serratia marcescens* NRRL B-486 is said to give better yields.

13.5.2   Biochemical mechanisms

It is thought that direct oxidation of glucose to 2-ketogluconate occurs.

13.5.3   Fermentation

A buffered aqueous medium containing 120 g glucose l^{-1}, calcium carbonate and inorganic nutrient salts is used, and a pregrown vegetative inoculum of *S. marcescens*. The fermentation is conducted at 30°C under aerated stirred conditions, and after 16 h almost quantitative conversion of the initial glucose to 2-ketogluconic acid is achieved. Further additions of glucose may be made during the fermentation to improve productivity and somewhat longer times are then required for complete conversion.

13.5.4   Product recovery

After removal of the cells, the calcium 2-ketogluconate is recovered by evaporation. The free acid can be obtained by treatment of the calcium salt with mineral acid.

13.6   Erythorbic acid

13.6.1   General

Erythorbic acid (also called D-araboascorbic acid or isoascorbic acid) is used as an antioxidant. It can be prepared chemically from 2-ketogluconic acid, but is also obtainable by direct fermentation of glucose. Erythorbic acid is produced by strains of species of *Penicillium*, particularly *P. notatum* and ultraviolet mutants thereof.

13.6.2   Biochemical mechanisms

According to Takahashi (1969: *Biotech. Bioengng* **11**, 1157) erythorbic acid is produced by the action of D-glucono-γ-lactone dehydrogenase on D-glucono-γ-lactone. Glucose is oxidized to D-glucono-δ-lactone by glucose oxidase, and interconversion of the δ- and γ-lactones is thought to be chemical in nature.

13.6.3   Fermentation

The fermentation medium consists of an aqueous solution of glucose $(80\text{--}120\,g\,l^{-1})$ containing calcium carbonate and inorganic nutrient salts.

A Cu concentration of 0.05 p.p.m. strongly inhibits and Fe must be controlled to near 0.3 p.p.m. A mixture of inorganic nitrogen sources was found to give the best yields. The fermentation is conducted at 28–32°C. Optimal pH is 3.8–4.5 and must not be allowed to rise above 5.5. Aeration/agitation must be controlled to avoid under- or over-supply of oxygen. After 5–7 days a 40–45% yield of erythorbic acid in broth is obtained.

13.6.4   Product recovery

The clarified broth, after action exchange treatment, is contacted with a weak base anion exchange column to absorb the erythorbic acid which is subsequently eluted with HCl in a multicolumn arrangement to avoid contamination of the product with HCl. Concentration *in vacuo* gives a recovery of 91% of crystalline erythorbic acid from the broth.

13.7   Tartaric acid

13.7.1   General

$$HO-CH-COOH$$
$$|$$
$$HO-CH-COOH$$

Tartaric acid exists in L(+) and D(−) optical enantiomers and an optically inactive *meso*-form. The racemate (an equimolecular mixture of the L(+) and D(−) forms) is produced by chemical synthesis.

The principal use of tartaric acid is as a food acidulant. For some industrial uses, such as in the retarding of setting of plaster, it is possible to employ the synthetic racemate, but this has a very low solubility, compared with either L(+) or D(−) tartaric acid, which renders it unsuitable for food uses.

13.7.2   Biochemical mechanisms

The racemic acid may be made by the reaction of maleic acid with hydrogen peroxide in the presence of a tungstic acid catalyst. The reaction proceeds by way of *cis*-epoxysuccinic acid, and can be stopped at this stage by temperature control. L(+) tartaric acid (the main article of commerce) is still made from the residues accumulating in wine vats. The fermentation method outlined below gives poor yields, and the method employing enzymic hydrolysis of *cis*-epoxysuccinate made synthetically, though developed up to pilot scale, is not at present economic.

13.7.3   Fermentation

Species of *Acetobacter* capable of producing 5-ketogluconate, form small amounts of tartaric acid under favourable conditions. More recently, Yamada and coworkers (see Reading List) have isolated mutants of *Gluconobacter suboxydans* able to produce L(+) tartaric in yields of the order of 30% from glucose.

The glucose is oxidized via gluconic acid to 5-ketogluconic acid. This is isomerized to 4-ketogluconic acid and thence by oxidative cleavage to tartaric and glycollic acids. No further developments of this process have been reported. One difficulty was the inhibitory effect of the byproduct glycollic acid.

13.7.4   Enzymic hydrolysis

A rather wide range of bacteria are capable of hydrolysing *cis*-epoxy-succinate sterospecifically to L(+) tartaric acid, and various detailed methods have been proposed in the patent literature. The organism may be grown in the presence of the *cis*-epoxysuccinate, or the organism may be grown and the resulting broth allowed to act on the substrate. Alternatively, intact washed cells, or an aqueous extract, or the extracted enzyme (*cis*-epoxysuccinate hydrolase) in an immobilized form may be used. The process may be carried out at pH values from 6–10, and at a temperature within the range 20–60°C. In many cases described the *cis*-epoxysuccinate is brought to the neutral to alkaline pH with sodium hydroxide. In one process, however, a suspension of the sparingly soluble calcium *cis*-epoxy-succinate is exposed to the action of the microorganism, thus producing the even less soluble calcium L(+) tartrate. The removal of the product from solution allows the equilibrium to proceed more completely in the direction of the tartrate (British Patent 1 532 041).

13.7.5   Product recovery

If the tartaric acid is produced in solution, cells or immobilized enzyme are removed. The tartrate is precipitated as the calcium salt which is filtered and washed, and then treated with sulphuric acid. The calcium sulphate is removed by filtration and the aqueous solution of tartaric acid is concentrated and crystallized. If the tartaric acid is produced as the insoluble calcium salt, the precipitate is collected, washed, and the tartaric acid released as above. Overall yields from maleic anhydride of about 75% are obtained.

13.8   Lactic acid

$$CH_3—CHOH—COOH$$

13.8.1   General

The production of lactic acid in the USA in 1881 is one of the earliest industrial applications of fermentation, and unlike any of the preceding organic acids it is produced by an anaerobic process. The acid exists in the L(+) AND D(−) forms, as well as in a racemic mixture.

Heating of concentrated lactic acid solutions gives rise to two types of self-condensation products; linear (e.g. lactoyllactic acid) and cyclic (dilactide). The formation of these products militates against the production of a pure crystalline acid.

Lactic acid is used as a food acidulant, and it also has a wide range of medical and industrial applications.

13.8.2   Biochemical mechanisms

Homolactic bacteria, e.g. *Lactobacillus bulgaricus* and *L. delbrückii*, have chiefly been employed for lactic acid production. The substrate may be lactose (whey), glucose (or glucose syrups), or sucrose, either pure or as beet molasses, and the fermentation is the classic example of an anaerobic process. Disaccharides are hydrolysed to hexoses, which are catabolized via the Embden–Meyerhof path to pyruvate, which is finally reduced to L(−)-lactic acid by lactate dehydrogenase. Under some conditions the DL-acid is produced, possibly by action of a racemase. The normal medium includes, besides the carbohydrate, inorganic nutrients to supply nitrogen, phosphate, potassium, etc. Organic nitrogen sources, e.g. corn steep liquor, malt sprouts, or yeast extract are often used, and additional vitamins may also be added. An actively growing suspension of *Lactobacillus* forms the first stage of a multistage vegetative inoculum.

13.8.3   Fermentation

L. delbrückii is used at temperatures of up to 50°C and *L. bulgaricus* up to about 44°C. The latter must be used to ferment whey because *L. delbrückii* cannot ferment lactose.

The use of a large inoculum and the relatively high temperature make rigorous sterilization of the medium unnecessary, and pH is maintained at 5.8–6.0 by addition of calcium carbonate. Although lactic acid is produced by an anaerobic process, small amounts of oxygen are not detrimental and precautions need not be extreme. The fermentation generally takes 6–7 days and yields in the range 80–90 g lactic acid per 100 g carbohydrate supplied (80–90% of theoretical) are normal. Large unagitated vessels, of steel or even wood, have been used for the process and a continuous variant has been described.

13.8.4   Product recovery

Several grades of lactic acid are marketed. Food grades need special purification methods unless a pure fermentation substrate is used in the first place.

In the presence of calcium carbonate, calcium lactate is formed in the fermentation. Sulphuric acid is added and the calcium sulphate so formed, together with the organism, is removed by filtration. In the simplest process the lactic acid solution from the filter is treated hot with active carbon, filtered and evaporated to a 50% solution. Where the fermentation substrate is very impure, e.g. beet molasses, ion exchange or solvent extraction is used.

Alternatively, the lactic acid is converted into a volatile ester which is fractionally distilled to effect purification. Hydrolysis of the ester, followed by evaporation, gives a water white product.

13.9   Acetic acid and other volatile fatty acids obtained by anaerobic fermentation

Much of the present world output of acetic acid comes from petroleum, while microbial oxidation of ethanol to acetic acid in the so-called quick vinegar process has long been known. Recently, newer processes have been under development for the manufacture of acetic acid and other volatile aliphatic acids (VFAs) by anaerobic fermentation of cellulosic and other wastes (Zeikus, 1980: *Ann. Rev. Microbiol.* **34**, 423–64).

13.9.1   Mechanism of VFA accumulation

The pathways involved are quite complex and are summarized in Fig. 13.3. Cellulosic wastes are hydrolysed to glucose by cellulases elaborated by anaerobic bacteria.

The production of butyric acid from glucose is seen to be less efficient than the production of acetic acid, where in practice the theoretical yield of three moles acetic acid from one mole of glucose is often approached, without the production of hydrogen and carbon dioxide which is a feature of the butyric fermentation.

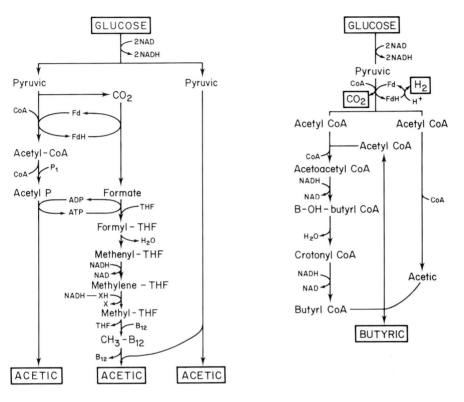

Fig. 13.3 Routes to VFA from glucose. [Reproduced, with permission, from Zeikus, J. G. (1980). *Ann. Rev. Microbiol.* **34**, 423–64. © 1980 by Annual Reviews Inc.]

The cultures used may be deliberate mixtures of anaerobic bacteria, e.g. *Clostridium thermocellum* with *Clostridium thermoaceticum*, or may be mixtures of organisms obtained by selection from sewage sludge. In the anaerobic digestion of sewage, for example, two types of organisms are present, namely those converting organics to VFAs and those converting the VFAs to methane and carbon dioxide. The action of heat on cultures from normal sewage sludge (80°C for 15 min) usually kills the methane-producing organisms, leaving anaerobic spore-forming bacteria which are not only able to digest cellulose yielding glucose, but also to ferment the glucose to acetic acid and other VFAs. Alternatively, the addition of 2-bromoethane sulphonic acid (BES) selectively inhibits the methanogenic organisms in whole sewage sludge.

The conditions in an anaerobic digester producing methane must favour both VFA producers and methanogens. Normally a pH of about 7.5 is aimed at and this is controlled by the rate of feeding the substrate. Overloading leads to a build-up of VFAs (because the methanogens metabolize more slowly than the VFA producers) and the concomitant fall in pH further favours the VFA producers. Methane production falls and VFAs accumulate. The addition of BES simplifies the operation of the digester when VFAs are required rather than methane.

13.9.2   Plant

Plant digesters are large closed tanks, often constructed of concrete. Air must be excluded rigorously, and mixing is by mechanical agitation, since the alternative means of using gas recirculation is not available when methane is not produced. The digester operates as a continuous fermenter and partial recirculation of organisms is an advantage, giving higher biomass concentrations and higher rates of conversion.

13.9.3   Product recovery

The manufacture of VFAs by anaerobic fermentation yields an aqueous solution in which the product concentration is rather low. Solvent extraction may be used to recover the product. Thus 20% trioctylphosphine oxide in kerosene may be used to extract acetic or propionic acids (Hydrosciences Inc.), while higher homologues (butyric acid upwards) can be extracted with kerosene. The VFAs can also be recovered using ion exchange.

13.10   L-Glutamic acid

13.10.1   General

$$HOOC-CH_2-CH_2-\underset{\underset{\displaystyle NH_2}{|}}{CH}-COOH$$

Glutamic acid (or monosodium glutamate) has been produced in Japan since 1908, initially by hydrolysis of wheat gluten or soya bean protein. The fermentation process was inaugurated by Kyowa Hakko Kogyo in 1957, while the Ajinomoto Company at one time made glutamic acid synthetically (e.g. from acrylonitrile) and isolated the L-glutamic acid by seeding a supersaturated solution of the racemate with crystals of L-glutamic acid. They now use a fermentation process exclusively.

The chief use of L-glutamic acid is in the form of the monosodium salt (MSG), which is very widely employed as a flavour enhancer.

The original organism of Kyowa was the so-called *Micrococcus glutamicus* (now renamed *Corynebacterium glutamicum*). Other organisms used are species of *Brevibacterium*, *Arthrobacter* and *Microbacterium*. A range of strains, each resistant to a particular type of bacteriophage, is required to enable phage infections to be countered.

Originally glucose, inorganic salts and biotin suboptimal for growth formed the medium. The biotin deficiency increased the permeability of the bacterial cell wall, allowing the glutamic acid to be excreted into the medium. Normally, an accumulation of more than 50 mg per g dry weight of glutamate in the intracellular pool would give rise to feedback inhibition. Additions of penicillin or of surface active agents such as Tween 60 or Tween 40 part of the way through the fermentation increase cell permeability at any level of biotin in the medium, and so allow cheaper biotin-rich substrates such as cane or beet molasses to be used. Processes based on substrates such as *n*-paraffins, acetic acid and ethanol have also been worked out and are used intermittently, according to raw material prices.

13.10.2   Mechanism of accumulation of L-glutamic acid

The metabolic pathways leading to L-glutamate formation in the commercial fermentation from sugars are as set out in Fig. 13.4, which again represents an incomplete form of the tricarboxylic acid cycle. A deficiency in α-keto-

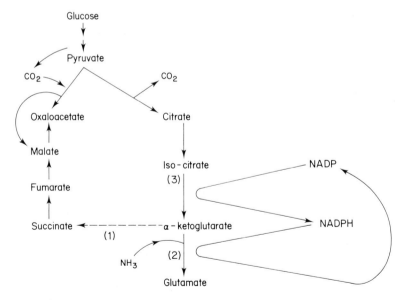

Fig. 13.4   Glutamic acid formation, (1) α-Ketoglutarate dehydrogenase (absent); (2) glutamate dehydrogenase (present); (3) isocitric dehydrogenase (present).

glutarate dehydrogenase and a nutritional requirement for biotin are characteristic features of glutamate overproducing strains. A lack of α-ketoglutarate dehydrogenase explains the flow of carbon to glutamate rather than completing the citric acid cycle.

The glutamate producers have a highly active glutamate dehydrogenase, for which the source of reducing power (NADPH) is mainly provided by the isocitrate dehydrogenase reaction. Carbon dioxide fixation, requiring biotin, is important for the formation from pyruvate of the C_4 dicarboxylic acids needed for the glutamate synthesis, and any impairment to this function leads to decreased yields. On the other hand, as noted above, a degree of biotin deficiency can be used to produce the necessary permeability of the bacterial envelope.

The basic carbohydrate source (cane or beet molasses) in the medium must be supplemented, where necessary, with phosphate, potassium, magnesium and manganese ions. Nitrogen is supplied as ammonium ion. Since nitrogen is contained in the product, large amounts are required, although high initial concentrations must be avoided. Substances conferring resistance to bacteriophage attack, which can cause complete lysis of the culture, are also added. The importance of biotin level in fermentations where penicillins or detergents are not added, has already been noted.

For inoculation, a culture of the bacteria in a shake flask is scaled up through several stages before transfer to the main fermenter.

13.10.3   Plant

The L-glutamic acid fermentation is carried out in stirred baffled tanks of stainless steel. Provision for cooling, dissolved oxygen measurement, and pH measurement and control, usually with ammonia, are required. A temperature between 30 and 35°C, and a pH between 7.0 and 8.0, are optimal. pH control is usually carried out using ammonia, which must be provided steadily as noted above. The oxygen transfer rate is fairly critical; a deficiency leads to poor glutamate yields, with lactic and succinic acids being formed instead, while an excess causes accumulation of α-ketoglutaric acid. Lactic acid is also produced when there is an excess of biotin, even at suitable aeration rates.

In media where excess biotin is present, penicillin (or surface active agents) must be added during the growth phase. A few units per ml of penicillin G, for example, are required; growth still continues but the cells are distorted.

The yield of L-glutamic acid obtained after 2–3 days fermentation is of the order of 50% by weight on the carbohydrate supplied. Generally speaking, yields from beet molasses, which already contains 3–5% of pyrrolidone-

5-carboxylic acid, the internal imide of glutamic acid, are higher than from other substrates.

13.10.4 Product recovery

In the standard process the cells are separated from the broth. The glutamic acid is crystallized by reducing the pH with hydrochloric acid to the iso-electric point (pH 3.2). It is of the greatest importance that this be done in such a way that the correct so-called α-type crystals are produced. One method is to add HCl and clarified broth separately to a suspension of the correct type of crystals at pH 3.2 at such a rate that the pH remains at this figure. When the correct type of crystals are obtained, filtration and washing are easily performed, yielding a product of high purity. Excessive levels of residual sugar in the broth lead to poor crystal forms, whereas the presence of certain other amino acids favours good crystallization. Monosodium glutamate is usually prepared from crystalline L-glutamic acid, which is dissolved in aqueous sodium hydroxide and the resulting solution decolorized, concentrated and crystallized. It is said (British Patent 1254972) that MSG may be obtained without the isolation of glutamic acid from the clarified broth, if the latter is treated with a strongly basic anion exchanger to remove organic acids other than amino acids.

13.11 L-Lysine

$$HOOC-\underset{\underset{NH_2}{|}}{CH}-CH_2-CH_2-CH_2-CH_2-NH_2$$

13.11.1 General

Lysine is an 'essential' amino acid which is very widely used to supplement cereal-based animal feeds.

13.11.2 Manufacture

A number of synthetic processes leading to DL-lysine are known. Since L-lysine is the form required in foodstuffs, various methods, either chemical or enzymic, for the topical resolution of either the product or some intermediate have been proposed. Only the process of Toray Industries (British Patents 1334970 and 1352860) is economic (see below). However, the bulk of the L-lysine production is by fermentation.

Organisms and mechanism of accumulation
The metabolic relation between lysine, its precursors and related amino acids is shown in Fig. 13.5. The accumulation of lysine by production strains

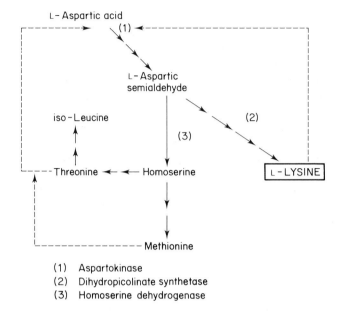

(1)   Aspartokinase
(2)   Dihydropicolinate synthetase
(3)   Homoserine dehydrogenase

Fig. 13.5   Mechanism of L-lysine accumulation. Dashed lines show inhibition.

depends upon their having been selected for defects in key synthetic steps and also for defects in the feedback control mechanisms which normally control amino acid synthesis.

The development of a lysine fermentation process was initiated by the discovery that homoserine auxotrophs of the glutamate overproducer *C. glutamicum* will produce lysine if grown under the appropriate conditions (Kinoshita *et al.*, 1958: *J. Gen. Microbiol.* **10**, 1975). Biotin must not be provided in limiting quantities, otherwise glutamate is produced. On the other hand, if too much biotin is available, lactate and succinate are produced. Penicillin addition induces glutamate production in the homoserine-less strains, as would be expected (see above).

The importance of homoserine deficiency can be appreciated from the figure (Fig. 13.5). In this species, the activity of aspartokinase, the enzyme which primarily governs lysine production, is controlled by multivalent feedback inhibition by lysine and threonine. Thus, homoserine auxotrophs (which lack homoserine dehydrogenase) will produce lysine because the threonine level can be kept low enough to avoid feedback inhibition of aspartokinase. Since both methionine and threonine are produced from homoserine (see Fig. 13.5), both amino acids must be added to cultures of the homoserine-deficient lysine producers. However, methionine helps to minimize inhibition of aspartokinase by threonine—presumably by competition for the allosteric binding site on the enzyme molecule.

Other bacteria like *C. coli* contain two aspartokinases, each subject to monovalent control by lysine and threonine, so that the multivalent control system in *C. glutamicum* must be considered a necessary feature of lysine-producing strains. Another feature of such strains is that their dihydropicolinate synthetase is insensitive to lysine. This is unexpected in that in branched pathways the first enzyme of a specific branch is usually subject to end-product inhibition. Similar behaviour is exhibited by mutant strains of several other types of bacteria (British Patent 1 304 067).

L-Lysine can also be accumulated by strains of *Brevibacterium*, *Bacillus* and *Corynebacterium*, selected for resistance to antimetabolites and antibiotics, and a genetically engineered lysine-accumulating strain of *E. coli* has been patented.

Fermentation

In the classical fermentation with *C. glutamicum* the medium consists of blackstrap molasses and hydrolysed soya bean meal. Patented strains are described for L-lysine accumulation from *n*-paraffins, aliphatic acids and alcohols. The fermentation is aerobic and is conducted in aerated stirred tank reactors. The pH is maintained in the neutral range by addition of ammonia, and temperature is controlled at 28°C. A rather small inoculum is used, grown up in the minimum number of steps from an agar slope culture. A major problem of the process is back mutation to prototrophs which generally outgrow the auxotrophic lysine-producing strains. This is why a small inoculum is desirable and why continuous processes have been unsuccessful. The addition of antibiotics, e.g. erythromycin, discourages the prototrophs. Alternatively, doubly auxotrophic lysine producers which are less prone to prototroph production may be used.

The fermentation lasts about sixty hours and yields of 40–45 g L-lysine l^{-1} are produced, starting with molasses concentration of 200 g l^{-1} (100 g sugar l^{-1}).

Product recovery

Although L-lysine hydrochloride may be directly crystallized from clarified broth after addition of HCl and evaporation provided certain conditions are fulfilled (US Patent 3 702 341), the usual process consists of acidifying the clarified broth with HCl and adsorption of the lysine on a cation exchange column in the ammonium form. The lysine is eluted from the column with aqueous ammonia, reacidified with HCl and the L-lysine hydrochloride crystallized.

13.11.3   Chemical synthesis with enzymic resolution

The Toray process is based on the sterospecific conversion of DL-α-amino caprolactam, which is itself produced chemically from cyclohexanone or cyclohexanol. Two enzymes bring about the selective hydrolysis of amino-caprolactam and the racemization of D-aminocaprolactam to the L-form. Both reactions take place in the same reaction vessel, where the yeast (*Cryptococcus laurentii*) and the bacteria (*Achromobacter obae*) are suspended. A glucose (1%) based medium is added, and the 'fermentation' takes 20 h at 50°C. The yield claimed in patents is 99.8%.

13.12   L-Aspartic acid

13.12.1   General

$$HOOC-\underset{\underset{NH_2}{|}}{CH}-CH_2-COOH$$

L-Aspartic acid has pharmaceutical uses and has potential as a raw material for the synthesis of peptide sweeteners. Although L-aspartic acid can be produced directly from glucose by fermentation (using mutant strains of *Brevibacterium* or *Corynebacterium* resistant to 6-dimethylamino purine), it is more usually prepared by the action of organisms containing aspartase on fumaric acid in the presence of ammonium ions:

$$HOOC-\underset{\underset{NH_4^+}{}}{\overset{\overset{CH-COOH}{||}}{CH}}\xrightarrow{\text{Aspartase}} HOOC-\underset{\underset{NH_2}{|}}{CH}-CH_2-COOH$$

Suitable organisms are said to include *E. coli*, *Bact. succinum*, *S. marcescens*, and *Ps. trifolii*. Strains of *Alcaligenes faecalis* also contain an isomerase which allows the cheaper *cis*-isomer maleic acid to be used as the raw material (US Patent 3 391 059).

13.12.2   Manufacture

Washed bacterial cells, with or without a detergent (e.g. cetylpyridinium bromide) to aid cell permeability, dried cells or immobilized washed cells (e.g. entrapped in a polyacrylamide gel) are allowed to act on buffered solutions of fumarate or maleate. Alternatively fumarate may be added to the broth of a fermentation which has progressed to the point where the aspartase content of the broth is maximal.

Conversion of fumarate using immobilized cells

An immobilized preparation of *E. coli* in an acrylamide gel is packed in a column and an aqueous solution of ammonium fumarate at pH 8.5 and 37°C is allowed to percolate through. If the flow rate is correctly adjusted, 100% conversion to L-aspartate can be obtained (US Patent 3 791 926).

Conversion of fumaric acid added to a fermentation broth

A strain of *Pseudomonas* is cultivated in a stirred aerated fermenter at 30°C using a medium based on beet or cane molasses at pH 7 and containing a small quantity of fumarate to induce the aspartase. After 11 h the temperature is raised to about 56°C and crystalline fumaric acid is added. The pH is kept at about 9.0 using ammonia. As the fumaric is converted to L-aspartic acid more fumarate is added so as to keep its concentration at $100 \, g \, l^{-1}$. Additions are decreased after 12 h and none is added after 15 h. Full conversion is reached after a further 5 h incubation (US Patent 3 933 586).

13.12.3 Product recovery

Hydrochloric acid is added to the L-aspartic acid solution (clarified where necessary to remove bacterial cells) until the pH reaches 2.8, the isoelectric point. The L-aspartic acid crystallizes.

13.13 L-Tryptophan

13.13.1 General

$$CH_2-\underset{\underset{H}{\overset{\displaystyle NH_2}{|}}}{CH}-COOH$$

Tryptophan is an essential amino acid. It is used medically and would be used as a feed additive if it could be made cheaply enough. Economic production by fermentation of sugars has not been reported, and commercial production is based on fermentation with added precursors, either anthranilic acid or indole (see Fig. 13.6).

13.13.2 Manufacture

Production from anthranilic acid

Strains of *Hansenula anomala* mutated to confer resistance to anthranilate are used and a suitable medium contains beet or cane molasses diluted

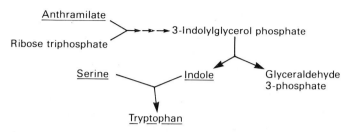

Fig. 13.6   Routes to tryptophan.

to 50 g sugar l⁻¹, NH₄NO₃, NaH₂PO₄, anthranilic acid and calcium carbonate. A culture of *H. anomala* is pregrown in a seed medium and added to the fermentation medium in an aerated/agitated vessel. Optimum pH is 6 and the pH must not fall below 5. This is ensured by the CaCO₃ in the medium. The temperature is maintained at about 28°C and the dissolved oxygen in the region of 10⁻⁴ M. Further additions of carbohydrate (or glycerol) and anthranilic acid may be made as the fermentation proceeds. Yields of tryptophan after 4 days are in the region of 2.5–3.5% based on sugar supplied.

Production from indole

Although a process has been devised using *Claviceps purpurea*, better yields are obtained with *H. anomala* strains. A medium containing glucose (50 g l⁻¹) calcium carbonate, indole, yeast extract and inorganic salts is used, and the fermentation is carried out at 27°C. Further quantities of indole in ethanol are added from time to time. Better yields may be obtained by using mixtures of indole and anthranilate.

13.14   D-Arylglycines

13.14.1   General

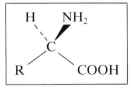

D-Arylglycines are not found naturally, and their technical importance arises from the fact that they may be coupled chemically with 6-aminopenicillanic acid to give semisynthetic penicillins.

As an example D-phenylglycine is considered. Chemical synthesis affords the racemic DL-phenylglycine amide as starting material, and selective hydrolysis of the racemic amide is carried out using a leucine amino-peptidase. The enzyme reaction may be carried out using an aqueous solution of the enzyme or with the enzyme in an immobilized form. The DL-phenylglycine amide is dissolved in an alkaline buffer (pH 9.0), with magnesium and/or manganese ions as activators.

A solution of the leucine aminopeptidase enzyme is added and the solution stirred at 20–40°C for several hours, when the L-phenylglycine amide is wholly hydrolysed to L-phenylglycine, Alternatively, the buffered solution of DL-phenylglycine amide is allowed to percolate down a column of glass beads to which the leucine amonopeptidase has been attached, using for example a 3-aminopropyl-triethoxysilyl compound. In this case the L-phenylglycine amide is hydrolysed during the passage of the solution through the column. In either case, the hydrolysate is passed through a column of IRC-50 ion exchange resin upon which the unreacted D-phenylglycine amide is absorbed, while the L-phenylglycine is washed from the column and effectively recovered by racemization to give further DL-phenylglycine.

The D-phenylglycine amide is eluted from the column with H_2SO_4, and after evaporation of the eluate *in vacuo* the amide is hydrolysed by further boiling yielding D-phenylglycine which crystallizes from the solution on neutralization with ammonia.

This process is an example of fixed enzyme resolution of racemic amino acids, and other amino acids may be resolved by this or similar methods. For instance, an immobilized acylase from *A. oryzae* can be used to hydrolyse *N*-acetyl-DL-methionine to give L-methionine.

Reading List

Amino acids
Johnson, J. C. (1978). 'Amino Acids Technology—Recent Developments'. Noyes Data Corporation: Park Ridge, New Jersey.
Yamada, K., Kinoshita, S., Tsunoda, T. and Aida, K. (Eds) (1972). 'The Microbial Production of Amino Acids'. Halsted Press: New York.

Citric acid
Kirk Othmer (Ed.). 'Encyclopaedia of Chemical Technology', 3rd Edn, Vol. 6, pp. 150–178, John Wiley & Sons: New York 1979.

Organic acids
Miall, L. M. (1977). *In* 'Economic Microbiology', Vol. 2, Primary Products of Metabolism (Ed. A. H. Rose), pp. 47–119, Academic Press Inc: London.

14 | Enzymes as Bulk Products

W. E. GOLDSTEIN

14.1 Introduction

Enzymes have been applied to practical uses for thousands of years, in the form of crude animal or plant preparations or as a consequence of permitted microbial development, and have been more consciously applied in operations of manufacture throughout the present century. However, the controlled production of microbially-derived enzymes is relatively more recent and is now a major sector of industrial biotechnology. The technology for producing and using commercially important enzyme products combines areas of microbiology, genetics, biochemistry and engineering. These disciplines have developed and matured through time both singly and in an interactive manner, leading to the evolution of techniques in enzyme manufacture. This evolutionary process has benefited enzyme applications because of increasing awareness of the relationships involved.

The nature of the methodology used in enzyme manufacture resembles that of other industries in that it is dependent on the final value and market volume of the product. Typically, the more expensive 'fine' enzymes are produced in lower volumes and at inherently greater unit manufacturing cost. The bulk or commodity enzymes are required in greater volume, but have an inherently lower unit value, so that they demand significantly lower manufacturing costs. A graph representing this aspect is shown in Fig. 14.1. Of necessity, therefore, the technologies for fine and for bulk enzymes have evolved in somewhat different directions. For example, enzymes for medical use must be refined to a high degree of purity, and within limits, cost is less of a factor than efficacy and safety in use (see Chapter 15). By contrast, bulk enzymes are only purified to the extent needed to meet actual customer requirements for activity and stability,

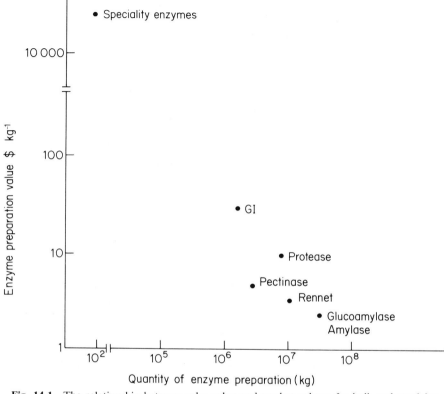

Fig. 14.1 The relationship between sales value and market volume for bulk and specialty enzymes.

though safety is still a requirement in terms of handling. In applications to food ingredients, the need to control costs is over-riding.

An illustration of this argument, useful in establishing a proper context, can be based on the use of bulk enzymes in starch processing, to first hydrolyse starch to a glucose-rich stream and then to isomerize the glucose to give, finally, a high-fructose syrup. Typical enzymes employed would be:

(1) thermostable α-amylase, an endo-amylase derived from bacteria (*Bacillus*);
(2) glucoamylase, predominantly hydrolysing 1,4-linked oligosaccharides, commonly derived from fungi (*Aspergillus*);
(3) glucose isomerase, converting glucose to an equilibrium mixture of glucose and fructose, and commonly obtained from bacteria, e.g. species of *Bacillus*, *Microbacterium**, *Streptomyces* or *Actinoplanes*.

The competitive basis for producing and selling these enzymes is achievement of desired product quality at an acceptable cost per unit of the final product—high-fructose syrup. We can therefore express the relevant factors in terms of this user cost to yield ratio expanded in the following manner:

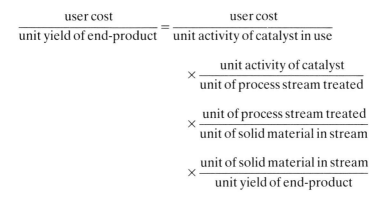

$$\frac{\text{user cost}}{\text{unit yield of end-product}} = \frac{\text{user cost}}{\text{unit activity of catalyst in use}}$$

$$\times \frac{\text{unit activity of catalyst}}{\text{unit of process stream treated}}$$

$$\times \frac{\text{unit of process stream treated}}{\text{unit of solid material in stream}}$$

$$\times \frac{\text{unit of solid material in stream}}{\text{unit yield of end-product}}$$

The components of this expanded cost equation suitably describe different factors which define bulk enzyme technology and stimulate advances in the field. The user cost per unit of catalyst activity is the selling price of the enzyme, made up of the manufacturing cost and the sales margin to the seller; this introduces the factor of the enzyme producer's control of sales margin and his need for technology which will reduce his costs in order to be competitive. The next two terms introduce the factor of the amount of catalyst needed to treat a given quantity of raw material and the actual concentration of the process stream being treated; the corresponding incentives are for the manufacturer to improve effective productivity (the inherent activity of the enzyme and its effective life) and to allow this to be expressed in as concentrated a stream as possible (thus recognizing the need to minimize requirements for water removal, water removal being a common cost to many enzyme applications). The final factor in the cost equation takes note of the product quality requirements; in our example, this could include obtaining the desired sugar mix, with maximal glucose/fructose content and minimization of undesirables such as di- and tri-saccharides with 1,6-linkages.

A further note on this equation is that it is only based on manufacturing costs. A new enzyme product requiring changes in capital equipment must also be cost effective in terms of improved product quality or reduced manufacturing costs to a sufficient extent to induce the customer to install the new equipment.

* Previously called Flavobacterium.

In the example we have chosen, the enzyme manufacturer has several courses to follow. Manufacturing cost per unit of enzyme activity can be reduced through selection of raw materials, by innovations making the best use of labour and utilities, by making the best use of capital-intensive facilities and by employing justifiable new technology in the design and use of new facilities and equipment; however, it is obvious that the most dramatic cost reductions will be obtained through yield improvement since this reduces all associated costs. Yield improvement is reached through changes in the organism, by optimizing the fermentation to give maximum enzyme yield, and in the design of downstream processing to isolate and sufficiently purify the enzyme. Thus the fermentation technology will be designed and operated so as to maximize the fermentation yield, and the processing technology will be correspondingly operated for maximum recovery.

The general pattern of fermentation yield improvement is that shown in Fig. 14.2, where the yield is plotted not against time but against expenditure. Process improvement by modification of the organism generally gives

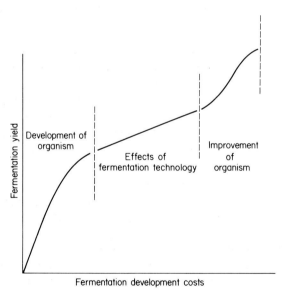

Fig. 14.2 The relative effectiveness of different technologies in improving fermentation yields, in relation to development costs.

the largest yield improvement in relation to costs incurred, while with an optimized organism, improvements in the fermentation technology provide further yield improvements but at possibly greater expense. Eventually the cycle is repeated as a new and better organism is introduced. In contrast, the improvement of product recovery by optimising processing, as shown in Fig. 14.3, usually follows an S-shaped curve; in the early stages, improve-

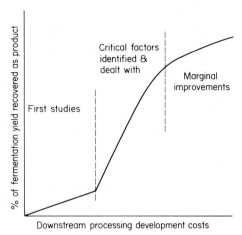

Fig. 14.3   Cost effectiveness of different stages in the development of product recovery technologies.

ments in recovery are hard to secure, but as critical process factors are identified enzyme yield will increase significantly as new methods are introduced. The next phase will be one of marginal improvements providing smaller increases in recovery for the cost incurred. The precise nature of the curve will of course vary from case to case, and the pattern will be repeated for each new enzyme-producing organism or strain introduced.

Factors pertinent to process design, for both fermentation and downstream stages, will therefore include:

(1) the degree of enzyme purity required:
   (a) what degree of microbial contamination is acceptable?
   (b) is the presence of other enzymes acceptable (e.g. proteases in an amylase)?
   (c) will additives be needed for chemical or microbial stabilization (e.g. calcium for amylase, sorbates as preservatives)?
   (d) are relevant materials being incorporated unintentionally (e.g. from the fermentation broth)?
(2) the required degree of control of temperature, pH, and ionic strength throughout the process;
(3) the required response to temperature and humidity during storage.

The successful development of fermentation and processing technology must take into account conditions for the use of the product as well as for its preparation. For example, the pH and temperature at which the enzyme is to be used need to be taken into account in selecting the producing organism. The nature of the substrate which is to be treated by the enzyme (e.g. the precise type of starch or protein to be hydrolysed) must be taken into account at all stages, as must the eventual process stream conditions and the eventual degree of substrate purity. Common influences, which are not always defined in advance, include additives intentionally introduced as activators or stabilizers (e.g. magnesium ion in glucose isomerase) and those which are present unintentionally. Examples of the latter might include unusual microbial contaminants and their metabolites (e.g. weak organic acids inducing pH changes in the manufacture of high-fructose syrup), contaminants that will destabilize an enzyme, such as heavy metal ions, or will serve as nutrients for undesirable microbial spoilage (e.g. sources of nitrogen or phosphate), or unusual by-products or breakdown products such as carbohydrates. The extent to which the customer's use conditions can influence research and development to produce new or improved enzymes is critically dependent on good communications between enzyme producers and users.

The most important commodity enzymes and their main applications are listed in Table 14.1, together with the producer organisms, the mode of use and the temperature and pH conditions this imposes, and notes of important activators, stabilizers, and inhibitors and inactivating entities. Evidently the list is quite a short one, and apart from glucose isomerase all the enzymes are hydrolases. Most applications are in food areas and particularly in starch processing. The brevity of the list reflects the difficulty of finding applications that can be met by the correct properties at an economic cost.

14.2   Organism sources and strain development

The source for a new or improved enzyme is guided by the intention that cultivation of the organism will be carried out in aerated stirred-tank reactors; there are detailed variants and many interesting new ideas, but practical enzyme manufacture is dominantly oriented towards submerged culture technology. An exception to this, based initially on tradition but since continued through intensive innovation, is the use of surface fermentations on solid or semisolid media for enzyme production in Japan.

The search for interesting and satisfactory enzyme-producing organisms can be either widely- or narrowly-based. If the end-use demands minimum costs combined with high assurance about product safety, the search is

Table 14.1 Commercially important enzymes

Application	Enzyme	Use form	Source organisms	pH and temperature (°C) for use	Activators, stabilizers	Inhibitors, destabilizers
Starch and carbohydrate processing						
Hydrolysis of 1,4-links in starch to give dextrans	α-amylase	Extracellular, soluble	Bacillus licheniformis	6.0–6.5; 100–140	Ca	Heavy metals
			B. amyloliquefaciens	6.0–6.5; 40–70		
Hydrolysis of dextrans to glucose	Glucoamylase	Extracellular, soluble	Aspergillus niger	3.5–4.5; 58–62		
Purified glucose equilibrated to glucose–fructose mixture	Xylose (glucose) isomerase	Immobilized whole cell	Bacillus coagulans, Streptomyces olivaceus, Microbacterium arborescens, Antinoplanes missouriensis	7.0–8.0; 55–65	Mg, Co	Ca, O₂, Heavy metals
Protein hydrolysis						
Protein breakdown in detergent formulations	Alkaline protease	Extracellular, soluble	B. licheniformis, B. subtilis	9.0–11.0; 40–60	Ca	
Protein breakdown in brewing substrates	Neutral protease	Extracellular, soluble	B. amyloliquefaciens, B. subtilis	6.0–9.0; 40–60	Ca	Zn
Milk coagulation and cheese flavour enhancement	Acid protease	Extracellular, soluble	A. niger, Mucor miehei, Mucor pusillus, Endothica parasitica	2.0–7.0; 40–50	Ca	
Pectin hydrolysis						
Fruit juice and wine processing	Pectinase, β-glucanase cellulase	Extracellular, soluble	A. niger, B. subtilis	4.0–7.0; 50–65	Ca	

likely to be confined to organisms which have already passed the very cost-intensive process of securing regulatory approval, or to their close relatives (compare Table 14.1). If, on the other hand, the search is for a really novel activity or combination of properties, it will have to be more widely based. In either case the search will be based first on culture collections, often built up within a company over a long period, and second on new isolates from especially relevant sources. For example, effluent disposal sites from the processing of starchy raw materials are suitable locations for collecting samples that may yield organisms with interesting starch conversion enzymes. Thermophilic organisms or ones active in conditions of extreme pH or salinity, will be sought in those specialized habitats where corresponding conditions have obtained naturally over long periods, such as volcanic hot springs or tropical salt lakes.

The basic strategy of culture collection (screening) and strain improvement for enzyme production is the same as that described for other microbial activities in Chapter 8. As explained there, the devising of satisfactory tests to select the desired cultures, whether new isolates or improved strains, is critical. Clearly the assay system will approach as closely as possible to one which measures the kind of enzyme activity being sought for the final application, with as much relevance to specific user requirements as can be built into the test system. In strain improvement, the full range of selective procedures will be added to random mutation; these will include selection for inhibitor resistance, insensitivity to catabolite repression, and so forth (see Chapter 8).

As explained in Chapter 2, the amounts of any enzyme produced by an organism are generally very closely controlled by its own regulatory mechanisms; specifically, by derepression or induction, by repression, and by catabolite repression, all operating against the background of breakdown or turnover mechanisms. To obtain a commercially useful level of production it is usually necessary to exploit and where necessary overcome these natural regulating mechanisms. In particular, the hydrolytic enzymes which are conspicuous in Table 14.1 are usually produced by organisms as a response to the depletion of more readily utilized nutrients in their environment; the limitation of essential nutrients promotes the release of catabolite repression and leads to the relevant enzyme synthesis. This mechanism may be combined with substrate-specific induction mechanisms; both catabolite repression and induction mechanisms can however be affected by mutation, and regulatory mutants are, in general, the end objective in strain improvement for enzyme production.

Overall, then, there will be a conflict between the need to produce a large amount of the organism, to obtain a good final yield of enzyme, and the need to restrict its nutrition so that the enzyme is synthesized.

In the research laboratory using well-defined media, the typical batch culture succession of growth, nutrient limitation, and enzyme production may be seen in a clear-cut way, but this is not necessarily the most efficient pattern for large-scale production. Commercial fermentations are almost always run on complex media, not only because the raw materials are more inexpensive but also because on complex media there is less separation between growth and nutrient limitation and the overall pattern of, say, enzyme synthesis is more efficient in terms of hours of plant use per unit product formed.

With a new organism, enzyme production is first studied using solid media (nutrient plates) and small-scale submerged cultures (multiple flasks on a vibratory shaker). Sufficient progress will then lead to studies in small-scale well-instrumented stirred fermentors. Some of the nutritional and environmental factors used in developing the main kinds of bulk enzymes are described in Table 14.2.

Table 14.2 Nutritional and environmental factors in development of commercially significant enzymes

Nitrogen sources	*Carbon sources*	*Elements and vitamins*
Soya flour and isolate	Corn, corn meal, wheat	Corn steep liquor
Fish meal	meal	Calcium, iron and sodium
Corn steep liquor	Starch from corn, wheat	salts of carbonate,
Yeast hydrolysate and	Corn steep liquor	sulphate, and citrate
extract	Barley	
Casein	Gum arabic	
Ammonia	Hydrolyzed corn starch	
Ammonium salts	Lactose	
	Glucose	
	Beet pulp	

Mechanical power input requirement mixing and gas dispersion per volume of broth	*Aeration requirements for mixing and oxygenation volumetric gas rate/volume of broth*
Bacteria and pelleted fungi	
1–$3 \, kW \, m^{-3}$	
	0.1–$1.0 \, m^3 \, m^{-3} \, min^{-1}$
Mycelial bacteria and fungi	
4–$6 \, kW \, m^{-3}$	

An optimization of medium components, preferably through statistical design, will be used to maximize formation of enzyme per unit of biomass and to increase biomass production itself so long as this does not impair the rate of enzyme synthesis. Apart from productivity, stability of the enzyme once formed will be a consideration, and where the final product is to be used in the form of an immobilized preparation of whole cells (e.g. glucose isomerase), the absence of cell lysis and retention of all the activity intracellularly will be an objective. Material balances for essential nutrients in the fermentation are desirable, particularly in order to be able to follow the exhaustion of particular nutrients and for maintaining appropriate ratios of others (e.g. C/N), but in practice the complexity of many media and the presence of solids may make such information very incomplete. Oxygen supply is usually critical, and the O_2 requirements are usually evaluated at the small fermentor scale (around 10 litres), since it is essential to avoid oxygen limitation throughout all phases of scale-up. The power requirements to maintain oxygen transfer become substantial as the scale of operation increases (see Table 14.2); as indicated, and as discussed fully in Chapter 5, the less viscous fermentations with bacteria and with fungal pellets require substantially less power input as compared with viscous mycelial growths.

14.3   Fermentation

14.3.1   Culture propagation

In typical enzyme fermentations, the organism is propagated through several stages of batch culture. Stock cultures from the research laboratories are generally preserved in freeze-dried ampoules, periodically sampled and tested. Inoculum is usually transferred from stage to stage in the proportion of 1–5% by volume, for example from seed culture flask through one or more seed tank stages and then to the main fermentation. The earlier propagation stages may use a very rich medium to give good growth, and later stages will then involve a process of adjustment to the production-scale medium; the more advantageous case where the final medium is the same as that used in the seed stages is unfortunately less common.

14.3.2   Instrumentation

Figure 14.4 shows a fermentor system schematically, in this case one at a pilot-plant scale used in the development stage of enzyme production studies. Units at laboratory scale (10–40 litre) or pilot-plant scale (100–200 and 1000–2000 litres) are normally well-provided with instrumentation as indicated. They may be fitted out for automated data monitoring and process control; this is more common at pilot-plant scale and is typically used by major manufacturers to assemble a proper data-base for running

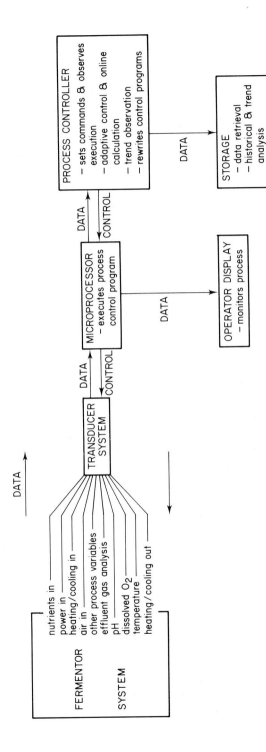

Fig. 14.4 A fully computerized data retrieval and process control system as applied to fermentation process control, and particularly as applied to process optimization at pilot plant scale.

production operations. Common measuring devices on fermentation units will include dissolved O_2, vessel and cooling-water temperature, off-gas monitoring of O_2, CO_2 or other gases by selective or general purpose instruments, and foam detection. The vessels will be equipped for top- or bottom-drive agitation and the introduction of air, and will be designed and run according to the best aseptic practice (see below). Laboratory units will be glass vessels equipped with small cooling coils while pilot-plant vessels have cooling jackets and are fabricated of high quality surface-finished stainless steel. The fermentation equipment is essentially the same as that used for antibiotics production, see Chapter 16.

14.3.3 Sterility

Cultures used for enzyme production often lack some of the self-protecting features (low pH; antimicrobial formation) that are helpful in other fermentations, so sterile operation is particularly critical; contamination can adversely affect both quantity and quality of products. To ensure that fully-effective procedures which meet the criteria set out in Chapter 17 are actually followed, it is desirable to establish fully standard practice for cleaning and for ensuring asepsis, possibly through the use of computer-controlled systems, since many problems of contamination are related to human error and to lack of attention to detail in repetitive operations.

14.3.4 Fermentation profiles

Fermentation procedures are worked out in detail on the pilot-plant scale (100–200 and 1000–2000 litres) before transferring to the production plant (typically 25–100 m³). Reasonable assessment of oxygen transfer needs can be made at this scale. The development of procedures for specific new strains is facilitated through extensive use of process automation (Fig. 14.4), and especially the capability to quantify CO_2 and O_2 balances through appropriate instrumentation.

Figure 14.5 provides a typical example of the results obtained in monitoring a particular enzyme fermentation, in which packed cell volume (a measure of biomass growth), dissolved O_2, CO_2 evolution, carbon source depletion, and the level of extracellular enzyme have all been measured. As indicated, biomass begins to increase early in the fermentation, with a concomitant increase in respiration and a fall in dissolved oxygen. Enzyme activity increases most significantly midway through the fermentation at about the same time as carbon source utilization accelerates. The end of

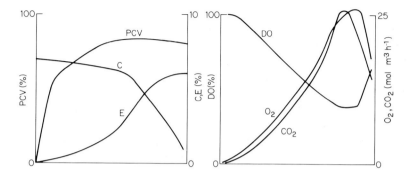

Fig. 14.5 Profiles of process variables through a typical enzyme-producing fermentation. PCV, packed cell volume (%); C, carbon source in the medium (kg m^{-3}); E, enzyme in medium (units/4000); DO, dissolved oxygen tension (% saturation); O$_2$, oxygen consumption rate (mol m^{-3} h^{-1}); CO$_2$, carbon dioxide production rate (mol m^{-3} h^{-1}).

the fermentation is marked by decreased respiration, a rise in dissolved oxygen, and the cessation of enzyme synthesis.

Such a set of data displays carbon source requirements, oxygen transfer needs, and the point at which the fermentation should be stopped and the broth discharged for processing. Power requirements consistent with obtaining these results can be measured only at what is defined here as pilot scale; the general geometry and the presence of many insertions in small laboratory-scale fermentors prohibit any useful interpretation of power input requirements on a smaller scale. Such laboratory units are used to establish the general feasibility of the fermentation and to provide small quantities of broth for testing processing.

14.4 Processing and purification

Once prepared, the fermentation broth must be processed without delays and with proper physical and chemical controls to maintain the catalytic effectiveness of the enzymes and to avoid contamination. A generalized flowsheet showing the sequences of operations appropriate to different forms of final product is shown in Fig. 14.6. As indicated, immobilized cell preparations, like glucose isomerase, involve isolation and treatment of the biomass itself as a product form. In other cases the enzyme itself must be isolated. Enzymes which are accumulated within the cells during fermentation, such as glucose oxidase, must be recovered by disrupting the cells to liberate the enzyme before subsequent purification, which can be varied according to the nature of the final application (e.g. for glucose oxidase, industrial or diagnostic grade). Processes for enzymes which are already extracellular involve concentration of the broth and some means of enzyme isolation before suitable final purification to whatever degree is necessary.

The general technologies for the unit operations listed in Fig. 14.6 are dealt with fully in Chapter 6. The specific requirements governing process choices for enzyme recovery are:

(1) restrictions as to temperature, ionic concentrations, metal ion con-tamination, etc., imposed by the need to conserve the integrity of the enzyme protein and its catalytic effectiveness;

(2) decisions as to the degree of purification and concentration which is specifically appropriate and economical for a product—the product marketed for a particular use, formulated in a given manner, and put into storage under specified conditions.

14.5   Examples: enzyme production

Continuing with illustrative examples based on the starch processing sequence already outlined (Section 14.4), we shall first consider the produc-tion of the three enzymes, bacterial α-amylase, fungal amyloglucosidase and immobilized glucose isomerase, and in Section 14.6 outline their practical use.

14.5.1   Alpha-amylase

An α-amylase (1,4-α-D-glucan glucanohydrolase) attacks in-chain linkages (endo-hydrolysis) in starch components having three or more linear α-1,4-glucan units; its depolymerizing action stops when fragments with two to six glucose units remain, typically molecules containing an α-1,4,6-linked branch-point residue. Its action on starch macromolecules causes a rapid reduction in molecular weight with an attendant drop in viscosity (thinning) which is of considerable practical significance and can be the basis for enzyme assay. The end-products of α-amylase attack on starch are dextrins, which as well as being substrates for further hydrolysis have their own uses as adhesives and as thickening agents in prepared foods.

For process reasons (see Section 14.6.1), an α-amylase must be used at high temperatures, and suitable enzymes are obtained from bacteria; the most heat stable, which can be used for long periods at 95°C and briefly at 105–110°C, are obtained from species such as *Bacillus licheniformis*.

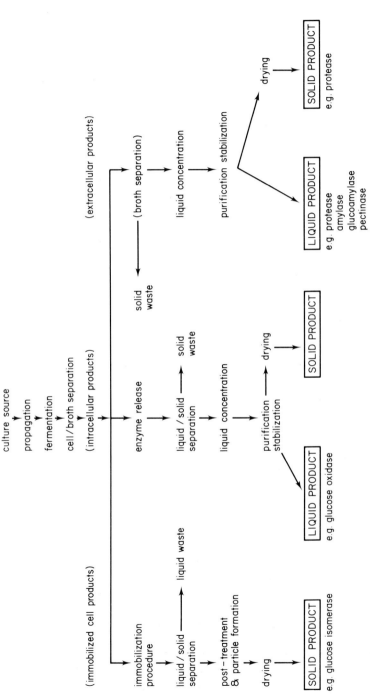

Fig. 14.6 Generalized scheme for enzyme production and downstream processing.

In batch culture on defined media, the extracellular enzyme is produced more rapidly as growth slows down. Free glucose represses enzyme synthesis, which is controlled by catabolite repression, and in many bacilli is associated with the early stages of spore formation. In commercial production, carefully selected strains of the organisms are grown on complex media based on corn or potato starch supplemented with a nitrogen source such as soya meal or corn-steep liquor. Typical media contain about 20% total dry matter and the final yield of enzyme protein, reached after about five days, will be a few per cent of this. The fermentation is carried out at about 40°C and neutral pH and requires effective aeration.

At the end of the fermentation the broth is chilled and the biomass and suspended solids removed by flocculation. The enzyme solution is then subjected to a degree of concentration and purification (see Fig. 14.6 and Chapter 6) that is carefully matched to user needs and storage requirements. Proteases as contaminants are undesirable in many applications, and also for storage stability; they are preferably eliminated from the strain through mutation. If present they can possibly be inactivated by heat treatment without damaging the thermostable amylase. The enzyme is always stabilized with calcium ions, and sold either as a liquid concentrate, in brine containing about 2% of enzyme protein, or as a somewhat more concentrated solid preparation.

The use of the enzyme for starch breakdown takes several practical forms. In addition to dextrin production and for the first stage in glucose manufacture, the enzyme is applied in brewing and in bakery work and for the removal of starch size in textiles manufacture. For each use there is a corresponding product specification and consequent variations in detailed manufacturing processes.

14.5.2   Amyloglucosidase

Amyloglucosidase, or glucoamylase, is an exo-hydrolase (exo-α-1,4-D-glucosidase) which removes terminal glucose residues stepwise from the ends of the shorter α-1,4-linked glucan chains present in dextrins. Commercially satisfactory preparations will also hydrolyse 1,6- and 1,3-linked residues (possibly at a slower rate), thus giving virtually complete conversion of the starch dextrins into glucose. Suitable enzymes are produced by fungi such as species of *Aspergillus* or *Rhizopus*, and some have a useful degree of thermotolerance permitting prolonged use at 55°C. Like many fungal enzymes they work best at rather acid pH (4–5.5).

The enzyme is produced on dextrin media and in wild-type strains its synthesis is repressed by free glucose; this characteristic is selected against in developing production strains. Selected fungal strains grown up from spore suspensions are inoculated into production-scale fermenters using a medium with a high concentration of starch (20% or more) which is treated with α-amylase during the sterilization cycle. The relatively viscous medium and the filamentous growth of the fungus mycelium combine to make aeration and agitation the limiting factor in reactor productivity. The batch fermentation is nitrogen-limited and lasts 4–5 days; pH is controlled above 4.5 to avoid enzyme inactivation.

At the end of the fermentation the mycelium is filtered off and the extracellular enzyme is concentrated to a liquid preparation containing about 5% of active enzyme, with specified levels of contaminants such as amylases and proteases.

14.5.3 Glucose isomerase

Fortuitously, enzymes with the physiological function of converting the common aldopentose, xylose, into the more readily metabolized keto-pentose, xylulose, will also catalyse the corresponding interconversions of hexoses, namely glucose and fructose. Thus commercial 'glucose isomerase' is actually D-xylose ketol-isomerase; the enzyme is produced by a wide range of bacteria and commercial preparations are from selected strains of both unicellular (*Bacillus, Microbacterium Arthrobacter*) and filamentous (*Streptomyces, Actinoplanes*) species. Parent strains produce the enzyme when grown on xylose or plant hemicelluloses as carbon source but some developed commercial strains produce the enzyme constitutively and even when grown on glucose-containing media. The organisms are grown in aerated batch culture for 2–3 days at 30°C and neutral pH.

At the end of fermentation the enzyme-containing biomass is separated, and although it is possible to extract and further purify the intracellular enzyme, most commercial preparations are based on whole cells treated in various ways so as to increase their catalytic activity and stability. The enzyme is most advantageously used in an immobilized form (see Section 14.6.3) and these are prepared by stabilizing the biomass into particulate form, for example by cross-linking with agents such as glutaraldehyde. The physical nature of the particles is very important for satisfactory reactor characteristics.

Some isomerase preparations require cobalt and magnesium for maximum activity and most need to be protected from oxygen both in storage and in use. For reasons of safety, syrup quality, and cost of cobalt removal, addition of cobalt is not a recommended commercial practice. Immobilization usually increases the temperature for optimum activity and shifts the pH optimum from neutral to slightly alkaline.

14.6   Examples: enzyme applications

14.6.1   Starch hydrolysis

Continuing with our set of illustrative examples, the first step in the hydrolysis of a native starch is to bring the polymer into solution, or more strictly to convert the solid starch granules into a gel. This is a high-temperature 'cooking' process which can be carried out continuously. Typically, a 40% starch slurry is heated to 105–110°C (or in some cases, 140°C) by direct steam injection. The resultant gel is very highly viscous and hard to handle, or even to cool, which is why the hydrolysis needs to be initiated immediately and requires a thermostable enzyme, which is added directly along with calcium salts for maximum activity. In typical practice, the partially hydrolysed starch is then subjected to further liquefaction at 90–100°C for extended periods of 1–2 h, additional enzyme being added. The process kinetics depend on both the temperature profile and the characteristic catalytic activity and stability over the temperature range used. Thus the holding time can be lengthened somewhat if less enzyme is used, but this means that the enzyme is used less efficiently. Further, conditions of enzyme addition and temperature in liquefaction must be such as to avoid excessive viscosity increase in gelling. Generally about 1 litre of the enzyme preparation (see Section 14.5.2) is used per tonne of starch transformed.

Most starch sources contain varying amounts of protein, and amino acids liberated by contaminating proteases can cause undesirable browning reactions. Process conditions can also lead to formation of undesirable sugars, and the high lipid content in some starches complexes some of the polysaccharide and may necessitate modified process conditions. The final slurry may or may not be reheated to about 120°C, often with added acid, to inactivate the enzyme, and the hydrolysate can then be passed on for further processing.

14.6.2   Production of glucose syrups

The dextrin solution from Section 14.6.1 can be further processed in a variety of ways; here we consider only its conversion into glucose. This is usually carried out as a batch process with relatively long reaction times. The starch hydrolysate, with 30–40% sugars, is cooled to 60°C and adjusted to about pH 4. The correct amount of enzyme is added and the mixture held at the optimum temperature for 2–3 days. If the temperature is too high it will be inactivated prematurely, and if it is too low there will be incomplete conversion plus the risk of bacterial contamination. Overlong exposure to the enzyme also leads to some reversal of hydrolysis, and formation, for example, of undesirable 1–6 disaccharides, and so the enzyme must be inactivated at the end of the batch by heat and/or acid. The overall conversion is easier if the sugar concentration is reduced, but this will increase the eventual cost of evaporation. There are trends toward use of immobilized enzymes to produce glucose syrups in much less time; attendant difficulties in conversion, asepsis, and practical economics must be addressed to prove the worth of this methodology.

14.6.3   Glucose isomerization

Conversion of glucose towards the equilibrium mixture (approximately 1 : 1) of glucose and fructose increases its sweetness and gives a product virtually identical with 'invert sugar' from beet or cane sucrose. The equilibration is what is catalysed by glucose isomerase and in practice about 95% of the equilibrium mixture is attained in a reasonable time. To produce fructose itself, further processing is required; the typical method of fructose enrichment is use of cation exchange resin to separate fructose from other sugars. Because the enzyme is relatively costly, the process is only economic if the catalyst can be re-used, i.e. as an immobilized preparation, while the process kinetics are most effectively arranged if the enzyme is used in plug flow fixed-bed (column) reactors.

The glucose syrup is first purified, usually by charcoal treatment followed by ion exchange and some evaporation, which also removes oxygen that would otherwise inactivate the enzyme. It is then passed, at about 40% sugar, pH 7–8, through single columns in parallel or possibly through two or more columns containing immobilized glucose isomerase arranged in series and held at about 60°C. As in dextran hydrolysis, higher temperatures give quicker reaction at the cost of shorter enzyme life, while lower temperatures mean that slower flow rates must be used. Although the enzyme is more stable, the throughput is less and the chance of bacterial infection is higher, unless the enzymic activity in the reactor is increased with the benefit of enhanced productive output from a more stable enzyme. The substrate is moderately viscous and temperature may markedly affect both bulk flow and intraparticle diffusion rates in the reactor. Under practical conditions a throughput rate of 0.25 to 1.0 volumes per bed volume per hour is optimal, decreasing as the enzyme decays and approaches the end of its useful life which is 2000–4000 hours. Enzyme longevity is dependent on process conditions such as operating temperature and exposure to inhibiting agents.

Successful operation of this process is critically dependent upon correct and effective planning of the process parameters in the first instance, and on sophisticated and precise control systems throughout. Catalytic effectiveness is a major reactor, and an advanced glucose isomerase preparation may have a productivity of 20 tonnes or more of product per kg of catalyst. The true value of the immobilized biocatalyst is, however, dependent on its cost as well as cumulative productivity. A biocatalyst with high productivity will only be of interest if the cost per unit of productivity is competitive—i.e. the purchase price is acceptable and the enzyme is able to withstand the demands of a production environment in a corn wet-milling plant.

14.7 Trends and perspectives

The use of recombinant DNA technology is not yet known to be widespread in enzyme manufacture. However, enzymes are relatively clear candidates for the methodology, being essentially single-gene products, and many of the situations and difficulties faced in enzyme fermentations, processing, and applications could conceivably be resolved through direct genetic modifications of suitable hosts. An example of this would be the cloning of the gene for calf rennin, a widely-used and desirable product, into *Bacillus subtilis* and into yeast (Harris *et al.*, 1982). Continual review and awareness of this field is needed, and much of the information must come from outside the normal scientific literature and through private contact. In a somewhat similar way, more widespread application of immobilization processes on a large scale may be justifiable, depending always on market circumstances and product form; for example (as noted previously), immobilized gluco-amylase is already a research topic and its practical application is a likely development. However, introduction of any such technology must be within the constraint of justifying capital requirements, balanced against the undoubted benefits, as previously noted.

Innovations in bioreactor design for enzyme production are noted throughout the literature, for example involving means of increasing the cell concentration and of operating in a continuous rather than a batch mode. Implementation of such designs for commercial enzymes may be less distant if the technique offers demonstrable practical gains in productivity and yield when tested on a production scale. Meanwhile there is a clear trend towards finding new production organisms in order to obtain enzymes with specifically desirable characteristics, such as stability to high temperatures or extreme pH; such research will allow innovation in fermentation to resolve difficulties in enzyme processing. However, the main accelerations in bulk enzyme technology will only follow from the intensive examination, exploration, and experimental proof of enzyme applications in areas where enzymology is not yet a significant operational technique.

Reading list

Books and reviews

Aunstrup, K., Andresen, O., Falch, E. O. and Kyaer Nilsen, T. (1979). Production of microbial enzymes. *In* 'Microbial Technology', Vol. 1, Ch. 9. Academic Press: New York and London.

Godfrey, T. and Reichelt, J. (1983). 'Industrial Enzymology'. Nature Press (Macmillan): London. [A comprehensive and wide-ranging account.]

Lambert, P. W. (1983). Industrial enzyme production and recovery from filamentous fungi. *In* 'Filamentous Fungi' (Eds J. E. Smith, D. R. Berry and B. Kristiansen), Vol. 4, Ch. 9. Edward Arnold: London.

Special topics

Aidoo, K. E., Hendry, R. and Wood, B. J. B. (1982). *In* 'Advances in Applied Microbiology' (Ed. A. I. Laskin), Vol. 28, pp. 201–233. Academic Press: New York and London. [Solid-state fermentations.]

Box, G. P., Hunter, W. G. and Hunter, J. S. (1978). 'Statistics for Experimenters, An Introduction to Design, Data Analysis, and Model Building'. John Wiley & Sons: New York. [Medium optimization.]

Harris, T. J. R., Lowe, R. A., Lyons, A., Thomas, P. G., Eaton, M. A. W., Millican, J. A., Patel, T. P., Bose, C. C., Carey, N. H. and Doel, M. T. (1982). *Nucleic Acids Res.* **10**, 2177–2187. [Genetic engineering.]

Ollis, D. F. (1982). *Phil. Trans. R. Soc. Lond.* **B297**, 617–629. [Genetic engineering.]

Toda, K. (1981). *J. Chem. Tech. Biotechnol.* **31**, 775–790. [Screening.]

15 The Production and Purification of Fine Enzymes

J. W. BREWER

15.1 Introduction

'Fine' enzymes are enzymes produced in relatively small amounts, but very considerable variety, and sold mainly for research purposes in biochemical and medical investigations. Increasingly they are also being marketed for analytical purposes, for example in clinical diagnostics systems, and to a small extent for therapeutic purposes.

Fine enzymes are more expensive because they are produced at lower volumes, thus creating a higher unit manufacturing cost. More time is entailed, and manufacturing these enzymes involves employing highly paid technical personnel. Sometimes, costly material and equipment are used, even though steps are taken to minimize costs. The intended application of the enzyme dictates the purity that is desired; for example pharmaceutical and medical (therapeutic) institutions apply high purity enzymes in their research programmes, while for therapeutic use very stringent standards must be met. In most applications of fine enzymes, batch-to-batch reproducibility is particularly important.

The enzymes may be of microbial, plant, animal, or in some instances, of human (blood) origin. Their diversity is so wide that only a few will be given as examples. Some of the high purity enzymes in most demand are RNA and DNA polymerases, restriction enzymes, lyases, ligases, and initiating and terminating enzymes used in recombinant DNA work. The competitive basis for producing and selling these enzymes is to produce enzymes of highest purity and yet be within an acceptable cost/volume range to be competitive. The user must be convinced of the superiority of the product produced by an enzyme manufacturer, in terms of its purity, effectiveness, and ease of reproducible use. Another very important selling factor is shelf life and/or stability in use. The manufacturers must be able to answer questions such as: is the enzyme stable under reasonable storage conditions? Is it easy to use in a normal facility? Does it require chemical or physical components that are difficult to obtain? If so, can a kit be provided that comprises these components? Does the kit include well-tested instructions and independent quality control data?

The yield of enzymes produced from microorganisms must be maintained and possibly increased, particularly to make the recovery process cost effective. These higher yields can be obtained by finding a mutant that overproduces the desired enzyme, or a metabolite inducer can be incorporated into the fermentation medium to force the microorganism to overproduce the enzyme, or genetic engineering can be used to modify the genetic mechanisms of the organism and cause it to overproduce the enzyme. Such methods for increased productivity in the fermentation will usually also lead to a recovered enzyme of greater purity and to lower production of undesirable material.

Most fine enzymes for research and therapeutic use are intracellular enzymes. This complicates matters, because after removing extracellular metabolites and debris the cells must be ruptured in such a way as to release the enzyme unharmed; the enzyme must then be separated from the interfering intracellular material. Such interfering substances include ribosomes, protease (unless protease is the desired enzyme), nucleic acids, nucleases (if nucleotide-forming enzymes are the desired enzymes), and cell wall or membrane debris.

15.2 Fermentation sources for enzymes

The general fermentation process will only be very briefly described, with the main emphasis on the yield and improved purity of enzymes in the fermentation stage. Enzyme activity throughout the fermentation should be assessed so as to define the point at which the highest recoverable yield of the designated enzyme is manufactured by the microorganisms. Some enzymes are produced at early, mid, or late log stages of bacterial growth. Further, the production of adverse, interfering enzymes or contaminating material that will interfere with the efficiency of the recovered enzyme must be taken into consideration before terminating the fermentation and harvesting the cells. Steps should be taken to reduce the production of these interfering substances by modifying the fermentation, if possible. Maximum yield of the desired enzyme may have to be sacrificed so that a stage where interfering material is reasonably low can be identified, to aid in ease of purification or isolation of the desired enzyme (Darbyshire, 1981); a typical example is provided in Fig. 15.1. Fermentations must also

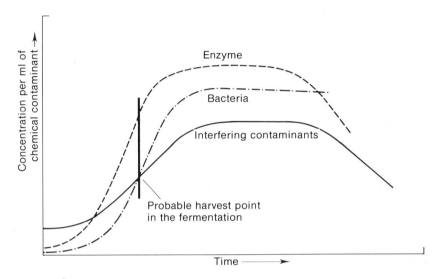

Fig. 15.1 Bacterial fermentation progress plotted against enzyme and interfering chemical contaminant production.

be continuously monitored to establish batch-to-batch consistency, so that unnecessary changes or modifications are not required in the purification process.

Modifying the medium by removing or adding a nutrient may well stimulate the organism to overproduce the desired enzyme (Davis & Blevins, 1979). Sometimes bacteria can be induced to produce enzymes in excess by growing the organism in the presence of a substrate specific for the enzyme in question (Dixon & Webb, 1979).

A microorganism can be modified by genetic engineering methods to aid in purification, aiming to produce the desired enzyme in such excess that the proportion of undesirable materials is minimized (Elander & Chang, 1979). A mutant that overproduces the needed enzyme can be selected as discussed by Elander in the present book (Chapter 8).

Microbial cells are usually harvested by centrifugation to concentrate the cells and free them from metabolites and medium constituents. At times, cells are washed, usually by suspending them in an appropriate buffer, and centrifuged again to concentrate the cells. The concentrated or 'packed' cells can either be frozen or used immediately for enzyme purification.

15.3   Extraction of intracellular enzymes

As already noted, most of the fine enzymes are intracellular. Plant cells and the cells of most microorganisms have cell walls, and all living cells have cellular membranes, that must be disrupted to release wanted enzymes. All extraction procedures are usually carried out in a 'cold' room or similar facility to maintain a temperature of about 5°C.

Most higher plant cells can be disrupted by maceration with a mortar and pestle, a high speed blender, or milling apparatus. For instance, fresh whole wheat germ (rich in enzymes) can be ground into a fine flour by using a mortar and pestle with ground glass or in a high speed mill. Another widely used method for cell disruption is by using high frequency sound (sonic disruption) generated electronically and transported through a metallic tip or cone into an appropriately concentrated suspension of cells. Sonic disruption is mostly used to rupture the cell walls and cellular membranes of bacteria and fungi (Penefsky & Tzagoloff, 1971).

Such cells can also be ruptured by exposing them to very high pressure (110 MPa or higher according to the manufacturers' specified limits) in a metal cylinder, using a piston to apply pressure, whereupon the cells are released into normal atmospheric pressure, causing the cells literally to explode. Care must be taken to keep this system cold, because heat is generated in producing the high pressures and in passing the cell mass through the press. Various systems are manufactured which use hand operation or electrical or pneumatic power to produce the required pressure; such systems can accommodate a wide variety of volume (Charm & Matteo, 1971), and some can operate continuously. Another well known method used to disrupt bacterial cells is the use of enzymes that attack the structure of the cell wall and thereby weaken or rupture it. In some instances two agents are used; one to weaken and destroy the cell wall leaving the cell with only its cellular membrane (spheroplast), and another to rupture the membrane completely. The destruction of the cell membrane can be completed either chemically or mechanically. For example, lysozyme attacks the structure of the cell wall of bacteria leaving them in the spheroplast form. A mild surfactant such as sodium deoxycholate is then added to destroy the cell membrane completely. The extracted material contains viscous polymers and nucleic acids which can be thinned using the shearing action of a blender at relatively high speed (Burgess & Jendrisak, 1975).

These methods can be used, for example, for multiple runs at 2 to 4 litres, pooling the batches for consecutive purification stages. The degree of cell rupture can be monitored quickly by centrifuging the cells and debris and determining dissolved nucleic acids by the ultraviolet absorption of the supernatant liquid at 260 nm (A_{260}) after dilution. Under normal disruption conditions, the light absorbancy at 260 nm will increase to a maximum, and level off when almost all the cells are ruptured. Treatment should not be prolonged far beyond the region of constant absorbance, as excess extraneous material may then be released, or under certain conditions the cellular debris will be rendered too minute, and difficult to remove. The extent of disruption can also be monitored by following the release of enzyme activity during the process.

Maceration of mammalian organs and tissue for the extraction of enzymes can be carried out using either an electrically or a manually operated meat grinder. Interfering tissue such as fatty tissue, blood vessels, and membranes must be removed before extraction so they will not interfere with further purification stages. The organs should be ground up finely to facilitate separation of a liquid extract from the solid residue by filtration or centrifugation (Chang & Bollum, 1970).

Cell debris or heavy cellular microbodies can be removed by high to very high speed centrifugation, by vacuum filtration through various grades of manufactured filter aids, or by gravity filtration. At this stage, clear filtrates are not necessary, but the liquid must be relatively free of particulate material.

15.4   Preliminary purification (preparation for column chromatography)

Preliminary purification of enzymes is usually intended to produce a semi-pure enzyme that can still contain impurities that will not interfere with the next stage of purification, usually column chromatography. This stage of purification should, for example, remove nucleic acids. Ammonium sulphate, streptomycin sulphate, sodium sulphate, high molecular weight polyamine polymers or liquid/liquid phase partitioning are some of the methods used in preliminary stages of enzyme purification. Figure 15.2 illustrates a typical case for the ammonium sulphate precipitation of an enzyme, demonstrating aspects of importance. As illustrated at A in Fig. 15.2, ammonium sulphate can be added at a concentration such that none of the desired enzyme is precipitated out of solution; at this point other precipitated (undesirable) material can now be discarded by either centrifugation or filtration. Ammonium sulphate is then added in greater quantities,

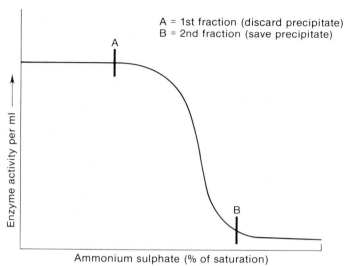

A = 1st fraction (discard precipitate)
B = 2nd fraction (save precipitate)

Fig. 15.2   Ammonium sulphate concentration plotted against enzyme activity of supernatant.

causing enzyme to precipitate; the point where the curve in the figure levels off indicates that precipitation of the desired enzyme is complete. The precipitate is then recovered by centrifugation or filtration and solubilized in an appropriate buffer solution, and supernatant liquid containing undesirable material is discarded. It should be understood that at this stage, the enzyme solution is still not very pure, and contains many impurities, in excess or in minor amounts (Dixon & Webb, 1979).

The use of polyamines provides an exception since the enzyme is precipitated in a polymeric solution containing a low concentration of salt ions, and is released when the precipitate is transferred to a solution containing a higher concentration of ions. Care must be taken not to have so high an ionic concentration in solution as to release excessive unwanted impurities, or so low as to release insufficient enzyme. Typical practice may allow a small amount of enzyme to be sacrificed to obtain an enzyme of comparable purity (Jendrisak, 1975). The actual definition of the effective range of these precipitants depends on the salt concentration, enzyme concentration, temperature, and pH (Dixon & Webb, 1979).

Liquid/liquid phase partitioning utilizes aqueous polymers (e.g. polyethylene glycol, dextrose, polyamines and sulphonated dextrans) of various molecular weights, charges, or affinity toward water to allow formation of separate liquid phases. The difference in molecular weights and/or affinity for water must be sufficient to cause phase separation. The charge or hydration differences must be enough to assure separation of the desired enzyme from enough of the soluble or insoluble impurities. This method can sometimes be used for direct extraction of whole fermentation broths without removal of the solid biomass suspension. The method of liquid/liquid phase partitioning locates the desired enzyme in either an upper or lower liquid phase, leaving impurities in the other remaining phase or phases. The enzyme/polymer mixture or complex is then separated in the subsequent stages of purification. As with salt or polymer precipitation, the ionic concentration and type, enzyme concentration, temperature, and pH determine the nature of enzyme partitioning and the concentration of each separate polymer to be used in forming partitions in the aqueous enzyme solution (Belter, 1979; Kula *et al.*, 1981). It may be helpful to add a surfactant to reduce denaturation effects at the liquid/liquid interface.

Care must be taken to ensure that conditions noted in reference literature describing the procedure remain invariant in purification efforts. Enzyme extracts from mammalian sources can often be clarified by adjusting to pH 5, and allowing to stand for a few minutes before centrifugation; this is a useful first step before the main purification begins. Very frequently, heat-labile impurity enzymes can be inactivated by heating the enzyme solution to temperatures at which the desired enzyme is stable for a sufficient period of time (Dixon & Webb, 1979); some enzymes are relatively heat stable and this can be a very useful method for their purification.

15.5 Chromatographic and other methods used in final purification

Types of column chromatography methods used in final steps of enzyme purification are: ion exchange, molecular sieve, gel filtration, and affinity chromatography. High speed centrifugation using gradient densities is also used to separate small quantities of enzymes. Fractions are usually collected from columns in tubes contained in racks that are automatically rotated or linearly manipulated based on time expired or drops collected. Each fraction is usually graphically monitored by the ultraviolet absorption at 280 nm, and by subsequent measurements of enzyme activity.

15.5.1 Ion exchange chromatography

In general ion exchange chromatography utilizes synthetic resins or natural matrices (e.g. cellulose) as the medium to which cations or anions are ionically bonded. These ions are either strongly or weakly attached. Matrices with weakly bound ions are preferred for enzymes, since media with strongly bound ions require extremes of pH and ionic strength to induce release of the enzymes; such extreme conditions are likely to be detrimental to the enzyme structure. When the enzyme in an appropriate buffer is brought into contact with the ion exchange column, the bound cation or anion is displaced (substitution reaction) by the enzyme. The cationic or anionic charges involved in the exchange are determined by the enzyme structure and by the pH and concentration of ions in the buffer. An appropriate buffer with a higher anionic or cationic concentration (or a different pH) is used to release the enzyme from the column. Ionic concentration, pH, or ionic concentration and pH combined, can be used appropriately to separate an enzyme from impurities using ion exchange column chromatography. Figures 15.3 and 15.4 exemplify the use of step and gradient elution of enzymes. General procedures are given in the literature; however, experiments must be performed to verify best conditions, first to bind the enzyme to an appropriate column and matrix then to release it. Figure 15.3 indicates that a different pH or higher salt concentration

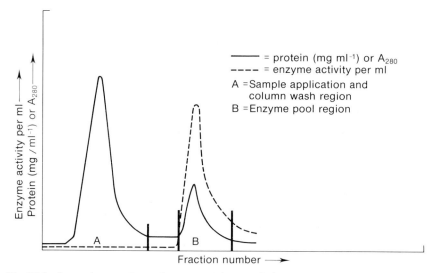

Fig. 15.3   Ion exchange column chromatography step elution pattern.

'step method' for enzyme elution can be used if the desired enzyme peak is widely separated from unwanted protein peaks at a certain buffer pH value or ion concentration. Alternatively, as indicated in Fig. 15.4, a pH or salt concentration gradient can be used if the two protein peaks can not be adequately resolved by step elution. The gradient can be made sharp or shallow as desired by the experimenter. Gradient control can be provided by adjusting the volumes of both gradient vessels, at the upper and lower pH or salt concentration gradient limits. With such a method, one may have to sacrifice enzyme quantity for enzyme purity. If a contaminant protein peak slightly overlaps the desired enzyme protein peak, the

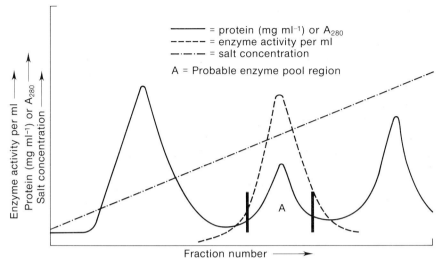

Fig. 15.4   Ion exchange column chromatography salt gradient pattern.

enzyme fraction must be pooled where the least amount of contaminate is present, without sacrificing too much. The techniques involved are at the discretion of the investigator (Conn & Stumpf, 1967; Pesce *et al.*, 1974).

15.5.2   Gel filtration

Gel filtration column chromatography separates protein molecules according to their molecular weight and/or size. The resin matrices for gel filtration columns are selectively prepared porous polymers manufactured to controlled average pore size. The pore size is such that the larger protein molecules pass around the beads and elute from the column rapidly. Smaller protein molecules enter the resin pores, and therefore move more slowly through the column. The ease with which molecules pass through the column resin pores determines the degree of exclusion and the time that the enzyme is collected from the bottom of the column. Figure 15.5 is an example of the use of this method to separate enzymes of widely different molecular weight and size. The protein peaks are located and selected as with other column chromatography methods (Dixon & Webb, 1979).

Either of these methods can be used separately or in combination. Generally it will be best to use the ion exchange chromatography method first, since the enzyme solution must be low in protein concentration but in fairly pure form in order to obtain acceptable enzyme and impurity separation by gel filtration.

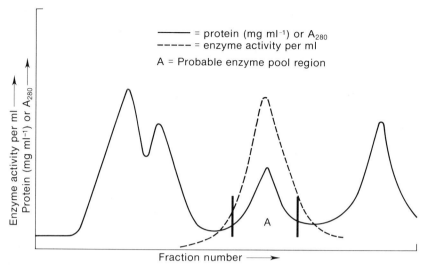

Fig. 15.5 Gel filtration column pattern.

The 'sieveorptive' method uses a resin combining both absorption and gel filtration chromatography properties, allowing separation of proteins by charge combined with separation by molecular size (Kirkegaard, 1976).

15.5.3 Affinity chromatography

Affinity chromatography methods involve use of a resin polymer modified chemically; the active form usually contains a bound substrate or inhibitor specific for the desired enzyme. This method uses the principle that the enzyme binds with its specific substrate or inhibitor in an intermediate reaction before the enzymatic reaction is complete. Because the enzyme binding is specific, only the enzyme will bind to the resin substrate or inhibitor complex. Impurities can then be washed out from the resin with a suitable buffer that will not disturb the bound enzyme. After the contaminants are removed, the ionic strength and/or pH can be changed to release the enzyme in a rather pure state from the ligand bound to the column (Cuatrecasas *et al.*, 1968).

15.5.4 Density gradient centrifugation

Density gradient centrifugation is used for small volumes of enzyme with low protein and contaminant concentrations. This method uses solutions of sucrose, glycerol, etc. that can produce stable density gradients, prepared in the centrifuge tube by layering more concentrated, dense solutions below less dense solutions, to partition the enzyme away from contaminants. The enzyme is layered on to the top of this gradient and centrifuged at relatively high gravitational force. The enzyme is partitioned separately away from the contaminants according to density under centrifugation, and remains stationary as a band at a certain density partition in the gradient. Usually the gradient is slowly drawn off from the bottom of the centrifuge tube as small fractions which are then pooled as appropriate (Blumenthal, 1979).

15.5.5   General notes

For effective absorption during ion exchange chromatography, the enzyme solution should be adjusted to the ionic strength and pH of the column buffer. The enzyme should be free of foreign material that will compete for binding sites on the column matrix, such as high quantities of contaminating protein molecules and nucleic acids. The enzyme concentration should not be too great, to avoid interference with the column flow or limit the availability of water of hydration. Column length should be sufficient to provide binding capacity necessary for all of the enzyme, and also extra length sufficient for enzyme fraction separation to occur when the enzyme is released from the matrix; since the enzyme usually binds to the very top portion of the column, the overall length must be adequate to facilitate good separation through the lower part of the column. The application flow rate of the enzyme should be slow enough to bind the enzyme to the topmost part of the column without excessive dispersion and back-mixing. The elution flow rate should be slow enough to facilitate good resolution and separation of the enzyme peak (Conn, 1967; Pesce *et al.*, 1974). In gel filtration, the enzyme must be sufficiently concentrated to form a thin band at the top of the column and thereby maintain sharp bands of proteins as they migrate down. The literature on the subject of column chromatography should be used to select the correct resin and the physical and chemical characteristics offering best conditions for column chromatography.

15.6   Preparation of enzymes for long-term storage

Enzymes are comparatively fragile substances, liable to undergo denaturation and inactivation under conditions promoting enzyme destabilization. Being proteins, they have both cationic carboxylic groups and anionic amino groups. Because of these charged centres the polarity of water exerts a stress or tension on the protein molecule, and the enzyme can easily be denatured, losing the specific active configuration of the macromolecule. Many enzymes are only active when their protein subunits are associated in a specific manner.

The water of hydration on an enzyme can be at least partially replaced by more hydrophobic molecules in high concentration (e.g. 50% v/v), to decrease such polarity effects. Suitable substances include propylene glycol or glycerol (Jarabak, 1966). In addition, enzymes can often be protected from stress and denaturation in the presence of their substrate, since for many enzymes, the presence of substrate aids in maintaining the configuration of the molecule at the active site (Dixon & Webb, 1979). An arguable disadvantage is that the presence of the substrate would lessen the purity of the enzyme and the substrate would eventually, over a longer time period, become the enzyme product. However, such effects are very slow, since enzymes are normally stored under very cold conditions (-20 to $-70°C$). Such conditions of themselves help to maintain the molecular configuration of the enzyme, slowing down molecular activity and reducing the extent of interaction of water dipoles with the charged enzyme molecule (Charm *et al.*, 1971).

Another additive often incorporated to stabilize enzymes is a reducing agent to present sulphydryl groups from forming inter- or intramolecular disulphides. Such reducing agents can be dithiothreitol, mercaptoethanol, and cysteine (Darbyshire, 1981). Similarly EDTA may be added to chelate destructive metal ions. Microbial contamination is controlled by preservation below $0°C$ particularly in the presence of hydrophobic substances such as glycerol.

Once frozen, most pure enzymes will not withstand repeated further freezing and thawing because of mechanical shearing of the enzyme by the stresses this causes. This is most dangerous at the transition stage because water is then present in both as solid and liquid, inducing a maximal stress on the enzyme molecules. Some enzymes that are deep-frozen are packaged in very small containers so that they can be discarded when an appropriate amount of enzyme is removed for use.

As a further precaution a non-active protein can be added to certain enzyme solutions at an appropriate concentration to give the enzyme a protective environment and to prevent denaturation effects when the enzyme is in dilute solution. Such additive proteins, for example, can be bovine serum albumin or gelatin.

15.7   Tests, assays and specifications

A material balance should be constructed for the whole purification process, illustrating the purity and per cent recovery of the enzyme at each stage (Dixon & Webb, 1979). Table 15.1 illustrates data normally collected during the purification of a sample enzyme.

Table 15.1 Sample enzyme material balance and purification table

Procedure	Volume (ml)	Enzyme concentration (units ml^{-1})	Total units $\times 10^{-6}$	Protein (mg ml^{-1})	Specific activity	Yield (%)	Fold purification	A_{280}/A_{260}
Homogenate	7500	1500	11.25	45.5	33.0	(100)	(1.0)	0.525
$\rightarrow$ (NH$_4$)$_2$SO$_4$ fraction	1000	10 690	10.69	106.9	100	95	3.03	0.825
$\rightarrow$ DEAE-cellulose eluate	500	18 000	9.00	18.0	1000	80	30.3	1.50
$\rightarrow$ (NH$_4$)$_2$SO$_4$ precipitation	50	168 750	8.44	30.7	5500	75	166.7	1.75
$\rightarrow$ Gel filtration column eluate	60	112 500	6.75	4.5	25 000	60	757.6	2.50

Enzyme assays can be obtained from published literature or developed by the investigator. The enzyme assay will require a specific substrate. If a new enzyme with an unknown substrate is being investigated then many substrates should be tested along with many possible inhibitors and activators (metal ions or other substances). Whenever possible, the substrate concentration should be adequate for saturation of the enzyme, so that the kinetics in the standard assay approach zero order. Where a suboptimal concentration of substrate must be used, both the Michaelis and maximal velocity constants should be determined from the observed rates. The principal elements which specify the initial velocity of the reaction are enzyme concentration, substrate concentration, pH, temperature and activators or inhibitors. The enzyme activity must be determined on the linear portion of the enzyme kinetic curve, usually with the substrate in excess, and occasionally where it is not, and the reaction rate order of the enzyme bears on the assay. Figure 15.6 provides an example of a zero order enzyme kinetics curve, but not every case produces such a curve.

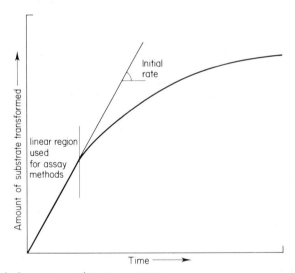

Fig. 15.6   Typical enzyme reaction progress curve.

Ideally, test procedures should be modified to eliminate any deviations, and allow use of an assay based on initial velocity. Various levels of enzyme concentration may be evaluated to determine the linear portion of the enzyme reaction progress curve and measure the initial velocity. Whether the disappearance of the substance, or the appearance of a product is measured will depend on resulting ease of measurement and accuracy (Dixon & Webb, 1979). Enzyme activity should be expressed in international units;

1.0 unit of enzyme activity causes transformation of 1.0 μmol of substrate per minute at 25°C under optimal conditions of measurement (Lehninger, 1970).

Protein quantities are usually determined on enzyme preparations throughout the purification process to determine the specific activity of the enzyme, defined as enzyme units per milligram of protein. The specific activity usually increases drastically as the enzyme goes through the purification stages, because of removal of protein impurities. Extent of purification is usually determined by dividing the specific activity at certain stages of the purification by the initial specific activity (Dixon & Webb, 1979). Enzyme activity recovery is usually determined by the following formula:

$$\% \text{ Recovery} = \frac{\text{Recovered enzyme activity}/\text{ml} \times \text{Volume (ml)} \times 100}{\text{Initial enzyme activity}/\text{ml} \times \text{Volume (ml)}}$$

Ultraviolet absorption at 260 and 280 nm is determined for the enzyme preparation to determine degree of absence of nucleic acids. Proteins absorb at 280 nm, mainly because of their tyrosine content, and nucleic acids absorb at 260 nm. A high A_{280}/A_{260} ratio indicates greater absence of nucleic acids from the enzyme preparation. If this ratio is sufficiently high, the protein content can be determined from the A_{280} reading using the purest enzyme available as a standard (Brewer, 1974a). Enzyme purity can also be assessed by gel electrophoresis. The two types of electrophoresis in use are native and SDS electrophoresis. Native electrophoresis is performed under conditions in which the enzyme protein is retained in its native state. The electrophoresis of a relatively pure enzyme using native electrophoresis usually yields a strong dark band along with other faint impurity bands.

The sodium dodecylsulfate (SDS) electrophoresis procedure uses SDS along with heat to denature the enzyme molecule by breaking the bonds holding the enzyme protein subunits. The electrophoresis containing SDS is performed to determine if the enzyme is completely intact and contains all its subunits and if the enzyme is actually the desired enzyme. By the analysis of extra bands, SDS electrophoresis is used to determine the purity of enzyme preparations (Brewer, 1974b; Gaal *et al.*, 1980). Gel electrophoresis is also used for enzyme assays to determine the activity of certain nuclease enzymes on synthetic or natural nucleic acid polymers, or to test for the presence of contaminating nucleases. This procedure is economically used for polymerases, ligases, nucleases, and restriction enzymes (Greene *et al.*, 1974).

Substrates can be purchased labelled with radioactive isotopes such as tritium (^{3}H), carbon-14 (^{14}C) and phosphorous-32 (^{32}P). The enzyme activity can be measured by determining loss of radioactivity through enzyme degradation of the substrate, or by the gain of radioactivity in the reaction product; the degraded substrate or resulting product can be collected on filters by precipitation or separated by other means such as by column chromatography. The use of radiolabelled substrates is a standard and rather general technique for enzyme activity assays (Conn *et al.*, 1967b; Pesce, 1974).

Specifications should be developed to meet needs of the user (e.g. purity) and to assure the user that the enzyme will perform under conditions of application. Sometimes the specifications are established by the producer, or based on results of marketing inquiry, or are in accord with requirements of regulatory agencies. Efforts in quality improvement often follow from difficulties in not meeting specifications and depend on the development of new or revised methods.

Most manufacturers of enzymes have quality assurance laboratories to assure consistent adherence of products to specifications.

15.8   Enzyme manufacturing credentials

Any manufacturer of enzymes should accompany the enzyme with a credential (see Fig. 15.7). In essence, the credential will give the scientific name of the enzyme and common name if it has one. A physical description of the enzyme (solid, solution) should be presented along with quantity per package. A brief description of the specification limits for the enzyme and other parameters, with a note of the methods of determination, should be included. The definition of enzyme activity units must also be included in the credential, followed by a description of the assay method. The purification procedure which has been used should be stated. References cited for preparation of the enzyme must be listed at the bottom of the credential to credit authors of procedures used. Storage temperature, precautionary statements concerning the enzyme, and any health hazards, must also be included.

15.9   Conditions for shipping enzymes

Some enzymes may need special precautionary notes for shipment (such as the temperature that the enzyme is to be kept at during shipment). The notice should then indicate that the container contains dry ice, wet ice or is shipped at ambient temperatures. Fine enzymes are usually packed in plastic vials tightly sealed with tape to ensure proper closure. Small vials in dry or wet ice must be packed tightly in an insulated container

Chemical credentials

Product: DNA polymerase (*M. luteus* 2.7.7.7)
Code no: 31-640
Lot no: 57A

This enzyme is purified from *M. luteus* by ammonium sulphate fractionation, DEAE cellulose and hydroxyapatite columns.

Specific activity: 560 units (mg protein)$^{-1}$
Enzyme concentration: 1892 units ml^{-1} in 0.05 M potassium phosphate,
pH 6.8, 0.01 M 2-mercaptoethanol, 35% ethylene
glycol
A_{280}/A_{260}: 1.10
Protein concentration: 3.38 mg ml^{-1} (based on A_{280}/A_{260})
Enzyme unit: Defined as the amount of enzyme required to convert
10 nmol of triphosphate deoxynucleotide to acid insoluble
material in 30 min in the presence of Poly(da-dT) at 37°C.

STORE AT −20°C   DO NOT FREEZE

Enzyme assay:	(0.03 ml)	
	Tris-HCl, pH 7.4	20 μmol
	$MgCl_2$	1 μmol
	2-mercaptoethanol	0.3 μmol
	dATP-[^{14}C]	20 nmol
	dTTP	20 nmol
	Poly(dA-dT)	0.2 A_{260} units
Incubation:	10 min at 37°C	
Enzyme diluent:	0.05 M Tris-HCl pH 7.5, 0.01 M 2-mercaptoethanol,	
	1 mg ml^{-1} Bovine Serum Albumin	
Reference:	Harwood, S. J. (1970). PhD Thesis, University of	
	Wisconsin, USA	

Fig. 15.7 Typical 'credential' labelling for a fine enzyme as shipped. Reprinted by permission of Miles Laboratories Inc., Research Products Division, 1127 Myrtle Street, Elkhart, IN 46514, USA.

to ensure that the vials are always in contact with the coolant. The coolant
and enzyme should be in an insulated container and the package must
be labelled as fragile, with any precautionary statements (relating to human
exposure) on all sides of the outer package.

424   *Chapter 15*

Reading List

Belter, P. A. (1979). *In* 'Microbial Technology, Fermentation Technology', 2nd Edn, Vol. II, (Eds H. J. Peppler and D. Perlman), pp. 413–418. Academic Press, New York, London.

Blumenthal, T. (1979). *In* 'Methods in Enzymology', Vol. LX, Part H, (Eds Kivie Moldave and Lawrence Grossman), pp. 628–638. Academic Press, New York, San Francisco, London.

Chang, T. M. S. (1980). *In* 'Enzymes, the Interface between Technology and Economics', (Eds James P. Danehy and Bernard Wolnak), pp. 123–130, Marcel Dekker, New York.

Charm, S. E. and Matteo, C. C. (1971a). *In* 'Methods in Enzymology', Vol. XXII, (Ed. William B. Jakoby), pp. 482–487, 528–530. Academic Press, New York, London.

Cuatrecasas, Pedro., Wilchek, Mier and Anfisen, Christian B. (1968). *National. Acad. Sci. (USA) Proc.* **61**, 636–643.

Darbyshire, J. (1981). *In* 'Topics in Enzyme and Fermentation Technology', **5**, pp. 150–151, Ellis Horwood Ltd. in Chichester, England, distributed by John Wiley and Sons, Halsted Press Division.

Davis, N. D. and Blevins, W. T. (1979). *In* 'Microbial Technology, Fermentation Technology', Vol. II, 2nd Ed., (Eds H. J. Peppler and D. Perlman), pp. 303–329, Academic Press, New York, London.

Dixon, M. and Webb, E. C. (1979). 'Enzymes', 3rd Ed., 1–37, Academic Press, New York, San Francisco.

Elander, R. P. and Chang, L. T. (1979). *In* 'Microbial Technology, Fermentation Technology', (Eds H. J. Peppler and D. Perlman), Vol. II, 2nd Ed., pp. 243–302, Academic Press, New York, London.

Gaal, O., Medgyesi, G. A. and Vereczkey, L. (1980). *In* 'Electrophoresis in the Separation of Biological Macromolecules', pp. 199–212, John Wiley and Son, Chichester and Akademiai kiado, Budapest.

Kirkegaard, L. H. (1976). *In* 'Methods of Protein Separation', Vol. 2, (Ed. Nicholas Catsimpoolas), pp. 279–319, Plenum Press, New York, London.

Kula, M. R., Kroner, K. H., Stach, W., Hustedt, H., Durelovic, A. and Gland, A. S. (1981). *In* 'Industrial Enzymes from Microbial Sources, Recent Advances', (Ed. M. G. Halperin), pp. 321–324, Noyes Data Corporation, Park Ridge, New Jersey.

Penefsky, Harvey S. and Tzagoloff, Alexander (1971). *In* 'Methods in Enzymology', Vol. 22, (Ed. William B. Jakoby), p. 210, Academic Press, Inc., New York, London.

16 | Production of Antibiotics

K. CORBETT

16.1 Introduction

Although 'penicillin' was discovered by Fleming in 1929, it remained an object of research interest until methods for its large-scale manufacture were devised in the early 1940s. Attempts to make use of the antimicrobial activity of penicillin were restricted to isolated incidents because insufficient quantities were available. Mainly as a result of the Oxford group led by Florey, the potential of penicillin was realized and spurred on by the necessity of developing a treatment for wartime casualties, a full-scale Anglo-American development was set up to turn this interesting research material into the first commercial antibiotic.

Viewed in the light of today's commonplace manufacture and use of antibiotics, it is all too easy to forget the problems encountered by the pioneers of this industry. The prospective newcomer would do well to read the account by Elder of the history of penicillin development and be thankful for the perseverance of these early workers. Initially penicillin was produced in milk bottles, simply because the technology already existed for handling and filling large numbers of these containers. It was not long, however, before the necessary technology was developed enabling deep tank fermenters to be used, resulting in a great leap forward in process efficiency and productivity.

The lesson, still not universally accepted, is that scientific discovery alone is insufficient to bring about radical change. Equally important is the development of suitable technology to enable the full commercial potential to be exploited. The successful manufacture and development of penicillin G demonstrated that microorganisms could produce medically useful compounds and also initiated the search for new antibiotics which has continued to the present day. In total, more than 5000 such compounds have been isolated, by far the largest proportion is produced by *Streptomycetes*. A relatively small number have found a role in therapy but hundreds of tonnes of the more popular products are produced each year, the total value of which approaches £750 million worldwide with penicillins and cephalosporins accounting for more than 70% of the market. These are used either in their natural form or as semisynthetic derivatives. The latter development was made possible by the isolation of the penicillin nucleus and its subsequent chemical modification. The search for new naturally occurring antibiotics is now accompanied by an equally determined chemical effort to produce new semisynthetic derivatives.

Antibiotic manufacturers are naturally jealous of the technology they use to produce their products; therefore, this chapter cannot, nor would it wish to, give a comprehensive account of all methods used to produce antibiotics, neither can it give a definitive account of any one manufacturer's process. The aim is to give an account of procedures which are well known and commonly used in the industry.

Every manufacturer uses a different process. The technology used will be derived from in-house research and development, from published methods, and from licensing arrangements with other manufacturers and consulting companies. Traditionally, each manufacturer has developed its own strains. However, since 1971, it has also been possible to buy the rights to use strains developed by independent organizations specifically set up to produce strains which are competitive with existing commercial strains.

The corollary of using dissimilar strains is the need for different operating conditions and media in order to obtain optimum productivity. The major components in the media used will probably be similar, but their relative proportions will vary and a variety of minor components may be used by different manufacturers. Almost all antibiotics are produced in mechanically agitated and aerated vessels, the design of which is of crucial importance to the process. The most important aspect for all commercial processes is the economics of the operation. Unless a profit can be made no manufacturer will stay in business for long, particularly in a competitive business like the antibiotic industry. Over the years the number of pencillin manufacturers has decreased as the drive to greater efficiency has intensified. This trend to greater efficiency has been particularly apparent since the dramatic increase in energy costs in the early 1970s. The emphasis on limiting process costs will be apparent in many of the aspects discussed in this chapter and should underline the overriding importance of this aspect of any biotechnological process.

16.1.1   Outline of the production process

A schematic representation of the key steps of a typical antibiotic production process is shown in Table 16.1. This illustrates the type and scale

Table 16.1   Key steps in a typical antibiotic production process

Scale and type of operation	Process step	Support work
Lyophilized Dried on soil or silica gel Slopes Preserved in liquid N_2	**Culture preservation** ↓	Strain improvement Method development
<10 L Flasks Steel inoculum cans	**Inoculum preparation** ↓	Test for contaminants Medium development
<20 m³ Stainless steel fermenter	**Seed stage** ↓	Medium development Process development Test for contaminants Biochemical and microbiological monitoring

10–300 m³
Stainless steel fermenter
	Final stage ↓	Medium development
		Process development
		Test for contaminants
		Biochemical and microbiological
		   monitoring
		Equipment design and
		   development
		Assay service

Rotary drum filtration Belt filtration Membrane filtration Centrifugation Whole broth extraction	**Harvest** ↓	Process development Equipment design and development Assay service
	Extraction and purification	

of each step, together with an indication of the associated analytical and development support necessary to achieve the optimum process performance.

A high yielding strain is the essential prerequisite of any good antibiotic-producing process. Continuous modifications of process conditions result in a steady improvement in efficiency, but the large steps forward in improved productivity usually come from the introduction of a superior strain. Thus, antibiotic manufacturers constantly strive to develop higher yielding strains. Having done so it is of paramount importance to retain the characteristics and viability over long storage periods. Although strain improvement departments would hope to provide a continuous supply of improved strains, provision must be made for the lean periods when improved strains do not emerge. The selected strain, therefore, must be preserved for possible long-term storage by methods which have been proved to maintain the desired properties. A full discussion of strain improvement and preservation methods can be found in Chapter 8.

The preserved culture is a valuable asset and as such it is used as sparingly as possible. Therefore, it is undesirable to inoculate a large production fermenter directly with the large quantities of preserved material that would be necessary to ensure satisfactory growth. Usually the smallest quantity possible is taken and the amount of material is amplified by growth on solid surfaces or in liquid culture (Fig. 16.1). The resultant inoculum is generally in the form of a spore suspension which may be prepared several days in advance. Occasionally with certain organisms it may be necessary to use a vegetative inoculum which would be prepared immediately prior to transfer into the fermenter.

In a commercial operation one aims to have the large final stage fermenters producing antibiotics for the maximum amount of time. However, at the start of the fermentation a certain amount of time is required for growth of the organism when very little antibiotic is produced. To optimize the productivity of the final stage fermenter it is usual to carry out part of the growth phase in a separate seed stage fermenter. Depending on the process and the strain, more than one seed stage may be necessary but in all cases the aim is the same; to produce the maximum amount of biomass in a suitable metabolic state to get the final stage off to a flying start.

Typically, the initial conditions for the final stage are designed to continue the rapid multiplication of the biomass which has been transferred from the seed stage. However, antibiotics are secondary metabolites; therefore to obtain maximum productivity it is necessary to limit this growth and switch the organism into secondary metabolism. This is usually achieved by designing the medium so that a key nutrient is exhausted at the appropriate time, producing the desired metabolic switch. The choice of limiting nutrient is important and varies between processes; glucose in the case of penicillin production and phosphate for a number of antibiotics produced by *Streptomycetes*. Figure 16.2 shows a typical profile for an antibiotic fermentation.

A simple batch fermentation as described above is the type of process that would be used initially for the production of a new product. In the case of some bacterial fermentations it has not been possible to improve on this simple procedure. However, in many fungal and streptomycete

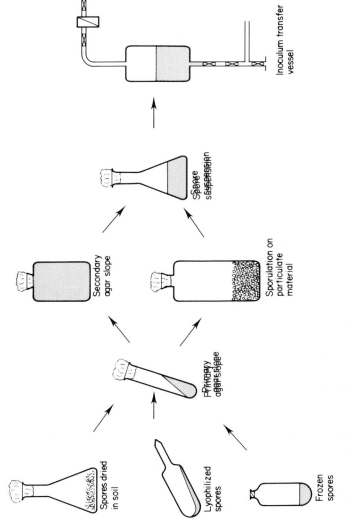

Fig. 16.1 Inoculation preparation.

Spores dried in soil

Lyophilized spores

Frozen spores

Primary agar slope

Secondary agar slope

Sporulation on particulate material

Spore suspension

Inoculum transfer vessel

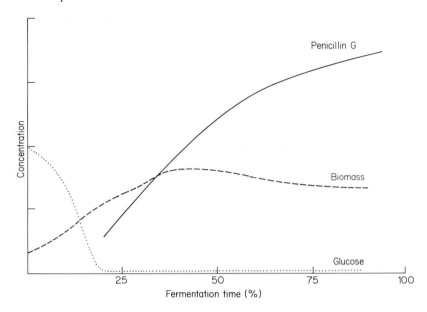

Fig. 16.2 A typical profile of an antibiotic production fermentation showing the limitation of an essential nutrient.

fermentations a fed-batch process is operated in which the productive phase is extended by feeding nutrients throughout the fermentation. This is carried out in a carefully controlled manner such that the maintenance requirements of the organism are provided but further growth is restricted. As far as is known no antibiotic is produced commercially by continuous culture, although this procedure is used extensively for experimental work in the industry.

At the end of the fermentation the broth contains not only antibiotic material and microorganisms, but also a large number of other chemicals. In fact, the antibiotic is only a minor component of the complete medium (between 3 and 35% of the total solutes in the broth). The initial step in their recovery and purification is usually separation of the liquid phase from the microorganisms by filtration or centrifugation. However, in certain cases it is preferred that the antibiotic is extracted directly from whole broth. Downstream techniques used for isolation and purification of antibiotics may include solvent extraction, ion exchange, ultrafiltration, reverse osmosis, precipitation, and crystallization (see Chapter 6). Finally, the bulk product is dried and used for preparation of derivatives or for the manufacture of pharmaceutical preparations.

The following section contains a more detailed account of the key steps and procedures involved in antibiotic production. Many of the examples used as an illustration are drawn from the production of penicillins.

The importance of pencillin fermentations cannot be overstated. Not only were they the first antibiotics to be produced commercially but they are still of prime importance today; the annual production of Penicillin G exceeds 10 000 tonnes. There has been continuing development of the fermentation processes for more than forty years so that they are among the most sophisticated of all antibiotic fermentations. As such they provide excellent examples of the generally important features of antibiotic production. Although in their natural form they are used to a limited extent, penicillin G is the raw material from which semisynthetic pencillins are produced and in some processes penicillin V is used as the starting material for ring expansion to cephalosporins.

16.2 Inoculum preparation

The preserved culture stocks are a valuable asset; for this reason the smallest quantity possible is taken and the amount of material is amplified by growth on solid surfaces or in liquid culture (Fig. 16.2). However, it should be remembered that the greater the number of stages between preserved material and the final stage the greater the chances of the organism losing productivity, particularly in the case of the highly mutated and often unstable production strains.

A sample of the preserved stock is usually spread on agar slopes made up from a medium specifically designed for this stage. The slopes are incubated under carefully controlled experimental conditions to encourage consistent growth and sporulation. In all stages of inoculum preparation, consistency is most important since a production process is run on a regular basis and any interruption due to inconsistent sporulation will have dramatic consequences on the schedule of the production department.

Having produced large quantities of spores an efficient harvest procedure is required. Spores from liquid culture may be collected by filtration through glass wool to remove vegetative material. The irrigating medium used to recover spores from solid surfaces does not have to support growth, therefore, a simple salts solution may be used. It may contain a surfactant to prevent clumping of spores, thus providing a consistent number of growth centres. Many surfactants can damage microorganisms, therefore great care is required in selecting one for this purpose.

Depending on the process, the spore suspension may have to be used within a few days; alternatively, it may be possible to prepare a bulk supply for use over a period of weeks. Stability on storage varies from strain to strain, some losing viability rapidly, others being hardly affected after six months or more. It is usual to impose a limit on the length of time they may be stored to avoid any risk of deterioration. Whichever method is used it is essential to ensure that the suspension is not contaminated before it is used in the production process. Consequently sterility plates and tubes are prepared, usually at 24, 30 and 37°C, and periodically a productivity test would be carried out in shaken flasks. If the suspension is stored for a long period, regular viability tests would be performed to ensure that no deterioration has occurred.

The final step of inoculum preparation is the transfer of the spore suspension into a container suitable for attachment to the fermenter. In the case of laboratory fermenters it is satisfactory to transfer from a glass flask carrying a side-arm. This is not the case with large production fermenters. Usually these are held above atmospheric pressure and there may be a danger of the glass vessel exploding. This can be avoided by depressurizing the fermenter, but the risk of contamination is then increased; it is therefore usual to use a metal container for the transfer process. Figure 16.3 shows

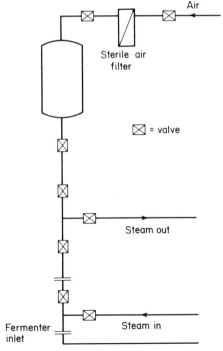

Fig. 16.3   Spore transfer vessel.

a typical spore transfer vessel. This type of vessel has the added advantage that the tube through which the spore inoculum passes may be steam sterilized after attachment of the container to the fermenter. The spore suspension may be blown into the fermenter using sterile air, or if the transfer tube is wide enough and the main fermenter depressurized, the spore suspension may be allowed to run in under gravity. Adequate time must be allowed for the tube to cool following steam sterilization otherwise damage to the spores may occur. If this technique is followed carefully the risk of contamination arising from the transfer to the fermenter is virtually eliminated.

16.3   Design of the fermentation plant

A typical antibiotic fermentation plant is illustrated in Fig. 16.4. Although batching (medium make-up) vessels and sterilizers are not essential to the operation, since medium may be batched and sterilized in the fermenter itself, they are usually included to improve the economics of the operation by ensuring optimum use of the fermenters. A single batching vessel and sterilizer can service a number of fermenters, thus allowing each fermenter to be used for the maximum time in producing the desired antibiotic. Shortening the time between successive fermentations allows a greater number to be performed each year, thus increasing the overall yield of antibiotic from each vessel.

The sterilizer may be either a batch sterilizer or a continuous sterilizer. Traditionally a simple batch sterilization vessel has been used but in more recent times continuous sterilizers have become fashionable, initially because they did less damage to the medium, as illustrated in Chapter 7, but even more importantly of late, substantially less energy is required to sterilize an equivalent volume of medium.

The seed and final stage fermenters must be capable of operating under aseptic conditions as must the feed vessels which are used to supply nutrient or precursor materials during the course of the fermentations. Harvesting of the antibiotic requires removal of the organism, which is achieved by use of a filter or a centrifuge. Recovery of antibiotics is a specialist topic which will not be discussed further in this chapter except to say that an efficient recovery process is as essential as an efficient fermentation process if the antibiotic is to be produced at an economically viable cost.

The most important vessel is the final stage fermenter itself. The potential productivity of the organism will only be achieved if the fermenter is correctly designed and constructed. There are a number of criteria which must be met when designing a fermenter. There must be both a high gas/liquid

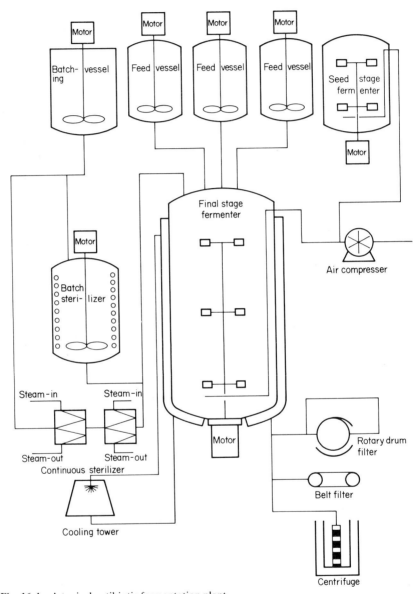

Fig. 16.4   A typical antibiotic fermentation plant.

mass transfer to ensure an adequate supply of oxygen, and a high inter-
phasial nutrient transfer to ensure that nutrients are rapidly distributed and
supplied to the organism. There must be good bulk mixing characteristics

to ensure homogeneity within the vessel. Fermentations produce heat and the larger the volume the greater the heat load; therefore the vessel must be a good heat exchanger. It is essential to ensure that unwanted organisms do not enter the vessel and in this era of genetic engineering and concern for health and safety, it is important to contain the organism within the vessel. To assist with aseptic running conditions it is usual to use a vessel which can be pressurized. Finally, it is desirable to design the vessel and its agitation system so that the organism is not unduly damaged during the fermentation.

Despite the proliferation of fermenter types, most commercial antibiotic fermentations are carried out in fermenters of traditional design (Fig. 16.5)

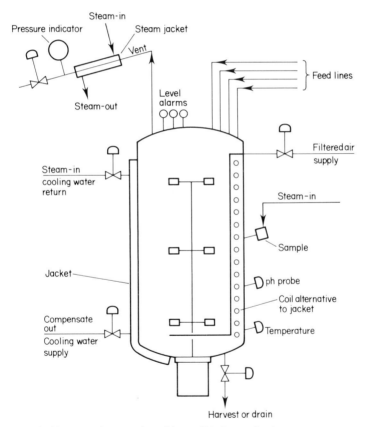

Fig. 16.5   Typical fermentation vessel used for antibiotics production.

capable of holding between 30 and 200 m³ of liquid. The vessels are usually constructed of stainless steel with a high quality smooth finish to the internal surfaces. To ensure that there are no crevices in which unwanted materials, including contaminating organisms can lurk, great attention is paid to points where pipework or probe holders enter the vessel and any welding marks must be polished to a flat smooth surface. The height of the vessel is between 2 and 4 times the vessel diameter and a number of vertical baffles (4–6), approximately 1/10th of the vessel diameter, are fitted to the vessel walls.

Agitation of the broth in most cases is via a series of flat-bladed (Rushton) impellers, although at least one penicillin manufacturer has used a simple air-lift system. To improve bulk mixing characteristics the top impeller might be of marine, skew-bladed or patented design, e.g. Sabre impeller. Great care must be taken in designing the sealing system around the impeller shaft to prevent the entry of contaminants, or in the case of a bottom drive system, the escape of broth. Sterile filtered compressed air is sparged into the vessel immediately below the bottom impeller at a rate between 0.3 and 1.5 vvm depending on the particular process requirements. The total power consumption of a large production fermenter of this type will be in the range of $1–4 \, W \, l^{-1}$.

Heat generated by the agitation system and by the organism as it metabolizes nutrients, must be rapidly dissipated if the fermentation is to be maintained at its optimum temperature, typically 24–26°C for most antibiotic-producing organisms. The heat produced is removed by the use of cooling water. The amount required depends on the rate of heat production and the temperature difference between fermentation broth and cooling water. Since large volumes of water are required and commercial organizations have to purchase water from a utility company, the water is recirculated via a cooling tower system reducing its temperature to between 15 and 20°C, depending on local conditions. Using water of this temperature it is possible to cool fermentations of up to approximately $100 \, m^3$ with a jacket or external limpet coil. For larger volumes the surface area available for heat transfer is insufficient unless very expensive chilled water (4–8°C) or chilled brine (0– −4°C) is used; therefore for larger vessels a more efficient internal cooling system is employed. Many smaller vessels may be supplied with an internal coil, particularly if they are used for *in situ* medium sterilization, since the temperature can be reduced more quickly to the normal working temperature following sterilization.

Entry or exit points from the vessel are potentially hazardous areas, particularly the air vent system, since in many cases it may not include a filter. Ideally it will contain a heated section to prevent organisms growing back along surfaces and will be designed to drain any condensate away from the vessel. It is usual for the vent system to include a control valve to maintain a constant head pressure within the vessel. Keeping the vessel above atmospheric pressure has the double advantage of impeding the entry of contaminating organisms from outside and increasing the oxygen transfer potential of the system. It is obvious that sample, transfer, harvest and drain lines must be fitted with appropriate valves to control inflow and outflow of materials as desired. These lines are provided with steam-in and steam-out points so that before operation of the valve the appropriate section of line can be sterilized. For high contamination risk systems it may be deemed prudent to maintain a continuously steam-sealed line throughout the course of the fermentation. This results in increased steam usage, and therefore higher fermentation costs, thus supplying an incentive for proper design and maintenance of valves and pipework.

The fermenters are fitted with ports allowing entry of various probes, traditionally for pH measurement and foam detection. Dissolved oxygen measuring probes have become common and penicillin probes have been reported. It is necessary to sterilize the probes before they are inserted into the vessel and several patented devices are available which also allow withdrawal and aseptic replacement during the fermentation.

Some antibiotic fermentations, notably penicillin production, require the addition of materials throughout the process. The long established method of addition is via a burette system which fills to a set volume and discharges into the vessel on a time basis. Although this is suitable for feeding nutrients it is not entirely satisfactory when adding acids or alkalis for pH control, and so other systems based on metering pumps or flow meters of various kinds have been introduced. Whichever system is preferred it must be capable of performing reliably over long periods of time, and most importantly, it must withstand sterilization on a routine basis.

Large fermenters also incorporate a number of small items specifically included to assist with routine operations. These include a manhole, allowing access to the vessel for cleaning or maintenance, and a sight glass for observing events within the vessel. In conjunction with the latter there will be some form of internal lighting since it might be necessary during a run to determine the progress of the fermentation by observation of the vessel contents. Finally, some means of measuring the volume of the vessel contents is needed. This may range from the simple expedient of using marks within the vessel, e.g. rungs of an internal ladder, notches on the baffle, etc., to the more sophisticated systems of internal differential pressure cells or mounting the vessel on external load cells.

The cost of installing a new fermentation plant is very high, and so having designed and installed the plant it is necessary to ensure that it operates for the maximum amount of time. For this to happen it is important to keep electromechanical failures and contaminations to a minimum. A specialist maintenance team operating a routine schedule should ensure that equipment failures are kept to a minimum and a suitable cleaning procedure should be adhered to between fermentations to reduce the chance of contaminations. This might entail washing the vessel first with an acid or alkaline cleaning fluid, then with water, steam sterilizing the empty vessel, cooling and pressurizing with air so that each valve and joint may be leak tested to ensure that the vessel is in a satisfactory state to receive medium for the subsequent fermentation.

16.4   Production media

A whole family of media are required for operating an antibiotic producing process (see Table 16.2); each specifically designed for a particular stage of the process. Reference has already been made to various media used in the strain selection and culture maintenance operation. Only small quantities of these media are required compared with the large volumes of seed and final stage medium used in the production plant itself. As the scale of the operation increases, greater emphasis is placed on controlling the cost of raw materials since this can significantly affect the price of the final product. All manufacturers spend considerable time and effort in developing media both for increasing productivity and for reducing costs. However, these are not the only criteria taken into account when selecting raw materials; all nutrients must be readily available throughout the year, they must be consistent in composition and they must be stable on storage.

A typical production medium is a complex heterogenous mixture, with a dry solids content of about 10% by weight. It must supply to the organism sources of energy, carbon, nitrogen, phosphate, trace elements and in some cases specific precursors for the biosynthesis of the antibiotic such as phenylacetic acid in the production of penicillin G. Although many commercial strains will grow and produce their specific antibiotics in a defined medium, productivity is much higher on complex media. The nitrogenous sources used are cheap protein-containing preparations such as cornsteep liquor, soya flour and fish meal. They may also supply other materials, often unknown, which stimulate growth or productivity. Additional cheap simple nitrogenous materials, e.g. ammonia, urea, nitrate, etc., may be used to supplement the protein source, but alone they do not result in satisfactory productivity. Technical grade glucose or starch-based products are usually used as the carbon and energy sources while phosphorus is supplied in the form of a salt of phosphoric acid. In many cases small amounts of yeast extracts or similar materials are added because they have been shown empirically to have a favourable effect on the fermentation.

Having developed suitable media for the various stages it is important to ensure consistency of preparation and sterilization on each and every

Table 16.2 Examples of typical medium ingredients for the various stages on a penicillin production process

Stage	Medium components	Percentage composition (w/v)
Storage of spores	Buffered salt solution	
	or glucose solution	10–20
	or glycerol solution	10–20
Growth and sporulation medium	Carbohydrate, e.g. glucose, glycerol, lactose, sucrose or molasses	0.5–2.0 either singly or in combination
	Nitrogen source, e.g. CSL, peptone, yeast extract, ammonium sulphate	0.2–1.0 either singly or in combination
	Chalk (calcium carbonate)	0–1.0
	Sodium chloride	0–0.5
	Phosphate	0–0.1
	Various salts	
	Solid support, e.g. maize grits, rice, perlite, bread or agar	
Removal of spores from solid surface	Water or buffer solution, usually containing a surfactant	
Shaken flask testing medium	Lactose	5.0–8.0
	CSL	5.0–8.0
	Chalk	0–1.6
	Ammonium sulphate	0–0.5
	Phenylacetic acid	Added daily
	or: Phenoxyacetic acid	0–1.0
Fermenter seed stage	Sugar, e.g. glucose, sucrose or lactose	3.0–5.0
	CSL	3.0–5.0 ⎱ either singly or
	Pharmamedia	2.0–5.0 ⎰ in combination
	Chalk	0–1.0
	Phosphate	0–0.1
	Ammonium sulphate	0–0.5
	Edible oil, e.g. lard oil	0–0.5
Fermenter final stage	Sugar, e.g. glucose, sucrose or lactose	5.0–10.0
	CSL	5.0–9.0 ⎱ either singly or
	Pharmamedia	3.0–5.0 ⎰ in combination
	Chalk	0–1.0
	Ammonium sulphate	0–1.0
	Phosphate	0–0.5
	Edible oil	0.3–0.8
	Phenylacetic acid or phenoxyacetic acid	0.3–0.6
	(Sugar, oil, PAA or POA and ammonium sulphate would be fed throughout the fermentation)	

occasion that a new batch of medium is prepared; this occurs at least once a day on a large production plant. Batch sterilization poses a particular problem. Heat treatment of any complex chemical mixture will inevitably result in reactions occurring between components and it is made worse by the considerable time required to heat up and cool down the large volume involved. It is essential to ensure that the process operating staff adhere closely to the standard operating conditions to ensure the minimum batch to batch variation.

16.5   Seed stage

The fermenter is usually smaller and simpler in design, in that it does not need to be equipped with feed systems. Depending on the duration of the production stage and the method of operation, one seed vessel may service between two and six final stage fermenters.

Seed stage medium is developed to encourage rapid growth but to prevent the production of antibiotics and also to prevent sporulation. Inoculum for the seed stage is carefully controlled and is either in the form of vegetative growth from flasks or spores from a solid surface, in which case a specific number of spores per unit volume will be introduced into the vessel. Incorrect levels of inoculum can lead to either inadequate growth or to an unsuitable type of growth.

Growth in the seed stage will be carefully monitored to ensure that it is transferred in the correct metabolic state for optimum performance in the final stage. In many cases this will be during logarithmic growth but this is not always the case, depending on the process and strain of organism being used. Assessments of growth and metabolic activity are many and varied and include viscosity, sediment, pH changes, time, oxygen utilization, carbon dioxide output, maximum oxygen demand, redox potential, ATP levels, DNA content, and assays of individual reactions. The most suitable system for each process must be determined empirically and once established, transfer conditions must be adhered to closely. When the seed stage attains the transfer criteria, a suitable volume, between 0.1 and 20% of the final stage volume, is transferred to the production fermenter to initiate the final stage.

In addition to monitoring growth of the desired organism, the seed stage must be carefully observed for growth of unwanted organisms. The use of heterogenous media increases the possibility of contaminating spores remaining viable after the sterilization process. The seed stage is more vulnerable to germination and growth of these contaminating organisms than the final stage which receives a heavy inoculum of a rapidly growing culture, producing an antibiotic which may kill off low levels of contaminating organisms. In the seed stage, conditions are designed to foster rapid growth and it is all too easy to find a contaminant outgrowing the desired organism. It is essential that such a contaminated seed is not transferred to the production stage medium. The loss of an occasional seed stage entails cost and time penalties, but these are small compared with the heavy loss which would result from a contaminated final stage. A high level of surveillance for possible contaminants is carried out throughout the seed stage by microscopic examination and by sterility tests on various solid and liquid phase media, usually at more than one temperature.

Although it may be possible to adversely effect productivity by poor operation of the final stage, it is rarely if ever possible to achieve good productivity if an unsuitable seed stage has been used to inoculate the final stage medium.

16.6   The production stage

The final fermentation step is the major cost centre of the antibiotic producing operation. A fermentation plant is not labour intensive and labour costs account for less than 15% of the overall costs. Energy costs for agitation and aeration and the cost of nutrients account for the major portion of the operating expenditure. While it is important to reduce production costs as much as possible, it should always be borne in mind that because antibiotics are expensive commodities, a small increase in productivity will give greater cost benefits than small savings in the use of power or raw materials. It is rare to undertake cost-saving measures that reduce productivity since any cost saving is rapidly lost if less product results.

All antibiotics are produced by batch or fed-batch fermentation methods and continuous culture has so far only been used as a tool in experimental studies. Fed-batch techniques are used to extend the duration and productivity of the fermentation. A nutrient, e.g. glucose, is added continuously throughout the fermentation, which may result in antibiotic production for up to 350 h. If all the glucose were batched into the medium at the start of the fermentation, the result would be massive growth leading to high viscosity which would adversely effect oxygen transfer and bulk mixing with resultant diminished productivity. The fed-batch method controls growth, and overcomes problems of catabolite repression.

Addition of nutrients may be made by metering pumps, flow meters or burettes. Traditionally, nutrient feeds have been programmed on a purely time basis with manual override to allow the workforce to modify the feed as required. The rate of feeding is not necessarily consistent throughout the fermentation but may change in accordance with a programmed profile. It is now possible to control feed rates of nutrient with the aid of microprocessors and several parameters have been used as the basis for such control. Thus changes in viscosity, pH, nutrient levels, dissolved oxygen and vent gas levels may be used for this purpose.

Fed-batch culture causes the broth volume to increase throughout the fermentation. Unless liquid is removed from the vessel the fermentation would have to terminate when the fermenter is full. In order to extend the fermentation beyond this time small volumes of broth may be withdrawn and the product recovered during the latter stages of such fermentations. Such mini-harvest procedures may also be used for partial replacement methods in which fresh medium is added to a fermenter, again to extend the period of productivity. As well as increasing the volume, and hence the amount of antibiotic, both techniques have the added advantage of maintaining the organism at its optimum growth rate for antibiotic production.

A characteristic of protein-containing media is their tendency to foam, particularly in the case of fermentations where the media is vigorously agitated and aerated. Maximum productivity can only be achieved if the full capacity of the fermenter is used. A thick head of foam on the media reduces the vessel capacity, therefore it is necessary to control the degree of foam formation. Foam control is usually achieved either by incorporation of specific chemical antifoam agents into the medium or addition of such agents during the course of the fermentation. The time at which a foaming problem manifests itself depends on the particular process. Some media tend to entrap air at the beginning of the fermentation before the protein is utilized; with these it is usual to incorporate an antifoam into the medium to permit the maximum volume to be held in the vessel from the inception of the fermentation. On the other hand, where a fed-batch process is operated and the vessel is only partially filled at the beginning, antifoam incorporation in the medium is not required. However, as the volume increases the need for foam control becomes more important, and an antifoam would be added at intervals either on a time basis or in response to a signal from the foam detecting probes. The antifoams used may be either a non-metabolizable type, e.g. silicones or metabolizable, e.g. vegetable oils, in which case they will also serve as alternative energy or carbon sources. As witnessed by the large number of antifoam agents available, there is no single antifoam that is effective under all conditions and the most effective material is a matter of empirical experimentation. While it is important to control foam formation effectively, the high cost of antifoams and the fact that they tend to deoxygenate the medium, reducing the dissolved oxygen level, dictates the rate of addition of antifoam. In order to limit antifoam usage a number of foam probes would be included in each fermenter and antifoam additions would only be made when it is apparent that the rate of increase of the foam level is such that, unless stopped, it will enter the vessel vent. It is essential to ensure that this does not occur since not only will it result in loss of valuable product, but also there is the added danger of contaminating the remaining vessel contents. In the dire situation that foam formation appears to be out of control the foam probes may cut off the air supply, stop the agitator and warn the operating staff that emergency action is required.

The escalating cost of energy has placed greater emphasis on reduction of electrical power consumption. Both aeration and agitation rates may be tailored to fit more closely the needs of the fermentation than was traditionally the case. Time profiled aeration and agitation programmes may be used or they may automatically be adjusted to maintain a specified dissolved oxygen concentration. If the latter technique is used the vessel will be fitted with electrodes measuring dissolved oxygen and possibly an oxygen analyser for the outlet gas. All the necessary calculations and control programmes are easily accommodated by the now ubiquitous microprocessor.

Accurate temperature control is important in obtaining maximum productivity in the final stage. Growth and productivity of the organism and stability of the final product are all sensitive to temperature. The optimum temperature for growth and productivity may be different and it has been suggested that the ideal situation would be to profile the temperature during the final stage. In practice the optimum temperature for the overall process is determined empirically and for most antibiotic fermentations it lies between 24 and 28°C. Once antibiotic accretion has ceased, the stability of the accumulated product is improved if the temperature of the broth is reduced, and it is common practice either to do this in the fermenter or to transfer the broth to a separate, simpler holding vessel which has been precooled. The second alternative is preferable, since the fermenter is free to be prepared for the next fermentation while the broth is being fed to the extraction equipment.

To enable the process to be correctly controlled, selected microbiological and chemical changes must be monitored during the fermentation. At its simplest this entails checking at regular intervals for the presence of contaminating organisms and the amount of product formed. Most manufacturers monitor a much wider range of parameters, particularly when a fed-batch fermentation is being performed with addition of a side-chain precursor, as in the case of penicillin manufacture. Addition of a suitable carbohydrate may be made until the required amount of growth is achieved, at which stage the carbohydrate level would be allowed to fall to the maintenance level required during antibiotic production. Monitoring at regular intervals allows feed rates to be adjusted, thus preventing a build-up of carbohydrate and the associated problems of catabolite repression. Nitrogen and sulphur sources are fed during penicillin production and it is necessary to monitor the broth constantly to ensure that correct levels of these compounds are maintained. The precursors of the side-chain of penicillin G and V, phenyl-acetic acid and phenoxyacetic acid must also be supplied continuously throughout the production phase. However, high levels of these compounds are toxic to the organism, making it essential that they are not allowed to accumulate or else antibiotic production will suffer. Both compounds are expensive, so it is undesirable to have excess quantities of unused precursors left in the broth at the end of the fermentation. Careful monitoring of their concentration allows the supply to match the demands of the organism.

Although the morphological characteristics of organisms used by individual manufacturers will differ from each other, each will produce a characteristic pattern of changes which will affect sedimentation properties of the organism and viscosity properties of the broth. Changes from the established pattern signifies deviation from optimum production conditions and should alert the operating staff to look for the cause of the change. Furthermore, by following changes in morphology it may be possible to predict when the fermentation should be terminated to ensure optimum conditions for harvesting the product.

It is obvious from the preceding discussion that careful control of the final stage is essential on a round-the-clock basis if optimum productivity is to be maintained. Traditionally analogue controllers have been used to maintain individual physical parameters at pre-set values. The operating staff, working on a shift rota, would monitor the values and make corrections and changes as appropriate to ensure they did not deviate from their set points. Data from each fermentation would be recorded either manually or directly on chart recorders.

The introduction of microprocessors and computers is having a dramatic effect on fermentation plant control systems. A computer may be used to supervise and adjust set points of the controlling analogue equipment much more frequently and more reliably than operating staff. Alternatively, using direct digital control it is possible to dispense with analogue controllers and use the computer directly to control individual parameters. This has the advantage of potential cost saving in equipment and greater flexibility in process control, but suffers from the disadvantage that a computer failure is more serious than with the supervisory system. When direct digital control is used it is common practice to install an emergency back-up system, at least on crucial parameters, to deal with a potential computer malfunction.

As well as being able to control precisely the fermentation and record accurately all parameters and process changes, the computer can perform more sophisticated functions. Sequences of events such as sterilization of the vessel or medium, can be controlled automatically. Moreover, it is capable of performing more intelligent functions than traditional equipment, for instance it can process, online, data from a number of physical and biochemical measurements to attain the optimum control of interacting parameters, e.g. aeration and agitation rates can be controlled by computed $k_L a$ or oxygen uptake rates. Ideally a mathematical model would be used to maintain all parameters in the optimum state; unfortunately antibiotic fermentations are as yet insufficiently understood for us to take advantage of this level of sophistication.

Coupled with improved process control the use of a computer can reduce labour, energy and raw materials usage, thus leading to a more efficient process. Realization of the potential cost benefits is resulting in the rapid introduction of computer control into antibiotic production plants.

16.7 Good manufacturing practice

It is self evident that a product meant for human consumption must be produced under clean, hygienic conditions but good manufacturing practice goes far beyond this stage. Good manufacturing practice is a whole philosophy governing the manner in which antibiotics are produced. Long before an antibiotic can be marketed it will have undergone rigorous toxicity tests. The material supplied for the tests will have been produced by a specific procedure. It is highly unlikely that the antibiotic will be 100% pure. It is therefore essential to know that not only is the antibiotic itself not toxic under the conditions of use, but also that the small amount of impurities present are safe to be administered to patients. Any deviations from the production procedure—a strain change, a raw materials change or a process operations change—could lead to differences in the final product, including the impurities. Consequently, it is extremely important to ensure that any change is carefully planned and consideration given to possible changes in the product and, if necessary, extra toxicity tests carried out before the material is approved for human use. To eliminate as far as possible the introduction of unscheduled changes it is usual to compile a standard operating procedure which must be closely adhered to by the operating staff. This would be in the form of a standard set of written instructions requiring process operators to sign the instruction sheet at the completion of each stage of the procedure, e.g. the weight of each raw material batched into the medium would be acknowledged in this way. These records would be checked by management and retained as evidence that each batch is carried out as intended. Not only is this necessary to satisfy regulatory authorities but also to ensure that the general public can have confidence in the antibiotics prescribed for their use.

16.8. Economics of antibiotic production

A detailed analysis of the economics of penicillin manufacture has been presented by Schwartz (1978). Although manufacturing figures are well-kept commercial secrets the article by Schwartz gives an indication of the distribution of costs between raw materials, labour, utilities and fixed costs. Table 16.3 indicates the typical cost ranges for the fermentation stages.

Table 16.3   Cost analysis of a typical antibiotic process

Process		% of total fermentation cost
Inoculum preparation	1–2	
Seed stage	1.5–6	
Materials		0.5–2
Labour		0.1–0.5
Steam		0.5–1.5
Electricity		0.5–2
Production stage	90–95	
Materials		45–65
Labour		5–10
Steam		5–10
Electricity		15–30

Inoculum preparation is conducted on a laboratory scale and as such contributes little to the overall cost which is dominated by the production vessel. Operating a fermentation plant is not labour intensive and this is reflected in the low labour charges throughout the operation. Materials and power costs, on the other hand, are high. Material costs range from 40% for fermentations requiring simple nutrients to 70% for those which require addition of expensive precursors.

Geographical location can significantly affect the cost of raw materials since the availability of local supplies, transportation costs and political decisions relating to importation levies will all have a bearing on the type and cost of material used. Energy costs are again subject to a number of factors, not the least of which are government policies on the cost of electricity and heating fuels to industry.

Table 16.4 shows the distribution of costs for the fermentation and the downstream processing stages. The fermentation step is the major cost factor in antibiotic production although the extraction process can contri-

Table 16.4 Cost analysis of fermentation and downstream processing

Process	% of total prime costs
Fermentation	50–75
Filtration	3–5
Extraction (solvent extraction ion exchange)	15–30
Crystallization	5–10
Drying	5–10

bute significantly, depending on the concentration of the antibiotic in the harvested broth and the complexity of the ensuing extraction methods.

The figures in the table make up the prime production cost of the antibiotic to which would be added various overheads in order to arrive at a total factory price. The overheads would include the cost of support services, maintenance engineers, analytical services, administration costs and depreciation of the capital investment in the form of buildings and equipment and research and development costs.

Even though a new antibiotic may be superior to existing products, it needs to be realistically priced in order to achieve a significant share of the antibiotic market. The ability to sell any antibiotic, therefore, depends on production at a cost which makes it competitive with other manufacturers products.

Although patent cover initially allows the manufacturer a clear field in which he can recoup monies spent on research and development of new products, the majority of antibiotics are either well established or use well established products as raw materials, e.g. semisynthetic pencillins and cephalosporins use penicillin G or V as the starting raw materials. The ability to sell well established materials depends on production at a cost which is competitive with other manufacturers. There is, therefore, a constant drive to improve the process, and over the years this has led to a steady reduction in unit costs of antibiotics. This constant need to improve means that there is always a need for scientific expertise to be applied to improving antibiotic production processes, so that for a long time to come there will be a ready market for engineers and scientists with an interest in biotechnology.

Reading list

Corbett, K. (1980). Preparation, sterilization and design. *In* 'Fungal Biotechnology' (Eds J. E. Smith, D. R. Berry and B. Kristiansen), pp. 25–42. Academic Press: London and New York.

Dobry, D. D. and Jost, J. L. (1977). Computer applications to fermentation operations. *Ann. Rep. Ferm. Processes* **1**, 95–114.

Elder, A. L. (1970). The history of pencillin production. *Chem. Eng. Prog. Symp. Ser.* **66**, 5–100. American Inst. of Chemical Engineers.

Hockenhull, D. J. (1980). Inoculum development with particular reference to *Aspergillus* and *Penicillium*. *In* 'Fungal Biotechnology' (Eds J. E. Smith, D. R. Berry and B. Kristiansen), pp. 1–24. Academic Press: London and New York.

Kirk, R. E. and Othmer, D. F. (1978). 'Encyclopaedia of Chemical Technology', Vol. 2, pp. 809–1036; Vol. 3, pp. 1–78. J. Wiley, New York.

Queener, S. and Schwartz, R. W. (1979). Penicillins: biosynthetic and semisynthetic. In 'Economic Microbiology', Vol. 3 (Ed. A. H. Rose), pp. 35–122. Academic Press: London and New York.

Sharp, J. R. (1983). 'Guide to Good Pharmaceutical Manufacturing Practice'. HMSO: London.

Schwartz, R. W. (1978). The use of economic analysis of Penicillin G manufacturing costs in establishing priorities for fermentation process improvement. *Ann. Rep. Ferm Processes* **3**, 75–110.

17 | Microbial Gums

G. W. PACE

17.1 Introduction

Many microorganisms produce large quantities of polysaccharides under a wide variety of conditions. Polysaccharides are usually thought to play specific roles, either as storage compounds, e.g. glycogen; structural compounds, e.g. chitin; or, in the case of extracellularly produced polysaccharides, as mediators in the interaction of the microorganisms with their environment. Some possible roles of exopolysaccharides include:

- protecting microorganisms against desiccation;
- acting as a barrier and preventing viruses and antibodies from attaching to specific sites on the cell wall;
- complexing and neutralizing charged toxins or toxic metal ions;
- acting as carbon and energy reserves;
- converting excess substrate into a foam which is less easily metabolized by other microorganisms;
- interacting with animal or plant cells in specific symbiotic or pathogenic relationships.

Commercially, extracellular microbial polysaccharides are of major interest as members of a class of water-soluble polymers or gums, which are used widely as thickening, gelling or suspending agents or protective colloids.

The production of gums for commercial use by fermentation compared to extraction from plants and seaweeds or chemical synthesis offers several potential advantages including:

- the wide diversity of polymers produced by microorganisms;

449

- the production of gum in reliable quantities and quality relatively independent of climatic conditions, by conducting well-controlled fermentations and using raw materials of constant quality;
- the manipulation of product composition and properties by altering fermentation conditions;
- as many microbial polysaccharides are extracellular the harsh techniques used in plant and algal gum extraction can be avoided, thus lessening product degradation during recovery.

However, the penetration of microbial polysaccharides into the water-soluble polymer market has been slow owing to high capital and energy costs that tend to make microbial gums less cost effective when compared to most other water-soluble polymers. The costs and times associated with obtaining approval for the use of novel gums in foods are also very high. In addition, the screening and selection of new microbial gums of commercial interest is difficult because of the poor understanding of the complex chemical and physical nature of polysaccharides and their relationships to the end use applications, and the marked effect of impurities on polymer performance. Thus, despite their potential advantages, only the extracellular polymers produced by *Xanthomonas campestris*, xanthan, and to a lesser extent dextran produced by *Leuconostoc mesenteroides*, are sold in significant quantities at present, although other microbial gums are attracting interest, see Table 17.1.

17.2   Structure and properties

The principle characteristic exhibited by gums is their ability to modify in particular ways the rheology or flow behaviour of solutions (Table 17.2). In addition, most water-soluble polymers are multifunctional, i.e. they exhibit a combination of properties which are essential to their effective performance in the end use application. For example, water-soluble gums used to make non-splash or cling type acid- or alkali-based cleaners, besides imparting suitable rheological properties, must also resist acid or alkali degradation. The properties of such a water-soluble polymer are determined by its chemical composition, molecular arrangement and bonding, and its average molecular weight and distribution. Microbial exopolysaccharides range from anionic through neutral to cationic polymers of monosaccharides or monosaccharide derivatives, and often contain side-groups such as acetate, pyruvate succinate, lipid-type components, organic nitrogen or inorganic ions. The degree of substitution by these side-groups have a marked effect on the polymer's properties. Structures of the more important (or potentially important) microbial polysaccharides are given

Table 17.1 Examples of typical microbial polysaccharides

Polysaccharide	Source organisms	Polymer type	Monomer units	Linkage types
Dextrans	Bacteria *Leuconostoc*, *Klebsiella*, etc.	Short-branched	D-glucopyranose	α1–6 (main chain) α1–3 (branch-points)
Scleroglucan	Fungi *Sclerotium* sp.	Short-branched	D-glucopyranose	β1–3 (main chain) β1–6 (branch-points)
Pullulan	Fungi *Aureobasidium* sp.	Linear block	D-glucopyranose	α1–4-linked trimers/tetramers linked β1–6
Alginic acid	Bacteria *Azotobacter*, etc.	Linear block poly-acid	D-mannuronic acid L-guluronic acid	β1–4 } α1–4 } in blocks
Xanthan	Fungi *Xanthomonas* sp.	Linear backbone with acidic trimer branches on alternate units	D-glucose (chain) branches: 6-acetyl-D-mannose D-glucuronic acid D-mannose-4,6-pyruvate ketal	β1–4 in chain attached α1–3 attached β1–2 attached β1

(Note: within each category the structures may be very variable, partly with the precise microbial strain and partly with culture conditions; the examples given are often merely illustrative.)

Table 17.2   Rheological properties exhibited by solutions of some water-soluble gums

Property	Description
Viscosity	A measure of the increase in the internal resistance to fluid motion.
Pseudoplasticity	Viscosity (apparent viscosity) is dependent on shear rate for fluid movement. At high rates of shear the apparent viscosity falls, i.e., the solution thins.
Thixotropy	At a fixed rate of shear the fluid becomes thinner with time.
Gel forming	The gum molecules form a network which causes the solution to set to a gel.
Yield stress	The presence of a weak gel at zero stress means that a yield stress must be applied to the solution before it will move. Solids can be suspended in such solutions.

in Table 17.1, which also indicates some of the variations within each structure type.

17.3 Biochemistry and physiology

17.3.1 Biosynthetic pathways

The biochemical pathways leading to extracellular polysaccharides on the one hand, and to dextran and levan production in bacteria on the other, are distinctly different. The synthesis of most exopolysaccharides takes the following generalized form:

    (a) synthesis, or transport into the cell, of the polymer's base sugar components as sugar 1-phosphate esters;

    (b) activation of these components by nucleotide triphosphate to form nucleotide diphosphate sugars;

    (c) where required, modification of nucleotide diphosphate sugars to other derivatives, such as activated sugar acids;

    (d) assembly on a long-chain isoprenoid lipid carrier of the activated sugars or derivatives to form the base polymer structure—the same isoprenoid lipid carrier also acts as the site for peptidoglycan (cell wall) and liposaccharide synthesis;

    (e) modification by addition of non-sugar groups to the polymer;

    (f) transport by and release from the lipid carrier to the external environment.

Several major control mechanisms can exist in these pathways. They include:

(a) substrate uptake controls;
(b) the size of the sugar nucleotide pool (determined by synthetic and hydrolase enzyme activities);
(c) the competing demands for substrates for the synthesis of other carbohydrates (such as intracellular and cell wall polymers);
(d) the availability of isoprenoid lipid carrier, with priority being given in the order peptidoglycan, lipopolysaccharide, and then exopolysaccharide synthesis;
(e) mechanisms controlling the level of variable substituents;
(f) chain elongation and release controls.

Knowledge of these controls can be useful in devising approaches to increasing both the conversion efficiency and productivity of the fermentation, as well as to altering the polymer's molecular weight and composition.

Biosynthesis of dextrans and levans is completely different and does not involve multienzyme systems, activated sugars, or lipid carrier intermediates. Their production is entirely extracellular and involves enzyme-catalysed polymerization of an oligosaccharide, which directly provides the necessary carbon and energy (Fig. 17.1). Control over the activity and mode of action of these polymerases can be exercised by manipulating the environment, and, for example, such mechanisms are exploited commercially in producing dextrans of various types using simple cell-free systems.

17.3.2 Physiology

With many microorganisms the kinetics and efficiency of polymer production, the molecular weight of the polymers, and their fine structure can all be affected by changes in growth conditions. However, the type of response observed often varies between microorganisms and probably reflects the variety of specific *in situ* roles played by different exopolysaccharides. Again, understanding and controlling the important environmental variables affecting polymer synthesis can be used to advantage in the design of an economic process; examples are given below to illustrate the types of effects encountered.

The kinetic relationships between growth and product formation are important in helping determine the most economic operating mode, such as high rate continuous culture, or alternatively batch fermentation in which the maximum cell concentration is reached rapidly, followed by slower growth with product accumulation, or some intermediate option. The specific rate of exopolysaccharide formation may be either independent of

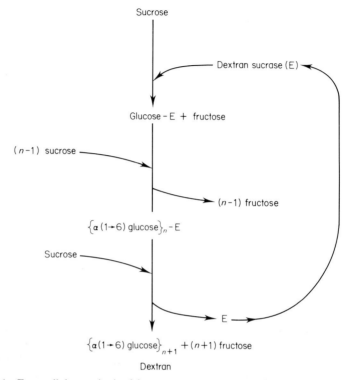

Fig. 17.1 Extracellular synthesis of dextran.

growth rate or it may increase with increasing growth rate, depending on the microorganism and the growth-limiting nutrient (Fig. 17.2).

In the microbial alginate and other exopolymer fermentations, both the conversion efficiency and the absolute value of the specific rate of product formation can be influenced by the nature of the growth-limiting nutrient. The ability of microorganisms to produce exopolysaccharide under carbon limitation also varies. For example, *X. campestris* produces copious amounts of xanthan under carbon limitation, whereas *Klebsiella aerogenes* produces no polymer.

The rate and efficiency of polymer production are also affected by other environmental variables such as dissolved oxygen, pH, and temperature, and there are some suggestions in the literature that high shear stresses can enhance polymer production by reducing the layer of polymer on the cell surface, thus improving diffusion to and from the cell.

The basic carbohydrate structure of most exopolysaccharides does not change with growth conditions, but the content of groups attached to the

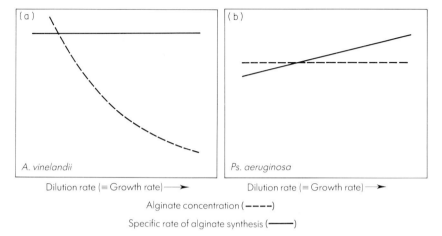

Fig. 17.2 Kinetics of exopolysaccharide synthesis. (a) *A. vinelandii*, specific rate of synthesis of alginate independent of growth rate; (b) *Pseudomonas aeruginosa*, specific rate of synthesis of alginate dependent on growth rate.

basic carbohydrate structure, such as acyl or ketal moieties, can vary widely. One interesting exception is the variation in the ratio of mannuronic to guluronic acids which make up the basic structure of microbial alginate (Table 17.1). The relative content of the uronic acids in this polymer depends on the activity of an extracellular epimerase, which catalyses the conversion of the mannuronic acid components of alginate to guluronic acid components after the alginate chain has been formed and released. The activity of this enzyme is markedly enhanced by the presence of high concentrations of calcium ions, and by controlling the calcium concentration, alginates of different compositions can be produced.

Variations in substituent groups such as acyl or ketal groups can have a dramatic effect on the rheological properties of the polymer and hence its effectiveness in various applications. For example, the pyruvate content of xanthan can be varied from almost zero to 8% (the maximum theoretical concentration) by altering the growth media. Thus it is essential in microbial gum fermentations that the design of the production media should take into account not only for the cost and desired rates of production and conversion efficiency, but also the effects on product quality. Another factor which is very important in determining polymer performance is molecular weight; the variables which control this are not understood, but the molecular weight can vary with growth conditions. For example, the intrinsic viscosity (a measure of the molecular weight) of xanthan, produced in batch fermentations on complex media, is considerably lower when the fermentation temperature is 35°C rather than 30°C. Typical ranges of molecular weight of various microbial gums are given in Table 17.3.

Table 17.3 Typical molecular weight ranges for industrially important polysaccharides

Polysaccharide	Molecular weight
Xanthan	2×10^6–5×10^7
Dextran	1×10^5–2×10^7
Pullulan	1×10^4–1×10^5
Scleroglucan	1.9×10^4–2.5×10^4
Alginate	5×10^5

17.4 Production processes

17.4.1 Fermentation

As previously indicated, the choice of growth medium, environmental conditions and growth pattern determines the product quality and the maximum attainable conversion efficiency, productivity and product concentration. However, industrial fermentation processes must also be designed such that the overall rate, efficiency and concentration of product are ultimately constrained by the heat and mass transfer capability of the fermenter, so as to make maximum use of the available energy. In exopolysaccharide fermentations the product concentration, through its effect on culture fluid rheology, has a strong feedback effect on the heat and mass transfer capability of the fermenter, and thus is of primary concern in scaling up (see Chapter 5 for more details).

Culture fluid rheology
The principal contributor to rheology of microbial gum cultures is the product dissolved in the continuous liquid phase. This contrasts with mycelial cultures, in which the main determinant of the rheology is the mycelium itself, which is discontinuous with the liquid phase. This can lead to marked differences in the effects of rheology on transport phenomena between these types of fermentations.

The main rheological feature exhibited by most polysaccharide culture fluids is extreme viscosity, at relatively low product concentrations, combined with shear thinning or pseudoplastic behaviour, which is modelled by the power law equation (Fig. 17.3). As the product concentration

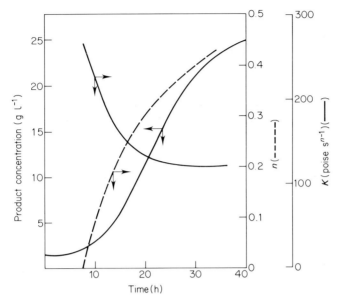

Fig. 17.3   Viscosity changes in a batch exopolysaccharide fermentation: the apparent viscosity μ_a is defined by the power law equation $\mu_a = K$ (shear rate)$^{n-1}$, where K is the consistency index and n is the flow behaviour index and is indicative of the pseudo-plastic nature of the culture fluid.

increases during batch growth both the viscosity and pseudoplastic nature of the broth increase. In some cases, such as the pullulan and *Azotobacter vinelandii* alginate fermentations, the culture fluid viscosity may decrease towards the end of the fermentation due to polymer lyase activity. Other types of rheological behaviour have also been observed; for example, first normal stress differences and yield stresses in xanthan fermentations, and viscoelastic behaviour in alginate culture fluids (particularly those containing high concentrations of calcium ions, which promote gel formation).

Fermenter design
The influence of rheology on transport phenomena in exopolysaccharide fermentations is not well understood and with a few notable exceptions, such as the prediction of power consumption in unaerated viscous pseudo-plastic fluids, proven design equations are not readily available or have not been established. Possible approaches are discussed in Chapter 5, but as this makes clear, most of the available literature centres on model system studies and their relevance to large-scale industrial fermentations is only

slowly emerging. Hence, the following discussion is limited to a qualitative description of the more significant identifiable effects.

Fermenters used for the commercial production of extracellular microbial polysaccharides are typically restricted to aerated, mechanically agitated vessels, which give the turbulence necessary for small bubble formation and mixing and hence good mass transfer (Chapter 5). The design of an efficient fermenter relies on supplying sufficient power to the impellers and correctly apportioning it between shear (or velocity head), which governs bubble formation, and impeller pumping capacity (or bulk flow), which determines the mixing and heat transfer in the fermenter. The geometry and speed of an impeller system mixing a solution of defined rheology determines the ratio of shear to impeller flow; hence impeller selection is critical. For example, in cultures exhibiting extreme rheology high-speed flat-bladed turbines can give sufficient small bubble formation, but they do not promote adequate bulk flow to distribute the bubbles throughout the vessel, or adequate bulk liquid turbulence to prevent bubble coalescence.

The reduced impeller flow or bulk mixing caused by increasing viscosity is compounded by pseudoplastic behaviour owing to the exponential increase in viscosity which occurs as the fluid slows down after it leaves the impeller. At the same time, the relatively low viscosity in the region near the impeller tends to promote channelling of gas up the centre of the vessel. In poorly mixed highly viscous solutions, which show viscoelastic behaviour, reverse flow (i.e. opposite to the normal direction) can be observed, and this may be important to transport phenomena where poorly mixed regions exist in the fermenter. The tendency to form poorly mixed or stagnant regions in agitated viscous systems is increased by the presence of a yield stress in the fluid (Fig. 17.4).

The rheology of polymer solutions partly affects mass transfer by influencing bubble interfacial area and the gas–liquid mass transfer coefficient (k_L), although it does not appear to markedly affect the diffusivity of oxygen. Interfacial area is determined by bubble shape, size and distribution, and hold-up. Bubble shape is affected by bubble velocity and liquid rheology; small slow-moving bubbles tend to be spherical or pear shaped if the fluid is viscoelastic, larger fast moving bubbles tending to be spherical cap shape. As discussed in Chapter 5, bubble size and distribution are determined by the balance between turbulence, surface tension and coalescence. The mechanism of bubble formation in aerated viscous fluids is similar to that in low viscosity systems with turbulent forces causing small bubbles to be broken off from large gas cavities attached to the trailing edges of the impeller blades. The turbulence in the vicinity of the cavity at the impeller tip is strongly affected by the local viscosity, which also influences the size

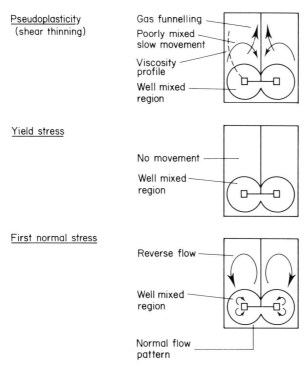

Pseudoplasticity
(shear thinning)

Gas funnelling

Poorly mixed
slow movement

Viscosity
profile

Well mixed
region

Yield stress

No movement

Well mixed
region

First normal stress

Reverse flow

Well mixed
region

Normal flow
pattern

Fig. 17.4   Effect of culture fluid rheology on fluid flow patterns.

and shape of the cavity. Bubble coalescence is promoted by increased fluid viscosity and pseudoplasticity, with trailing bubbles being caught and accelerated in the longer wake which exists behind bubbles moving in these fluids. The extent of the subsequent break-up of the coalesced bubbles is determined by the turbulence in the bulk liquid.

Rheology also influences the mass transfer coefficient (k_L) through its effect on bubble rise velocity and the local velocity profile of the liquid surrounding the bubble. The motion of small slow-moving, rigid-interfaced bubbles is affected by viscosity, viscoelasticity, pseudoplasticity and yield stress, whereas the movement of large fast-moving, free-interfaced bubbles is insensitive to changes in rheology.

17.4.2   Recovery

The recovery and additional processing of microbial exopolysaccharides closely follows the technology used in the processing of many other water-soluble polymers such as the plant and algal gums. The number of techniques available to achieve the isolation and purification of gums is large, and the actual method used depends on the objective of the processing and its cost. The usual unit operations used for the recovery of microbial polysaccharides often result in simultaneous purification and concentration. A typical scheme for the recovery of a microbial exopolysaccharide involves cell removal, precipitation, solids separation, mechanical dewatering and drying (Fig. 17.5).

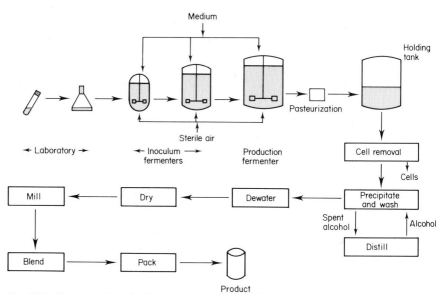

Fig. 17.5   Process outline for the production of microbial gums.

Cells may be physically removed using centrifugation or filtration, although the solution may have to be diluted or heated to lessen the effect of viscosity on the separation. Alternatively, chemical or enzymatic methods may be used to degrade cells which can then be washed out in later purification/concentration operations. Cell removal is carried out when the presence of the cells may affect the product performance, cosmetic appearance in solution or its toxicology. For example, the presence of *Azotobacter* cells in microbial alginate used in printing past formations can result in the formation of a heavy precipitate. Interestingly, commercial processes for the production of food grade xanthan do not involve the removal of the *Xanthomonas* cells from the gum, but the bacteria are rendered non-viable by pasteurization.

The favoured method for the primary recovery of microbial gums is by precipitation using water-miscible non-solvents, such as isopropanol, ethanol, methanol or acetone. In some cases, e.g. the neutral polysaccharide dextran, the addition of alcohol results in the formation of a

fluid coascervate containing about 30% gum. The Food and Drug Administration (USA) regulations on food grade xanthan gum prescribe the use
of isopropanol for its precipitation.

The major variable cost in the recovery of xanthan, as with most alcohol-
precipitated polymers, is in the cost of distilling spent alcohol from the
precipitation stage and solvent-handling losses. The amount of alcohol
necessary for the precipitation of xanthan is dependent on ionic strength
but virtually independent of gum concentration. Thus it follows that the
production cost of xanthan is very sensitive to the product concentration
attained in the fermentation. For example, doubling the product concentration will approximately halve the variable cost of alcohol recovery per
weight of product. Thus, improved heat and mass transfer to allow the
fermentation to proceed to higher product concentrations, or concentration
prior to precipitation, should reduce recovery costs.

The amount of solvent required for complete precipitation is also a function of the composition of the gum; for example, as the pyruvate concentration in xanthan increases, the amount of alcohol required to precipitate
the gum increases. Solvent precipitation also results in partial purification
of the gum by removal of alcohol-soluble components.

Several alternative methods for the primary recovery of microbial gums
are reported in the literature. These mainly centre on producing an insoluble
form of the gum by addition of certain salts or adjusting pH. For example,
microbial alginate can be recovered by forming the insoluble calcium salt
or by acidifying to precipitate alginic acid.

After precipitation, the solid is separated and where necessary re-con-
verted to a soluble form by solid or alcoholic titration with base. The product
is then mechanically dewatered and dried. Alternatively, the aqueous
extract containing the gum may be dried directly. However, because only
water is removed the resultant product is relatively impure compared with
that harvested using a precipitation method. The dried product is milled
and then packed into containers with a low permeability to water. The
particle size of the milled product has a marked effect on its dispersibility
and hydration rate.

Throughout the recovery process, exposure to heat and mechanical
stresses may result in product modification or degradation. For example,
rapid drying at high temperatures may result in a product that has poor
solubility or poor solution rheology, as can occur with microbial alginate.
Contrastingly, controlled heating of xanthan can result in a product with
improved solution rheology.

The polysaccharide may be physically or chemically treated or derivatized
during recovery to change its purity, cosmetic, rheological or other properties. Some examples of such treatments include reaction of xanthan with

glyoxal to give an easily dispersible form; treatment with propylene oxide to destroy the cellulase activity in xanthan gum; controlled degradation using polysaccharases; and formation of a copolymer by reaction of xanthan with polyacrylic acid.

Reading list

Blanch, H. W. and Bhavaraju, S. M. (1976). Non-Newtonian fermentation broths: rheology and mass transfer. *Biotechnol. Bioengng* **18,** 745–790.

Kang, K. S. and Cottrell, I. W. (1979). Polysaccharides. *In* 'Microbial Technology', 2nd Edn, Vol. 1, pp. 417–481. Academic Press: New York and London.

Lawson, C. J. and Sutherland, I. W. (1978). Polysaccharides. *In* 'Economic Microbiology', Vol. 2 (Ed. A. H. Rose), pp. 327–392. Academic Press: New York and London.

Margaritis, A. and Zajic, J. E. (1978). Mixing, mass transfer and scale-up of polysaccharide fermentations. *Biotechnol. Bioengng* **20,** 939–1001.

Pace, G. W. and Rhigelato, R. C. (1980). Production of extracellular microbial polysaccharides. *Adv. Biochem. Engng* **15,** 41–70.

Sandford, P. A. (1979). Exocellular microbial polysaccharides. *Adv. Carbohydr. Chem. Biochem.* **36,** 265–313.

Smith, I. H. and Pace, G. W. (1982). Recovery of microbial polysaccharides. *J. Chem. Technol. Biotechnol.* **32,** 119–129.

Sutherland, I. W. (Ed.) (1977). 'Surface Carbohydrates of the Procaryotic Cell'. Academic Press: New York and London.

18 | Biotransformations

CLAUDE VÉZINA

18.1 Introduction

Biotransformation is a biological process whereby an organic compound is modified into a recoverable product by simple, chemically defined reactions catalysed by enzymes contained in the cells. Biotransformation differs from fermentation in which the products, such as antibiotics, enzymes, amino acids, organic acids and solvents, result from the complex biosynthetic machinery of primary or secondary metabolism. Microbial, plant and animal cells can supply the enzymes for transformation, but microbes surpass plant and animal cells in several respects. Their high surface–volume ratio confers rapid growth and high rates of metabolism leading to efficient transformation of the substrate added. Moreover the microbial world, rich in species, provides a varied assortment of enzymes for a tremendous variety of reactions on many classes of compounds, and with the facility to adapt to the artificial environment imposed by technical and economic requirements. The substrate to be transformed can be added to growing cells, resting cells, spores, dried cells or to enzymes isolated from these cells; the transforming agents can be suspended into a medium for use in agitated aerated vessels, or packed in columns for continuous operation; they can be 'immobilized' on solid support for increased stability and prolonged operation.

Plant cells possess a repertory of enzymes for transformation (see Chapter 21 and the review by Kurz and Constabel, 1979) and can even surpass microbes in effecting certain reactions, such as the hydroxylation of cardenolides. However, plant cells grow slowly in artificial culture and the transformation of a natural or unnatural precursor into a product may require several weeks. In the past twenty years, significant advances in plant cell culture have been achieved, and industrial processes can be expected when the desired reaction is exclusive to plant cells, and the product has a high value or is required in limited quantity.

463

Animal cell cultures and organ perfusion preparations may be useful in transforming organic compounds (see Chapter 20), but are complex to operate even on the laboratory scale. The 11β-hydroxylation of deoxycorticosterone to corticosterone (Fig. 18.1) led to the first industrial process

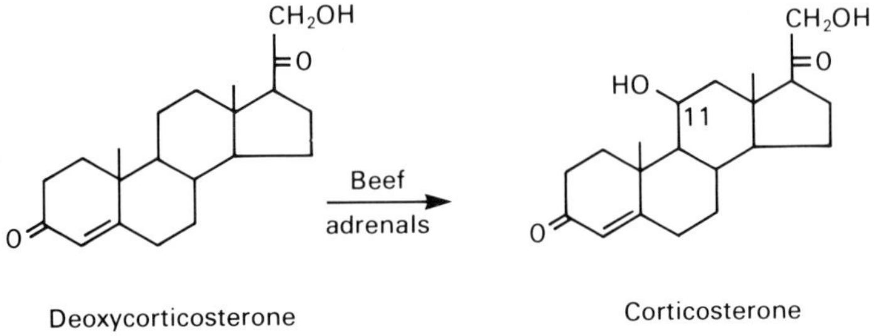

Deoxycorticosterone   Corticosterone

Fig. 18.1   Hydroxylation of deoxycorticosterone to corticosterone.

for the biological preparation of hydrocortisone (cortisol) from Reichstein's compound S, but this observation also prompted microbiologists to search for microbes capable of 11β-hydroxylating steroids, and the process with adrenals was soon replaced by the much more efficient microbial process (see below).

18.2   The world of biotransformations

Biological systems obey the laws of chemistry! Nevertheless, biotransformations often are preferred to chemical processes when high specificity is required, to attack a specific site on the substrate and to prepare a single isomer of the product. This accounts for the high yields typical of biological conversions, which often exceed 90% with microbial cells. Biotransformations proceed at ambient temperatures (20–40°C) and pressures (0–34 kPa), an advantage over chemical processes which often call for significant input of energy, and are generally run in aqueous media which may pollute less than the organic solvents used in chemical conversions. On the other hand the dilutions at which biotransformations are operated constitute a real disadvantage.

18.2.1   Preparation of intermediates and end-products

The souring of wine into vinegar was practised in Babylon by 5000 B.C., but it was only in 1864 that the role of microbes in acetification was perceived (Pasteur). In the latter part of the nineteenth century, a systematic approach to microbial transformations was initiated by Bertrand (1896), and by the turn of the century, the following reactions had been found:

Oxidation:	*Reduction:*
Ethanol to acetic acid	Malic acid to succinic acid
Glucose to gluconic acid	Fructose to mannitol
Polyhydric alcohols to	
corresponding sugars	
Hydrolysis	*Resolution* of racemic mixtures of:
Tannin to gallic acid	Tartaric acid (famous experiment of
Di- and tri-saccharides to	Pasteur)
constituent monosaccharides	Lactic, mandelic and glyceric acids

Several transformations were soon added, including the oxidations of iso-propanol to acetone, glucose to 2-ketogluconic acid, glycerol to dihydroxy-acetone, and D-sorbitol to L-sorbose. The observation by Mamoli and Vercellone (1937) that fermenting yeast can reduce 17-ketosteroids to 17β-hydroxysteroids (Fig. 18.2) led 15 years later to the most dramatic applica-

Fig. 18.2 Reduction of 17-ketosteroids to 17β-hydroxysteroids, applied in the manufacture of steroid hormones.

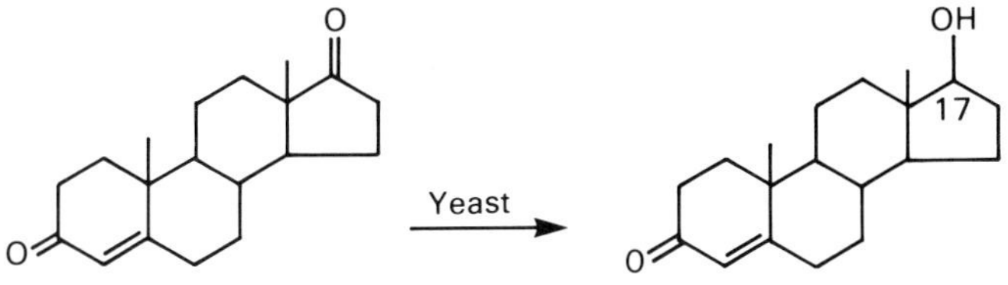

tion of biotransformation, the manufacture of steroid hormones.

Nowadays, microbes are used as chemical reagents for the preparation of key intermediates needed in organic synthesis (e.g. 6-aminopenicillanic acid by microbial deacylation of penicillins G and V for use in the synthesis of semisynthetic penicillins) and for the economic production of end-products, especially steroid hormones. The successful biotransformations of steroids provided an impetus to the development of many transform-ations that have operated or are still operating in industry: antibiotics [see Chapter 16 and review by Sebek (1981)], natural and synthetic antitumour agents, resolution of racemic amino acid mixtures and preparation of organic acids (see Chapter 13).

18.2.2 Degradation of organic compounds

Another facet of microbial transformation embraces the degradation of fungicides, insecticides, and herbicides. Here the objective is not to prepare useful, recoverable products, but to seed contaminated areas with microbes capable of converting pollutants into innocuous compounds, sometimes carbon dioxide and water, in an attempt to improve the environment. Degradation will not be discussed in this chapter.

18.2.3   Other applications

Microbes serve as a source of enzymes for use in analytical systems, and microbial transformation plays an important role in the metabolism studies that must accompany the clinical evaluation of drugs. Microbial models are useful in predicting, and sometimes necessary for preparing, the metabolites of drugs administered to animals and to man.

This chapter outlines, as specific examples, the microbial transformation of D-sorbitol into L-sorbose and the role of microbes in the preparation of steroid hormones. A brief account of microbial models of mammalian metabolism is also given.

18.3   Biotransformation of D-sorbitol to L-sorbose

Studies on the oxidation of polyhydric alcohols to sugars led to the formulation of the 'Bertrand rule', which states that only polyhydric alcohols with *cis* secondary hydroxyl groups adjacent to a primary alcohol can be oxidized to corresponding sugars by *Acetobacter* species. Among several alcohols that conform to the rule only D-sorbitol and glycerol are used on an industrial scale: D-sorbitol is transformed into L-sorbose, using *Acetobacter suboxydans* (Fig. 18.3), and glycerol to dihydroxyacetone, using *A. suboxydans*, or, better, *Gluconobacter melanogenus*. L-Sorbose is produced at the rate of 35 000 tonnes annually as the key intermediate in the chemical synthesis of L-ascorbic acid (Vitamin C), while dihydroxyacetone is mainly used as a suntanning agent.

The current industrial process for L-sorbose is probably based on the procedure first decribed by Wells and coworkers in 1937. A medium comprising glucose, yeast extract and a slight excess of calcium carbonate is supplemented with 15–30% of D-sorbitol, inoculated with a suspension of active cells of *A. suboxydans*, and incubated at 30°C. Dry yeast extract can be replaced by corn steep liquor, provided octadecanol is used to control foaming. A fermentor with vigorous aeration and agitation is crucial for maximum efficiency, which is assisted by using oxygen-enriched air or by operating the vessel under excess pressure. Yields of 90–95% can be obtained in one or two days from D-sorbitol concentrations of 20–30%, while on a laboratory scale a continuous transformation, using immobilized

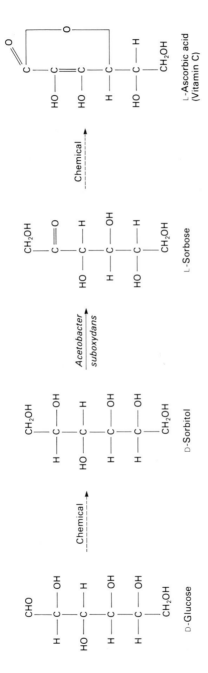

Fig. 18.3 Transformation of sorbitol to sorbose in the manufacture of Vitamin C.

cells, was reported to lead to higher yields in shorter process times. The L-sorbose is recovered by filtering the beer and concentrating the filtrate under vacuum to a syrup, which crystallizes on cooling. The crystals are washed with ice water and dried, and a second crop can be obtained from the mother liquor. The overall recovery is about 65% based on the starting material.

18.4   Microbial transformation of steroids and sterols

18.4.1   The necessity of biotransformation

The crucial role of microbes in steroid synthesis has been in connection with the synthesis of the adrenocortical hormones, corticosterone (Fig. 18.1), cortisone and hydrocortisone (Fig. 18.4), and their therapeutically

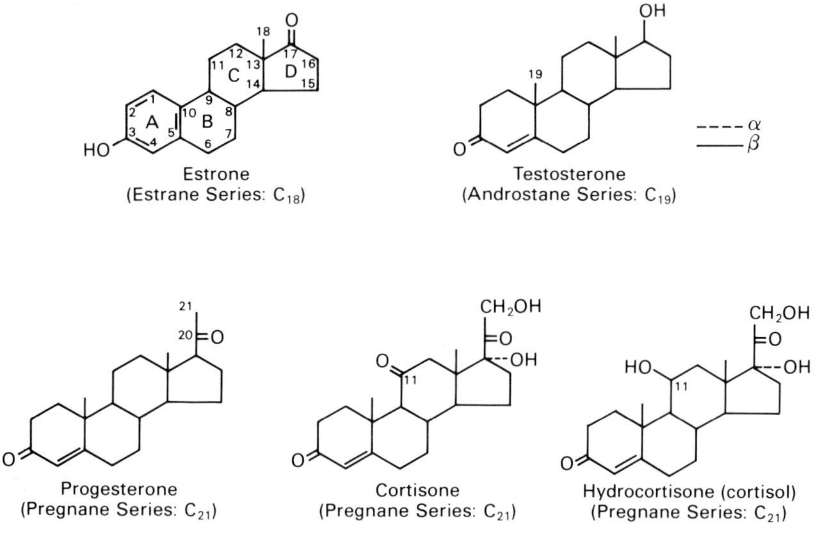

Fig. 18.4   Steroid hormones of estrane, androstane and pregnane series.

superior derivatives, such as prednisone, prednisolone, triamcinolone, etc. It centres on the introduction of an oxygen atom at carbon 11 of the steroid molecule. Following the isolation of cortisone by E. C. Kendall and T. Reichstein in the 1930s, Hench and coworkers at the Mayo Clinic announced in 1949 that the administration of cortisone could relieve the pain of patients with rheumatoid arthritis. This created a tremendous incentive to provide cortisone by a synthetic route more efficient than extraction from ox adrenals. L. Sarett had developed (1946) a chemical synthesis from deoxycholic acid, but the process involved no less than 31 steps and was uneconomic. Inspired by the work of Mamoli and Vercellone (Fig. 18.2) and the results of O. Hechter and his collaborators (Fig. 18.1), D. H. Peterson and H. C. Murray attacked the problem microbiologically and reported in 1952 the 11α-hydroxylation of progesterone in a single microbial step using *Rhizopus arrhizus*. This reaction was decisive for the economic synthesis of adrenocortical hormones and afforded vast possibilities for the preparation of derivatives. Microbes have also proved to be useful, though not essential, in the preparation of other steroid hormones (Fig. 18.4), the estrogens or female hormones (estrone, estradiol) and the androgens or male hormones (testosterone). They can extend the range of useful raw materials to include abundant sterols, such as β-sitosterol, which cannot serve as substrates for chemical processes alone, and they facilitate the preparation of key intermediates in the synthesis of useful derivatives.

Today, the steroid industry consumes 2000 tonnes of diosgenin equivalent to make, by combined chemical and microbiological processes, products with a market value of well over one billion US dollars. Corticoids and contraceptives represent respectively 85 and 5% of the market in volume, or 55 and 35% in value. The diuretic spironolactone, the conjugated estrogens (from pregnant mare urine) and the antibiotic fusidic acid have the remaining share. Steroid biostransformation is second only to antibiotics production as a vital contribution of microbes to the production of pharmaceuticals.

18.4.2 Substrates for synthesis of steroid hormones and their derivatives

The substrates in Fig. 18.5 are preferred for their abundance in nature and for their ease of transformation into essential intermediates. Stigmasterol, from soya beans, has a double bond between C_{22} and C_{23} and is easily converted by 4 chemical steps into progesterone. Diosgenin, a sapogenin from the Mexican barbasco yam (*Dioscorea composita*), is a versatile substrate which can be transformed in high yields into progesterone and Reichstein's compound S (see structures in Fig. 18.8), as well as into C_{19} intermediates for the synthesis of androgens, estrogens and various 19-nor steroids; however, trends to increasing costs have made diosgenin less attractive. Meanwhile the older process for converting deoxycholic acid, from ox bile, has been improved and is suitable for the preparation of estrogens, progestagens and 19-nor-progestagens. β-Sitosterol, the most ubiquitous of plant sterols, and campesterol (not shown), both from soya beans, are biotransformed in a single step into two key intermediates, androstenedione and androstadienedione (see below). Cholesterol, which

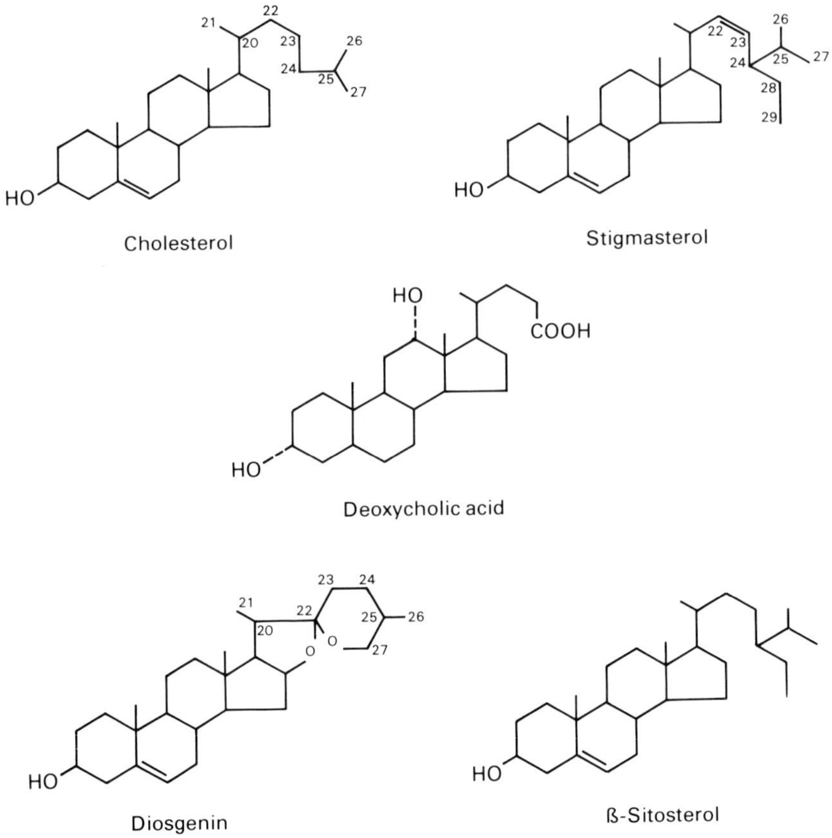

Cholesterol

Stigmasterol

Deoxycholic acid

Diosgenin

ß-Sitosterol

Fig. 18.5   Biotransformation substrates for steroid hormone manufacture.

is the natural substrate for the biosynthesis of all the steroid hormones in mammals, can be obtained from wool grease and biotransformed into the same intermediates. Hecogenin (not shown), a sapogenin from the African sisal, is used on a limited scale for preparing 16-methylated corticoids.

18.4.3   Classes of microbial transformation

Four transformations presently used on an industrial scale are illustrated in Fig. 18.6. The 11α-hydroxylation of progesterone, referred to above, is now performed with *Rhizopus nigricans* and yields more than 85% of the desired 11α-hydroxyprogesterone. *R. arrhizus* also has a strong 6-hydroxylase and on prolonged incubation it transforms the 11α-hydroxy derivative into the undesirable 6β,11α-dihydroxyprogesterone. Many other fungi

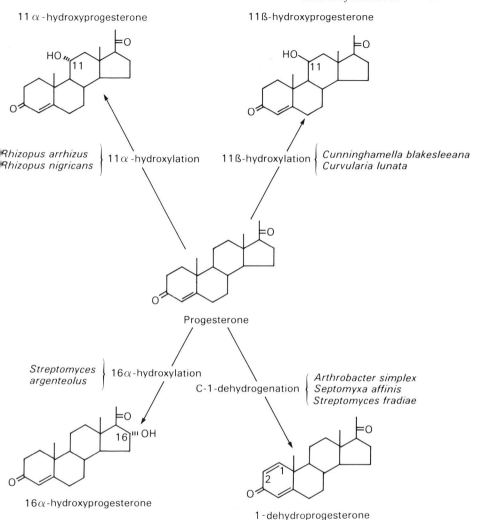

11 α -hydroxyprogesterone

11 ß-hydroxyprogesterone

Rhizopus arrhizus
Rhizopus nigricans } 11α -hydroxylation   11ß-hydroxylation { *Cunninghamella blakesleeana*
Curvularia lunata

Progesterone

Streptomyces argenteolus } 16α-hydroxylation   C-1-dehydrogenation { *Arthrobacter simplex*
Septomyxa affinis
Streptomyces fradiae

16α-hydroxyprogesterone

1-dehydroprogesterone

Fig. 18.6   Transformation schemes with progesterone.

can also 11α-hydroxylate; the conidia of *Aspergillus ochraceus* are used on a limited scale and will transform 95% of added progesterone into the 11α-hydroxy derivative. The 16α-hydroxylation of steroids is effected by *Streptomyces argenteolus* and is mainly useful to prepare triamcinolone (see below); conidia are active. The 11β-hydroxylation reaction was first described in *Cunninghamella blakesleeana* and *Curvularia lunata* (1953). Compound S can be 11β-hydroxylated into hydrocortisone in 60–70% yields using *C. lunata*; much higher yields are obtained when the substrates are acetylated at positions 17α and 21. 11β-Hydroxylation by spores is very rare, the only known example being the transformation of progesterone into 11β-hydroxyprogesterone with conidia of *Stachylidium theobromae*. Progesterone, shown in Fig. 18.6, is by no means the unique substrate for these hydroxylations; Reichstein's compound S and a large variety of 3-keto-Δ^4-steroids are also hydroxylated. Steroids without the 3-keto-Δ^4-conjugation are sometimes transformed, but more slowly; highly substituted substrates may be transformed more slowly or not at all. Acetylated substrates are hydroxylated as efficiently as the corresponding alcohols, but are often hydrolysed before they are hydroxylated; this is always the case with *A. ochraceus*.

The broad substrate specificity of hydroxylase systems is a great advantage for these processes, allowing selection of the optimal intermediate for hydroxylation in a sequence of chemical and microbial steps to a desired end-product. Microbial hydroxylations all involve direct replacement of the hydrogen atom on a given carbon. The oxygen atom in the hydroxyl group is derived from molecular (gaseous) oxygen, not from water, and the hydroxyl group thus formed always retains the stereochemical configuration of the hydrogen atom that has been replaced. Hydroxylases are extremely sensitive enzymes and few successful cell-free studies have been reported. The enzymes are inducible and both NADPH- and O_2-dependent. The requirement for an exogenous carbon source (glucose) in *A. ochraceus* conidia has been related to the regeneration of NADPH.

C-1-Dehydrogenation (introduction of a double bond) has been observed in *Cylindrocarpon radicicola*, *Streptomyces lavendulae* and *S. fradiae* and in *Fusarium solani* and *F. caucasicum*, but these organisms also degraded the side-chain (when present) extensively. *Septomyxa affinis* and *Arthrobacter* (*Corynebacterium*) *simplex* are now preferred, giving higher yields of the C-1-dehydrogenated product and smaller amounts of side-chain cleavage products. C-1-Dehydrogenases are also inducible enzymes, and several have been purified. Oxygen is not required, but a hydrogen acceptor, such as phenazine methosulphate, is generally added to cell-free extracts for maximal activity. In *S. affinis* conidia, C-1-dehydrogenase is not inducible, but the reaction is repressed or inhibited by glucose.

Ring A aromatization is a consequence of C-1-dehydrogenation (not shown) and takes place when the substrate has no methyl group at carbon 10 (19-nor) or is suitably substituted at carbon 19 (19-hydroxy, 19-oxo). The C-1-dehydro-derivative then formed is unstable and undergoes spontaneous non-enzymatic rearrangement to ring A aromatic products, such as estrone. The side-chain of pregnenes and sterols is often cleaved by organisms that carry out this dehydrogenation. Thus 19-hydroxycholesterol and 19-hydroxy-β-sitosterol are converted (by aromatization and side-chain cleavage) to estrone in high yields with *Nocardia restricta*, the side-chain being degraded by a stepwise β-oxidation to the 17-keto compound.

When the substrate carries a 19-methyl group, the steroid nucleus can be degraded by *N. restricta* and many other C-1-dehydrogenators through the introduction of a double bond at C_1 and a hydroxyl group at 9α (or vice versa). The 9α-hydroxylase reaction can be blocked by chelating agents, divalent cations or redox dyes, but a definitive solution to this problem was given by Marsheck *et al.* (1972). They prepared a mutant of *Mycobacterium* sp. which transformed cholesterol, β-sitosterol, stigmasterol and campesterol into androstadienedione, an important intermediate for the synthesis of estrogens (estrone). A further mutant was derived which produced as the main product androstenedione, a key intermediate in the preparation of the diuretic, spironolactone. The sequence of reactions is given in Fig. 18.7.

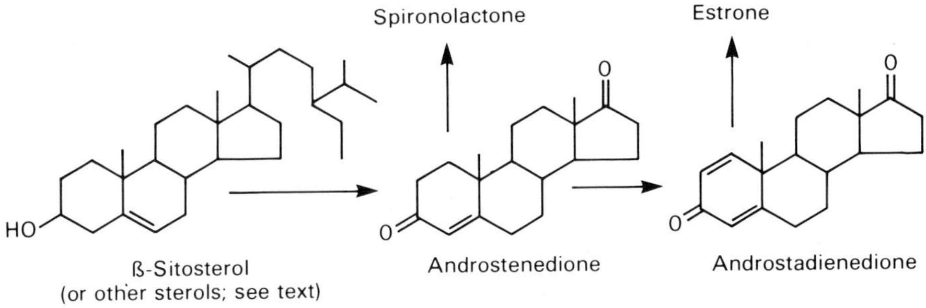

Fig. 18.7   Transformation of sterols to androstadienedione.

These are the main biotransformations of economic importance, although many other reactions have been reported (Table 18.1) and thousands of steroids, sterols and related compounds biotransformed by one to several of the reactions listed.

18.4.4   Synthetic routes to clinically important steroids

The contribution made by microbial transformations to the overall preparation of both natural and synthetic corticoids, using the reactions described in Section 18.4.3, is illustrated in Figs 18.8 and 18.9. Progesterone (Fig. 18.8), obtained by the chemical conversion of stigmasterol, is transformed by *R. nigricans* into 11α-hydroxyprogesterone, a key intermediate, which is chemically modified to hydrocortisone and cortisone, or dehydrogenated at C-1 (*S. affinis*), after chemical modification, to prednisolone and prednisone (which afford 4–10 times more anti-inflammatory activity than C-1 saturated hydrocortisone and cortisone). Alternatively, diosgenin is chemically converted to progesterone or Reichstein's compound S. Compound

Table 18.1   Microbial transformations of steroids

Oxidation				*Oxidation (cont'd)*
1. Hydroxylation				8. Side-chain cleavage with:

Oxidation

1. Hydroxylation

$1\alpha,\beta$	$6\alpha,\beta$	$11\alpha,\beta$	$17\alpha,\beta$
$2\alpha,\beta$	$7\alpha,\beta$	$12\alpha,\beta$	18
$3\alpha,\beta$	8β	$14\alpha,\beta$	19
4β	$9\alpha,\beta$	$15\alpha,\beta$	21
$5\alpha,\beta$	10β	$16\alpha,\beta$	26

2. Dehydrogenation (insertion of double bond)

Δ^1	Δ^7	Δ^{16}
Δ^4	$\Delta^{9(11)}$	
$\Delta^{1,4}$	Δ^{14}	

3. Epoxidation

4. Oxidation to ketone through hydroxylation

5. Aromatization of ring A without degradation:
   Through Δ^1 and enolization
   Through Δ^1, enolization and reverse aldolization
   Through 19-hydroxylation and reverse aldolization

6. Degradation of the steroid nucleus (sequence of reactions)

7. Oxidation of alcohol to ketone:
   3β-OH to 3-CO (with isomerization of Δ^5 to Δ^4)
   11β-OH to 11-CO
   17β-OH to 17-CO

Oxidation (cont'd)

8. Side-chain cleavage with:
   Formation of 17β-OH or 17-CO
   Ring D expansion and lactone formation

9. Oxidation of sulphide to sulphoxide, amine to ketone

Reduction
Double bond: Δ^1, Δ^4, Δ^5
Acid, aldehyde and ketone to alcohol

Hydrolysis
Ester to alcohol
Oxide to diol

Isomerization
Δ^5 to Δ^4
$\Delta^{5(11)}$ to Δ^4

Resolution of racemic mixtures

Other reactions
Amination
Enolization of carbonyl compounds
Esterification
Halogenation
D-Homoannulation
Michael addition
Reverse aldol rearrangement
Wagner–Meerwein rearrangement

S is 11β-hydroxylated (*C. lunata*) into hydrocortisone, which is then C-1-dehydrogenated into prednisolone (*A. simplex*); also, cortisone can be C-1-dehydrogenated into prednisone.

In the synthesis of triamcinolone (Fig. 18.9), the substrate can be either hydrocortisone, obtained by one of the pathways shown in Fig. 18.8 or 11α-hydroxy-compound S (not shown). The substrate is converted chemically into 9α-fluorohydrocortisone, which is 16α-hydroxylated (*S. argenteolus*), then C-1-dehydrogenated (*A. simplex*) into triamcinolone. The presence of the fluorine atom at the 9α position enhances anti-inflammatory activity, but increases the undesirable salt-retention effect; the 16α-hydroxy

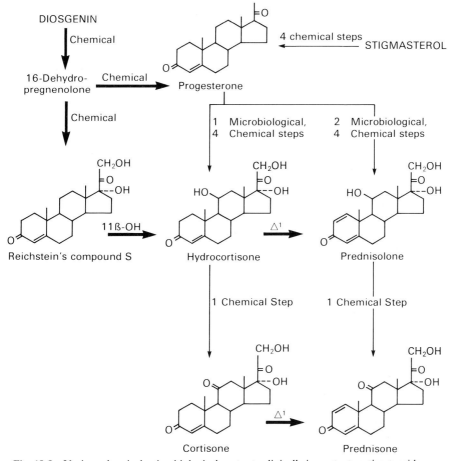

Fig. 18.8 Various chemical–microbiological routes to clinically important corticosteroids.

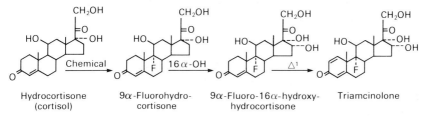

Fig. 18.9 Reactions involved in the chemical–microbial preparation of triamcinolone.

group in triamcinolone counteracts this. The preparation of triamcinolone incorporates more microbiological manipulation than that of any other steroid presently in clinical use.

These examples suffice to illustrate the role of biotransformations in the synthesis of clinically important steroids. Many other drugs presently used in the clinic, not only in the steroid field, have been made by similar sequences of chemical and microbial modifications. The collaboration of chemists and microbiologists in elaborating these complex sequences must be emphasized.

18.5   Technology of biotransformation

The methods illustrated in Fig. 18.10 are currently used in the microbial transformation of steroids and can be adapted to the biotransformation of many other classes of organic compounds. The reader is referred to the excellent recent review by Goodhue (1982; see Reading list).

18.5.1   Selection and maintenance of microbes

The conventional approach rests on the screening of microbes isolated from natural sources, or available in culture collections, and the selection of a strain that catalyses a reaction of interest. In early days, the selection was random; today, it can be based on prior observations (see Reading list) that certain microbes perform the desired reaction with a related compound. One difficulty with steroids is the lack of selective methods or specific reagents to recognize colonies with desired activity. The plate assay developed by Vézina *et al.* (1969) can be used to select organisms which can aromatize 19-nor and 19-substituted steroids and sterols into equilin and related estrogens; these ring A aromatic products (with a double bond in ring B) react specifically with *para*-nitrobenzene diazonium fluoborate to yield an intense red colour. Colonies developing on a solid medium (which contains a suitable steroid substrate) are replicated before the reagent is sprayed; a red ring develops around active colonies. The method has been invaluable in screening large culture collections and has accelerated strain improvement.

A modified enrichment method can be useful to isolate mutants blocked in substrate dissimilation: a steroid substrate is incorporated as the sole carbon source in a mineral medium seeded with soil dilutions; cells that can degrade the substrate will grow, and are transferred to the same medium enriched with another carbon source, such as glucose. Mutants may be present that are blocked at various steps in the degradation of the steroid substrate, but can utilize glucose as the carbon source. Intermediates

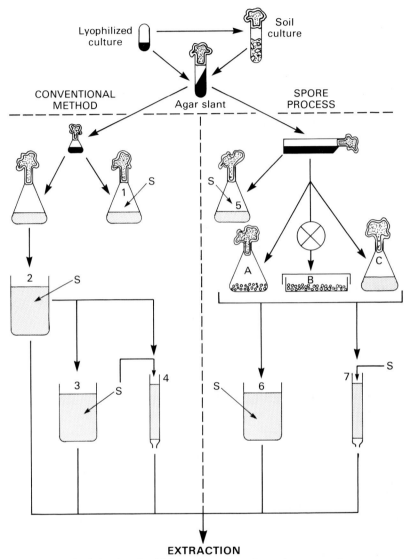

EXTRACTION

Fig. 18.10   Methods for the microbial transformation of organic compounds.

(before the metabolic block) can accumulate, and the mutants bearing the lesion can be isolated. (For further details, see Abbott, 1979.) Mutants have also been isolated which cannot accumulate an undesirable compound; thus, a mutant of *A. ochraceus* blocked in the transformation of 11α-hydroxy-progesterone into 6β,11α-dihydroxyprogesterone (undesirable reaction) was prepared which could yield only the 11α-hydroxy derivative of progesterone.

Microbes for transformation belong to the eubacteria, streptomycetes and fungi (yeasts and molds) and are maintained according to standard methods (above the horizontal dotted line in Fig. 18.10); frozen cultures (-20 to $-170°C$) are also suitable.

18.5.2   Batch and continuous processes

The conventional method is illustrated in Fig. 18.10 (left of the vertical dotted line). A suspension of vegetative cells or spores from agar slants is used to inoculate a small shake flask containing a suitable medium. After 1–2 days of incubation at the optimum temperature on a gyrotory or reciprocating shaker, the vegetative growth (cells or mycelium) is used to inoculate shake flasks (marked 1 in Fig. 18.10), which are convenient for screening and for studying factors of transformation (medium, temperature, pH, carrier solvent, substrate concentration, incubation time). The steroid substrate is added toward the end of the growth phase (18–36 h) and the transformation allowed to proceed; samples are taken at time intervals and analysed by TLC, paper chromatography, gas chromatography or HPLC, to follow the course of transformation (1–5 days). The structures of the transformed products are elucidated by classic chemical methods.

On an industrial scale, an inoculum is built up and used to inoculate a fermenter (2). The organism is grown under optimal conditions until a desired concentration of active cells is reached. The substrate is added ($1–20 \, g \, l^{-1}$), and transformation is allowed to proceed under the same conditions until a maximum is reached. Steroids are insoluble or sparingly soluble in water, and the technique of addition to the aqueous medium is important. The substrate (S in Fig. 18.10) is often dissolved in a suitable water-miscible solvent (propylene glycol, ethanol, methanol, acetone, dimethylsulphoxide, dimethylformamide) and the solution added slowly (sometimes continuously) with vigorous agitation for maximal formation of fine crystals and maximal efficiency of transformation. The concentration of some organic solvents must be kept to a minimum to prevent inhibition of growth and metabolism, while the steroid concentration can range from $1–20 \, g \, l^{-1}$, depending on the substrate and the organism. Alternatively the steroid substrate can be suspended in a surfactant, such as Tween 80, and 'sonicated' to fine particles before it is added to the beer. In a process called 'pseudo-crystallofermentation', as much as $500 \, g \, l^{-1}$ of finely powered solid hydrocortisone, added to *A. simplex* culture, can be transformed in high yields ($>93\%$) into prednisolone in 5 days. The efficiency of this process matches that of chemical processes.

Finally, the transformation products are recovered by extraction with methylene chloride, chloroform or ethyl acetate and purified by column chromatography and other methods.

In a modified process, the cell or mycelium in the fermenter (2) are separated from the medium by filtration or centrifugation and resuspended in buffer (3) before the steroid is added; the efficiency of transformation is not diminished, and extraction and purification are simplified. The cells and the mycelium (or their enzymes) can also be 'immobilized' on a solid support and the mixture packed in a column (4) for continuous addition of the substrate and recovery of the product. Yet another approach to the biotransformation of water-insoluble substrates, such as steroids, is the use of a reverse-phase system whereby the reaction is conducted in organic solvents. The potential advantages are the rapid transport of substrates and products between the organic and aqueous (cells) phases and the greater solubility of oxygen in organic solvents than in water for oxygen-dependent biotransformations. For further details on these modified processes, see Abbott (1979).

18.5.3 Transformation with spores

The process described by Vézina *et al.* (1969), used on a small industrial scale, is illustrated in Fig. 18.10 (right of the dotted vertical line). Spores are prepared in Roux bottles and suspended in a non-nutrient medium for transformation in shake-flasks (5), or propagated on a larger scale in Fernbach flasks (A), trays (B) or in submerged culture (C). The spore suspensions are washed and resuspended ($2–5 \times 10^8$ spores ml^{-1}) in a buffer (6); for hydroxylations, glucose must be added and the pH continuously readjusted to 6 during transformation. Spores are generally stable, and can be stored or transported as a concentrated biochemical catalyst which is readily available when required; they give high, reproducible conversions of substrate and minimal amounts of undesirable products. The medium is simple (water or buffer), foaming is minimal and extraction of transformation products efficient. This is a process of choice when the substrate is expensive and large-scale fermentation equipment for aseptic operation is not available. Spores can also be immobilized for continuous operation (7).

18.6 Microbial models of mammalian metabolism

The development of any new drug requires an understanding of its metabolism in several mammals, including man. In the determination of absorption, distribution and excretion patterns it is often observed that the administered drug is modified by mammalian enzymes into derivatives that circulate in blood and/or are excreted in urine. One difficulty is the small amount of these derivatives in animal tissues and urine and the inability of most animal systems (cell culture, organ perfusion) to produce them in sufficient quantity for structure elucidation and pharmacological–toxicological testing. In an early but typical study, Sehgal and Vézina (1967) screened several microbes for their ability to transform the anticonvulsant, 5-H-dibenzo(a,d)cycloheptene-5 carboxamide (cyheptamide) (compound

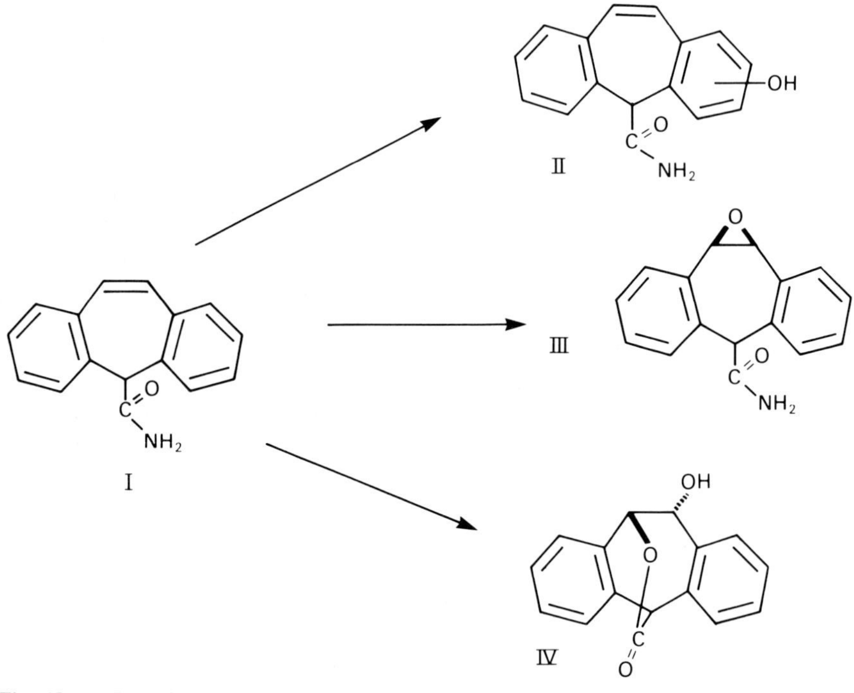

Fig. 18.11 Transformation of cyheptamide (*T*) into metabolites II–IV in the dog and by *Streptomyces*.

I in Fig. 18.11), and selected *Streptomyces lavendulae* to obtain significant quantities of compounds II, III and IV. Compounds II and IV were identical to metabolites previously found in dog's urine. Compound III had not been detected in dog's urine, and it was suggested that this epoxide could be the precursor of compound IV. Further metabolism studies established that compound III was present in very small quantities in dog's urine. The metabolites were found to be less active and less toxic than the parent drug.

Microbial transformation is now a routine procedure in drug metabolism studies. It is performed to predict the presence and nature of metabolites, to prepare large quantities for structure elucidation and biological evaluation, and, sometimes, to find new compounds which can be active *per se* or serve as intermediates for synthetic modification. The methodology is the same, *mutatis mutandis*, as that practised for steroid biotransformations. In the past decade, drug biotransformation was developed extensively by J. P. Rosazza and his collaborators who coined the term 'microbial models of mammalian metabolism', and the interested reader should refer to the review by Smith and Rosazza (1982).

18.7 Conclusion

Biotransformation offers a vast repertoire of reactions, a few of which have emerged as important industrial processes. For the manufacture of organic chemicals, the choice of the industrialist oscillates between organic synthesis and biotransformation, and his decision is purely economic. Newly introduced microbial processes often require important investment to secure the equipment necessary to handle large volumes of biological reagents under aseptic conditions. Low concentrations of substrates and products and relatively slow reactions are typical of biological processes. Nevertheless, microbial transformations will be preferred when they replace several chemical steps, e.g the 11-hydroxylation of steroids, or when they give abundant access to new complex substrates which defeat organic manipulation, e.g. the substitution of diosgenin by plant sterols which are degraded microbiologically into useful intermediates. Similar shifts from organic synthesis to biotransformation are more frequent as steroid biotechnology is applied to other classes of compounds.

The field of biotransformation is evolving as new methods are designed to replace random screening procedures, new analytical tools (such as HPLC) are available to compensate for the lack of selective methods, and new concepts are developed to increase the efficiency of the transformation process: immobilization of vegetative cells, spores and enzymes for increased stability and continuous operation, reverse-phase systems for transforming water-insoluble substrates. The contribution of microbial genetics to biotransformation has so far been modest. It is expected that recombinant DNA technology will be exploited to clone in a single organism the genes that code for desirable reactions in different organisms. Such a clone could perform a sequence of reactions in a single step and improve the economic nature of biotransformation.

Reading list

Abbott, B. J. (1979). Some new approaches to biotransformations. *In* 'Developments in Industrial Microbiology', Vol. 20 (Ed. L. A. Underkofler), pp. 345–365. Society for Industrial Microbiology: Arlington VA.

Charney, W. and Herzog, H. L. (1967). 'Microbial Transformations of Steroids—A Handbook'. Academic Press: New York and London.

Goodhue, C. T. (1982). The methodology of microbial transformation of organic compounds. *In* 'Microbial Transformations of Bioactive Compounds', Vol. 1 (Ed. J. P. Rosazza), pp. 9–44. CRC Press Inc.: Boca Raton, FL.

Kurz, W. G. W. and Constabel, F. (1979). Plant cell cultures. *Adv. Appl. Microbiol.* **25,** 209–240.

Marsheck, W. J., Kraychy, S. and Muir, R. D. (1972). Microbial degradation of sterols. *Appl. Microbiol.* **23,** 72–77.

Perlman, D. (1979). Ketogenic fermentation processes. *In* 'Microbial Technology', 2nd Edn, Vol. 2 (Eds H. J. Peppler and D. Perlman), pp. 173–177. Academic Press: New York and London.

Sebek, O. K. (1981). Microbial transformations of antibiotics. *In* 'Economic Microbiology', Vol. 6 (Ed. A. H. Rose), pp. 575–611. Academic Press: London and New York.

Sebek, O. K. and Perlman, D. (1979). Microbial transformation of steroids and sterols. *In* 'Microbial Technology', 2nd Edn, Vol. 1 (Eds H. J. Peppler and D. Perlman), pp. 483–496. Academic Press: New York and London.

Smith, R. V. and Rosazza, J. P. (1982). Microbial transformations as a means of preparing mammalian drug metabolites. *In* 'Microbial Transformations of Bioactive Compounds', Vol. 2 (Ed. J. P. Rosazza), pp. 1–42. CRC Press Inc.: Boca Raton, FL.

Vézina, C. and Singh, K. (1975). Transformation of organic compounds by fungal spores. *In* 'The Filamentous Fungi', Vol. 1 (Eds J. E. Smith and D. R. Berry), pp. 158–192. Edward Arnold: London.

Vézina, C., Rakhit, S. and Médawar, G. (1984). Microbial transformation of steroids. *In* 'Handbook of Microbiology', 2nd Edn, Vol. 3 (Eds A. I. Laskin and H. A. Lechevalier), pp. 65–466. CRC Press Inc.: Boca Raton, FL.

19 Genetic Engineering and its Applications

K. MURRAY

19.1 Introduction

Today's vast opportunities for the innovative exploitation of biological systems result from several decades of painstaking work and careful observation in a variety of interrelated scientific disciplines, amongst which the relatively recent combination of microbiology and biochemistry has developed into the exciting subject of molecular genetics.

The basis of molecular genetics is the process of genetic recombination, the breakage and rejoining of DNA molecules, which is of fundamental importance to all living organisms as a mechanism for adaptation and variation. Its study has provided an understanding of inheritance and contributed to ideas on evolution, and has already been exploited to great advantage by plant and animal breeders, often quite empirically. Since 1960, the manipulation and analysis of genetic crosses in bacteria and their viruses (or phage) have been major factors in the development of molecular biology, to the point that we can now construct DNA molecules containing genes and controlling elements from any source and can clone and propagate these new molecules in suitable host cells, particularly bacteria. These procedures of molecular cloning enable us to by-pass the normal sexual constraints between species and to construct and disseminate quite novel genetic combinations.

This manipulation of genes by both biochemical and genetic methods, together with their transfer, replication and expression constitutes genetic engineering.

19.1.1 Gene structure and function

A gene is a region of a chromosome responsible for some observable phenotype or characteristic of an organism. The chemical structure of genes and

the basic mechanisms of their replication and expression are well understood (see Chapter 2). Genes are nucleic acids, usually segments of DNA,
though some viruses use RNA as their genetic constituent. The DNA molecules are generally present in the well-known double-stranded form proposed by Watson and Crick in 1953, but some viruses have a single-stranded
DNA molecule as their genome; this is usually circular and has a double-
stranded form as its replicative intermediate. Biologically, the important
aspect of the Watson–Crick structure is that the DNA molecule comprises
two polynucleotide strands that are aligned antiparallel and held together
by the highly specific (but physically weak) interactions between purine
and pyrimidine bases which result in one strand being complementary to
the other. This structure (Fig. 19.1) immediately offered a basic mechanism

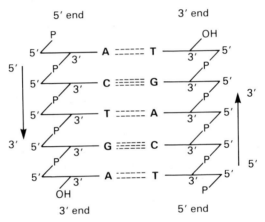

Fig. 19.1   Segment of a DNA duplex showing antiparallel orientation of the complementary
chains.

for faithful replication of the DNA molecule, and hence the store of genetic
information, and much is now known of the complex biochemical reactions
involved in this process. Expression of the information carried in genes
has also been studied intensively, especially in bacteria, by both genetic
and biochemical methods; the two principal steps, transcription and translation, are illustrated diagrammatically in Fig. 19.2.

   Transcription of the gene by RNA polymerase gives a messenger RNA
molecule (mRNA) from which the sequence of nucleotides is decoded in
the *translation* process with synthesis of the polypeptide chain. Transcription begins with the interaction of RNA polymerase at a specific site, the
promoter, and the protein continues along the DNA molecule until a specific termination sequence is encountered (Fig. 19.2). Ribosomes bind to
a particular site (r.b.s.) on the mRNA in the presence of appropriate

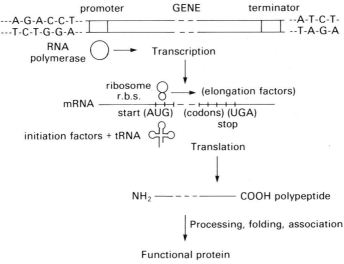

Fig. 19.2   Gene expression and its regulation in bacterial systems.

cofactors and at the correct trinucleotide sequence (AUG), translation is initiated by interaction with specific proteins and a tRNA molecule that has a methionine residue attached at its 3′ terminus. In the presence of elongation factors, translation of the triplet code continues until a termination codon (e.g. UGA) is encountered. The polypeptide chain then folds to give a functional protein, which in some cases may require the correct association of a number of polypeptide chains.

Several complex interactions between proteins and both RNA and DNA are involved, and a large number of highly specific recognition processes between these macromolecules are necessary for the correct expression of the biological properties specified in genes from a particular organism. If genes are to be transferred effectively from one organism to another, the fidelity of recognition processes through these macromolecular interactions must be maintained or it must be possible to make whatever changes are necessary to accommodate them.

19.1.2   Transfer of genetic information

Three basic processes may be distinguished for the transfer of DNA between different cells. The first is *transformation*, first recognized over fifty years ago as a morphological change in *Pneumococci* in infected animals after injection with killed cells of a different type. It is now known that this arises because DNA itself can be infectious. Direct uptake of DNA by the cell is readily demonstrated in bacteria by the change in phenotype after adsorption of DNA prepared from the wild-type (prototrophic) organism by cells of a mutant (e.g. auxotrophic) strain that have been treated in such a way as to make them 'competent' for the uptake and utilization of the DNA. In some of the recipient cells the donor DNA becomes incorporated (recombined) into the chromosome; this is recognized because the auxotrophic requirement is permanently relieved and the cell becomes phenotypically wild-type.

The second process of gene transfer is *transduction*, which is mediated by a virus and can occur by two versions of the same basic mechanism. One is *generalized* transduction in which any DNA within the infected cell, including some of its chromosomal DNA, may become incorporated into a virus particle and then transferred by the normal infection process to another cell after lysis of the original host. This is characteristic of those viruses, such as coliphage P1, that package a headful of DNA in a non-specific fashion. Other viruses, such as bacteriophage λ, are able to integrate their own chromosome (or genome) into that of their host, an essential part of the phenomenon of lysogenization, by a process that is normally reversible. However, the reverse of integration, i.e. excision, sometimes occurs aberrantly so that the DNA molecule excised from the lysogenic host chromosome carries with it some of the host genome (Fig. 19.3). This DNA molecule may then be packaged into a virus particle which subsequently infects another host cell giving rise to gene transfer by the process of *specialized* transduction.

Conjugation, a sexual process that may occur between bacteria of the same or different species, is the third means by which genetic material, even large plasmids or an entire bacterial chromosome, may be exchanged between individual cells. Where plasmids are involved, they may sometimes be integrated into the recipient cell chromosome, by mechanisms analogous to those in specialized transduction.

These three natural processes for the transfer of genetic information vary appreciably in their range and specificity; the new biochemical methods for the construction of various combinations of DNA sequences *in vitro* broaden and extend their range enormously.

19.1.3   Restriction and modification

The efficiency with which DNA transferred into a new cellular environment is utilized is very variable. One important factor is the ability of many cells to recognize and destroy foreign DNA. This phenomenon of host-controlled 'restriction' is perhaps best illustrated by a specific example. If a lysate containing phage λ that has been grown on a certain strain of *Escherichia coli* (e.g. *E. coli* strain C, in which case the phage is described

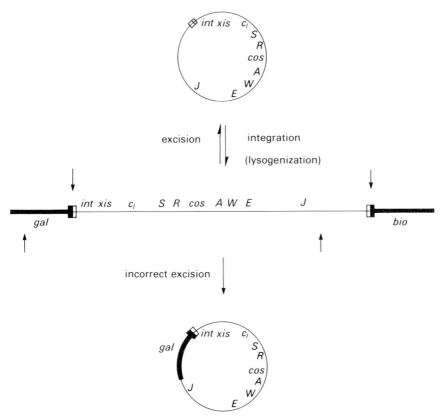

Fig. 19.3 Lysogenization and the formation of specialized transducing phages. The heavy line denotes part of the host chromosome and the short vertical arrows indicate the positions of interaction of the enzymes responsible for integration or excision of the phage chromosome, the boxes showing the position (*att*) at which these reactions normally occur. The approximate position of some of the phage genes is shown and *cos* represents the position of the cohesive ends of the linear molecule formed as the phage DNA is packaged into the virion.

as λC) is used to infect a second strain (*E. coli* strain K, for example), the efficiency of phage growth on this second strain may often be several orders of magnitude lower than that of the first. However, whereas if the surviving phage in the new lysate (λK in our example) are used directly to infect a further culture of the second strain (*E. coli* K) they will grow quite normally upon it, but if they are first passed through their original host (*E. coli* C) (where they may or may not grow normally) they will again grow very poorly upon the second strain (*E. coli* K). This *restriction* of growth of the phage is thus dependent upon the host strain on which the phage were last propagated. Such host-controlled restriction results from degradation of the infecting bacteriophage DNA by an endonuclease specific to the particular host strain; these enzymes are called restriction endonucleases. The host cell protects its own DNA against the action of its own restriction enzymes by methylating certain bases (usually adenine or cytosine) located within the specific nucleotide sequence recognized by the restriction enzyme. This process is known as host-controlled *modification* and explains why those phages that survive attack by the restriction enzyme subsequently grow normally upon the host strain; their DNA has been replicated in the presence of the modifying methylase so that the appropriate bases within the phage DNA targets for the restriction enzyme have been methylated and hence protected. Many examples of restriction and modification systems have now been found in numerous species of bacteria. In some cases the genes for restriction and modification enzymes are chromosomal and in others they are carried on plasmids (autonomously replicating extra-chromosomal elements, usually double-stranded circular DNA molecules, often quite large). The systems found in different species of bacteria, and in different strains of a given species, exhibit distinct specificities for unique oligonucleotide sequences (usually tetranucleotides or hexanucleotides) that constitute the targets for the methylation or restriction enzymes. As a result, we can deploy a large range of highly specific endonucleases as reagents for the fragmentation of DNA into defined segments for the construction of new genetic combinations.

19.1.4   Construction of recombinant DNA molecules *in vitro*

The average size of DNA fragments generated by the action of a restriction enzyme is determined by the length of its target sequence. The frequency of any particular nucleotide sequence in a polynucleotide of random base composition is 1 in 4^n, where n is the number of bases in that sequence. Thus, a given tetranucleotide sequence would occur once in 256 base pairs and a specific hexanucleotide once in 4096 base pairs. DNA molecules are not random polymers, and in practice there is a very broad spread around these average values. The critical point is that the DNA fragments produced are of a size such that many will encompass a gene, or sometimes several genes (given that each amino acid of protein is defined by a sequence of three nucleotides).

The target sequences recognized by most (but not all) restriction enzymes have a two-fold axis of symmetry and the restriction enzyme breaks phosphodiester bonds in both DNA strands within the target sequence (Fig. 19.1). With some enzymes the phosphodiester bonds immediately opposite each other in the DNA duplex are broken (even breaks) leaving fragments with 'blunt' ends, while in other cases the enzyme breaks bonds a few base pairs apart in opposite strands of the duplex (staggered breaks) to give fragments with short, single-stranded projections of complementary base sequence, or 'cohesive' ends. The latter type of DNA fragments readily reassociate by normal base pairing, and the action of DNA ligase (poly-nucleotide ligase) then joins them together covalently. Restriction enzymes producing staggered breaks found immediate application in the construction of new types of DNA molecules, but DNA fragments resulting from even breaks may also be joined together covalently through the action of DNA ligase of bacteriophage T4. This reaction, together with the use of short synthetic oligonucleotides of defined sequence for linking together fragments with projecting ends of differing sequences, extends the range of biochemical methods so that the possible combinations of DNA sequences that can be constructed are virtually unlimited.

19.2 Molecular cloning

19.2.1 Purposes of molecular cloning

Since the only requirement for cleavage of a DNA molecule by a restriction endonuclease is the presence of an unmodified target sequence, DNA from any source can be fragmented by the action of one or more restriction enzymes. The mixture of fragments will often be very complex, but a number of methods are available for their fractionation and in simple cases, such as certain digests of viral DNA or plasmids, for the recovery of individual fragments. As the complexity of an organism increases its genome becomes larger and therefore the number of fragments generated by a given restriction enzyme increases, so that even the simpler higher organisms may give a million or more DNA fragments in such digests.

Figure 19.4 illustrates the effects of the size of the target sequence for the restriction enzyme and the complexity of the DNA preparation upon the size and number of the fragments obtained when enzyme digests are separated by electrophoresis on agarose gel. Electrophoretic mobility of a DNA fragment is inversely proportional to the logarithm of its relative molecular mass. Tracks (a) to (d) contain samples of phage λDNA, a relatively simple molecule with a molecular mass of 3.2×10^7, before and after digestion with various enzymes. The individual bands are well separated. Tracks (e) to (g) contain digests of *E. coli* DNA, which is about 100 times the size of phage λDNA; resolution of the individual bands is still possible, at least partially, at the analytical level. The last three tracks (h–j) contain digests of human liver DNA with the same enzymes, and here the complexity of the DNA is so great that the resolution of individual fragments is lost, although bands arising from the breakage of highly reiterated

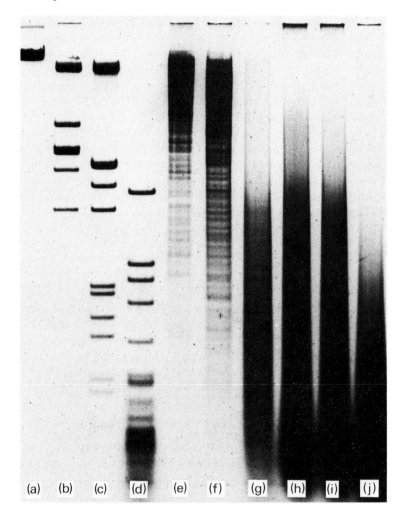

Fig. 19.4   The separation of DNA fragments in restriction enzyme digests by electrophoresis in agarose gel. (a–d) Samples of phage λDNA (a) before digestion, and after digestion with (b) R.*Eco*RI, (c) a mixture of R.*Eco*RI and R.*Hin*dIII (both of which recognize hexanucleotide sequences), and (d) R.*Hae*III (which recognizes a tetranucleotide sequence). (e–g) Corresponding digests of *E. coli* DNA. (h–j) Digests of human liver DNA with the same enzymes. 0.5–3.0 μg DNA was digested and after electrophoresis the gel was stained with ethidium bromide and photographed under ultraviolet light.

sequences may often be discerned. Although physical fractionation of such complex mixtures is not practical, the DNA fragments could still, in principle, be used to transform bacteria. In the absence of an obvious biological indicator, such as the change from an auxotrophic to wild-type phenotype as described earlier, the difficulty is to recognize cells that have acquired DNA fragments.

Molecular cloning methods enable these difficulties to be overcome. DNA fragments as a mixed population are joined by incubation with DNA ligase to simple carrier molecules which can be replicated in the bacterial cell and which contain a genetic marker such that only transformed cells are capable of growth in a selective medium. If each transformed cell has taken up a single DNA molecule, populations of cells containing identical copies of the newly constructed molecule may then be propagated and the recombinant DNA molecules recovered. Individual segments of the genome of virtually any organism may thus be cloned and, by growing cultures of the transformed bacteria, prepared in quantity for detailed biochemical (sequence) and genetic analysis. The molecular cloning procedures thus provide the means for isolating and working with individual components of a mixture present in less than one part per million. The carrier DNA, or vectors, used are either small plasmids or viral genomes, such as that of phage λ, and their necessary characteristics are listed in Table 19.1. A wide range of vectors is now available, and choice depends

Table 19.1   The *Desiderata* of a DNA molecule that is to serve as a vector for DNA fragments in *in vitro* recombination experiments

1. The molecule must be capable of autonomous replication in an appropriate host cell.
2. The site for insertion of a DNA fragment must be such that the insertion does not destroy an *essential* function.
3. The vector must carry a means for selection of transformed cells, such as a drug resistance determinant, the ability to confer immunity (for example to a colicin or to a phage), or the production of a phage plaque in a lawn of bacterial cells.
4. A means for distinguishing, or preferably selection of, recombinant DNA molecules from the parent vector DNA.

upon the particular experiment. Both plasmid and phage vectors have their advantages and limitations, and in some cases vectors constructed from parts of both plasmids and phage may be preferred.

Once a particular gene has been cloned (and identified, see below) it will often be possible by further biochemical manipulations to increase the level of expression in a bacterial or other cell, sometimes by more than a hundred-fold. In favourable cases, a single gene product may then constitute as much as 30% of the soluble protein of the cell.

19.2.2 Recognition of cells carrying recombinants

Several methods are available for the detection of specific recombinants. Where applicable the simplest rely upon complementation of an auxotrophic host strain. For example, if the objective is the isolation of the gene for a particular bacterial enzyme a population of recombinants made from the DNA of that bacterium (wild-type) would be used to transform an *E. coli* strain with a conditional mutation (temperature-sensitive or suppressible) in the gene for that enzyme; transformants that carry the cloned gene concerned will grow under the non-permissive conditions. Alternatively, transformation leading to the production of particular enzymes or their specific metabolic products may sometimes be detectable by simple colour tests. To continue with the enzyme example, introduction of recombinants that contain the *lac* gene of *E. coli* into a *lac⁻* strain permits synthesis of β-galactosidase by cells that cannot otherwise make this enzyme, which is not essential for cell growth (except on lactose). Convenient and sensitive colour tests such as the use of MacConkey lactose-agar plates (red colonies) or 5-bromo-2-chloro-indolyl-β-D-galactoside (blue colonies or phage plaques) then enable cells that produce β-galactosidase to be readily distinguished from those that cannot (colourless colonies or plaques). Cells carrying recombinant phage as a prophage (lysogens) can be detected in an equivalent manner; strains unable to hydrolyse glutamine, for example, when lysogenized with a phage carrying a glutaminase gene, may be easily detected on appropriate indicator plates by virtue of the pH change attending the release of ammonia. Recombinant phage made by the *in vitro* methods are often themselves useful as vectors since restriction of the phage DNA allows replacement of the inserted fragment by a different DNA fragment. If the fragment being replaced contains a readily detectable, or, preferably, a selectable genetic marker this provides a convenient distinction between the parent and new recombinant phages.

In other instances methods using fluorescent antibodies or solid phase radioimmunoassay procedures have been adapted for screening populations of transformed cells or phage for recombinants expressing a particular gene product. These have the advantage that fusion products (polypeptides resulting from the direct linkage of two genes either in whole or in part; (see Fig. 19.10)) and non-functional or defective gene products can often be detected because they nevertheless react with suitably specific antibodies (see Fig. 19.11).

Direct analysis of cloned gene products by gel electrophoresis is another useful method for the study of expression of cloned genes, but is only effective with small populations of recombinants or when a specific gene has been identified. It is often useful to make small-scale preparations of recombinant plasmids or phage DNA for direct analysis by restriction enzyme digestion and gel electrophoresis.

The methods described depend upon expression of the cloned genes, but genes which are not expressed in the new host cells may nevertheless be identified by nucleic acid hybridization methods, applied directly to bacterial colonies or phage plaques. A purified mRNA or cDNA copy labelled to a high specific radioactivity with ^{32}P provides an extremely powerful means for identification of the corresponding recombinant in a large population. When combined with the packaging of phage genomes *in vitro* the method permits very large numbers of recombinants to be screened very efficiently and detection of one specific sequence in a population of a million cloned fragments is not unduly difficult.

19.2.3 Analysis of complex genomes

The isolation of individual segments of the complex genomes of higher organisms by molecular cloning in phage λ vectors, and propagation of the recombinant phage, provided the means for detailed analysis of the organization of genes in these systems. This analysis is done at two levels. Initially, digestion of the cloned fragment with a range of restriction enzymes gives a series of smaller fragments which can be readily aligned to give a map, termed a *restriction map*, on which regions that are transcribed can be identified from the fragments that hybridize with mRNA for a given gene product. Determination of the nucleotide sequence of the restriction fragments then gives a detailed picture of the linear structure of the gene and, in many cases, of the relative positions of several genes within the cloned segment of the original chromosome. Clones of adjacent fragments of the chromosome can be located and it is sometimes possible to overlap several such fragments to give a detailed picture of very large stretches of a chromosome.

Several higher organism genes have now been examined in this way and detailed pictures of the structure of genes for α-amylase, globins, immunoglobulins, interferons, ovalbumin, and a number of hormones, are amongst the more prominent examples. One of the most remarkable features revealed by this analysis is that in many cases the coding information for a eukaryotic gene product is not a continuous linear sequence of nucleotides, but may contain one, or sometimes several interruptions which can often be very long and which have been termed *introns*; the coding regions of the gene are called *exons*. This phenomenon clearly has important consequences for gene expression and its regulation. In eukaryotic cells the gene is transcribed to give a large primary RNA transcript which is then processed to give an mRNA by elimination of the intron regions, a process

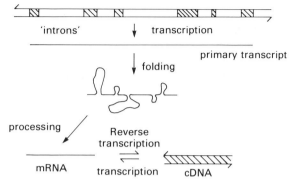

Fig. 19.5 The organization and expression of some eukaryotic genes. The sequence coding for a polypeptide, represented by the hatched area, is interrupted by additional non-coding sequences called introns, which may be quite long. The primary transcript must be processed (spliced) to give a mature mRNA molecule, which is translated in the normal way (Fig. 19.2). The action of reverse transcriptase upon a mature mRNA gives a DNA molecule containing uninterrupted coding sequences analogous to gene sequences in bacteria.

referred to as splicing (Fig. 19.5), and the mRNA is then translated in the anticipated way. Bacteria do not have to perform splicing operations and are not equipped to handle the direct transcripts of eukaryotic genes that contain introns, so such genes will not be expressed in bacteria. However, the enzyme reverse transcriptase can be used to make DNA copies (cDNA) of mRNA preparations (usually separated from other RNA species by virtue of the polyA sequences at their 3′ termini) and these DNA copy preparations may then be cloned in bacterial vectors, frequently as mixed populations, where they can be made to function as a prokaryotic gene.

19.2.4   Gene expression and its amplification

Once a prokaryotic gene (or a cDNA copy of a eukaryotic gene) has been cloned in a phage or plasmid vector in a configuration such that it is expressed, either as the natural gene product or fused to another polypeptide (Fig. 19.10), the yield of the gene product can often be greatly increased. Several factors affect the efficiency of gene expression, some of which are summarized in Table 19.2.

Table 19.2   Factors governing cellular yields of gene products

1. Number of gene copies.
2. Efficiency of transcription, usually dependent upon the promoter used.
3. Stability and secondary structure of the mRNA.
4. Efficiency of translation, dependent upon a strong ribosome binding site and its distance from the initiation triplet.
5. Stability of the polypeptide in its novel cellular environment.

Both phage and plasmid systems, or hybrids of the two, may be used to amplify gene expression. Phage λ has been studied intensively and the position of many of the virus genes as well as their control elements is known with precision; some of the more important control systems are shown in Fig. 19.6. At the 5′ ends of the DNA are single-stranded projec-

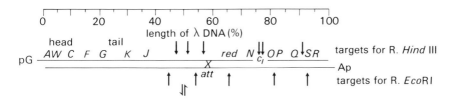

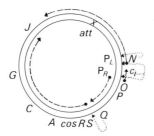

Fig. 19.6 A simplified version of the map of the bacteriophage genome showing some of the more important control systems. The mature phage contains a single linear duplex DNA molecule with a relative molecular mass of 3.2×10^7.

tions of 12 bases with complementary sequences. Upon injection into the host cell, the DNA circularizes by base pairing of these cohesive ends. Chromosomes of lambdoid phages are normally drawn in the linear form; here genes are located at the positions corresponding to the percentage of the length of the wild-type phage DNA. Genes on the left of the linear map code for head and tail proteins of the phage. Much of the central region is inessential and can be deleted without seriously impairing phage growth. *Red* represents the phage recombination system, *O* and *P* are concerned with replication of the phage DNA, and *S* and *R* code for proteins that lyse the host cell when the phage products have been assembled into infectious particles. The c_I gene codes for a repressor protein that interacts at the sites shown by the dotted arrows in the lower part of the figure to prevent expression of the phage genes. Removal of the repressor permits expression in both directions from P_L and P_R, as shown by the broken arrows. *N* and *Q* are positive regulatory genes, the products of which interact at the positions shown by dotted arrows. *Q* is necessary for the expression of genes S and R and genes to the left of these, as indicated by the long broken arrow inside the circle. Thus, after circularization of the chromosome, gene Q activates the expression of genes A, C, E, etc. (i.e. those on the left of the linear map) as well as genes R and S. Also shown in the figure is the attachment site by which the phage chromosome may be inserted into its host chromosome (where it may be stably replicated along with the host), and the positions of targets for the restriction enzymes R.*Hind*III and R.*Eco*RI in the wild-type chromosome. Some of these targets must be removed by deletion or mutation in order to make phage derivatives that can be used as receptors (or vectors).

The amplification of T4 DNA ligase, an enzyme that is particularly useful in recombinant DNA methodology, provides an example of the way in which the basic genetics of λ can be exploited. DNA from phage T4 was digested with restriction enzymes, the digestion products inserted into an appropriate λ vector, and recombinants recovered by transfection of a suitable host strain. The population of recombinants was then screened for phage that complemented a ligase-deficient (temperature-sensitive) strain of *E. coli*; phage that by this criterion expressed the ligase gene were purified and used to study the level of the enzyme in infected host cells. In Fig. 19.7 the amount of ligase obtained from such cells is compared with that from an equivalent quantity of *E. coli* B infected with a phage T4 strain in the normal way.

Preparative experiments are best performed with thermo-inducible lysogens of the recombinant phage and various aspects of λ can be exploited to optimize expression of the inserted gene and isolation of its product. Use of the efficient λ promoters increases the level of transcription of the incorporated genes, and an amber mutation in gene S prevents lysis of the host so that gene products are contained within the cells. (This is useful for laboratory-scale preparations, but would not be chosen for large-scale operations.) For T4 DNA ligase an important additional factor is the genetic purification afforded by transfer of the T4 genes to λ, whereby the ligase gene is isolated from other T4 genes, many of which code for potent nucleases that would interfere with the subsequent purification and assay of the enzyme. More detailed manipulation of the recombinant phage gave further amplification of the enzyme on induction of appropriate lysogens. The best relied-upon transcription towards the right from the late λ promoter P'_R under the positive control of gene Q, in a phage with amber mutations in gene S (necessary for cell lysis) and gene E the product of which is required for production of phage heads. Induction of the prophage permits replication of the phage DNA to give many gene copies which, in the absence of the E gene product, cannot be encapsidated and so remain available for continued transcription over prolonged periods, while the

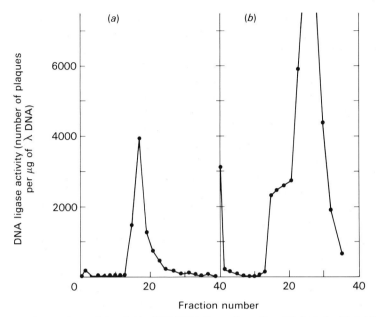

Fig. 19.7 Comparison of the yield of T4 DNA ligase from *E. coli* infected with (a) T4 and (b) a λ-T4 hybrid. Similar masses of the two cell pastes were extracted and the figure shows chromatographic fractionations on DEAE-Sephadex A50. The assay measures the restoration of infectivity of a phage λDNA cleaved into two fragments by a restrictive enzyme.

translation products of these transcripts accumulate within the cells since lysis is blocked. In this way, cells have been obtained with a ligase content of about 2% of the soluble cellular protein.

Genes for several other bacterial enzymes have been manipulated in similar ways, often giving amplification several hundred-fold above the normal level. They include DNA polymerase and DNA ligase from *E. coli*, as well as some restriction and modification enzymes. However, problems sometimes arise due to the lethal effects of high levels of a particular gene product. In such cases it is useful to work with cells lysogenized with a phage that can be induced at the appropriate stage of the growth cycle so that the recombinant is propagated as a single gene copy per bacterial chromosome and then induced to give many functional gene copies at a stage when the cells can be sacrificed.

Several examples also exist where addition of a simple inducer compound at the appropriate stage of growth of the culture results in derepression of genes cloned in plasmids and hence higher levels of transcription. The *trp* operon of *E. coli*, which can be induced by addition of indole acetic acid, has been harnessed effectively for this purpose and in some cases it has been claimed that the product of genes placed under this regulatable promoter constituted over 20% of the soluble protein of the cell.

Another useful and widely used system for amplification of gene expression, combines a thermosensitive repressor gene of phage λ (usually c_I) in plasmids of the *colE* type which exist in a reasonably high copy number in the bacterial cell, but this copy number can be greatly increased if a particular mutant is used.

For gene expression, the importance of mRNA structure and stability, as well as the efficiency and position with respect to the initiating codon of the nucleotide sequence responsible for ribosome binding, are increasingly apparent as more examples of eukaryotic genes (or their cDNA equivalents) are studied in bacterial environments. These aspects are encountered in some of the specific examples described below.

Stability of the newly synthesized gene product within its cellular environment is obviously important, but is difficult to predict; a number of eukaryotic gene products when made as the native polypeptide, or as fusion products to other (prokaryotic) peptides, prove to be quite unstable. Use of host strains deficient in proteases sometimes helps, but more often it is necessary to make alternative gene structures such as an arbitrary range of deletion mutants, or to consider a different microbial host species. Synthetic or semisynthetic oligonucleotides are often useful in such experiments.

19.3   Some examples of applications

The following examples illustrate the type of problem where the general procedures described above have been used on both prokaryotic and eukaryotic systems to improve yields of particular products, to provide alternative routes to high value products such as hormones that cannot be obtained in large quantities by conventional methods, and to provide a safer and more convenient way of handling pathogenic organisms such as some viruses.

19.3.1   Adaptation of biosynthetic pathways

Not all microorganisms use the same metabolic pathway to a particular compound, and what is less efficient in one context may be more desirable in another. Uptake of ammonia may be effected through the use of glutamate dehydrogenase (GDH) to convert 2-oxoglutarate into glutamic acid; the pathway is relatively inefficient in terms of ammonia but uses little energy. The organism *Methylophilus methylotrophus*, a commercial source of single cell protein (see Chapter 10), uses a two-stage pathway involving glutamine synthetase (GS), and glutamate synthase (GOGAT) to effect the same conversion; ammonia uptake is very efficient but the pathway utilizes an additional molecule of ATP for each glutamate molecule formed (Fig. 19.8); this reduces the biomass yield from the cost-determining sub-

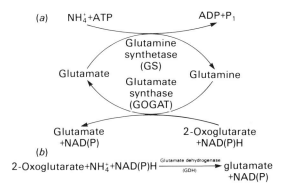

Fig. 19.8 Pathways of ammonia assimilation by bacteria. (a) GS/GOGAT; (b) GDH.

strate, methanol. Transfer of the gene for GDH from *E. coli* into a strain of *M. methylotrophus* lacking GOGAT through the use of plasmids having an extended host range reduced the ATP requirement for ammonia assimilation by the organism, with a consequent improvement in the conversion of the substrate. These elegant experiments, from the research laboratories of ICI Limited, provide an excellent example of the application of genetic engineering methods to yield improvement of a commercial bulk product (see Chapter 10).

19.3.2 Bacterial enterotoxins

With many pathogenic bacteria infection begins with a specific adhesion of the bacteria to epithelial cells; enterotoxigenic strains of *E. coli* adhere specifically to the epithelial cells of their host's small intestine and secrete toxins that cause diarrhoea, often severe. The adhesive capacity of *E. coli* strains is host-specific and associated with characteristic proteinaceous antigens on the bacterial fimbria (or pili) which are encoded on large plasmids. A genetic analysis of this system in *E. coli* strain K88, carried out with the aid of plasmid cloning methods by Kehoe *et al.* (1981), showed that four genes were involved in expression of the adhesion (*adh*) system, three of which were expressed as a single operon. The polypeptides corresponding to these four genes were identified and from an analysis of a series of deletion and insertion mutants a scheme was advanced for their role in the formation of pili. The product of one gene (*adhC*) appeared to be a positive regulator for expression of the *adhD* product which was identified as the subunit of the fimbria, and hence the major antigen; a large polypeptide (70 K) encoded by *adhA* is probably necessary (perhaps together with the *adhB* product) for attachment of the fimbria to the bacterial cell wall. This picture provides the background necessary for manipulation of the *adh* gene cluster for the production of vaccines against K88; strains carrying a plasmid engineered for high levels of expression of the *adhD* product on the cell surface would be candidates for oral vaccines. Purification of the K88 antigen in large quantities would be greatly simplified by construction of a mutant defective in *adhA* which could secrete large quantities of the K88 antigen into the culture medium.

19.3.3   Viral antigen genes and their expression in *E coli*

Expression of human virus genes in *E. coli* has been achieved with hepatitis B virus (HBV), the causative agent of serum hepatitis which constitutes a serious and world-wide problem in public health. Fundamental studies of this virus have been seriously impaired because it cannot be obtained in large quantities and cannot be grown in cells in tissue culture. The virion, or Dane particle (Fig. 19.9), is roughly spherical with a diameter of 42 nm.

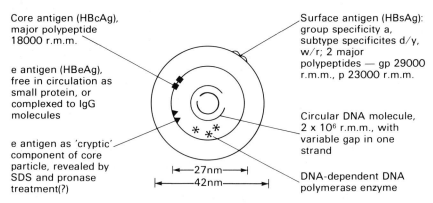

Core antigen (HBcAg), major polypeptide 18000 r.m.m.

e antigen (HBeAg), free in circulation as small protein, or complexed to IgG molecules

e antigen as 'cryptic' component of core particle, revealed by SDS and pronase treatment(?)

Surface antigen (HBsAg): group specificity a, subtype specificites d/y, w/r; 2 major polypeptides — gp 29000 r.m.m., p 23000 r.m.m.

Circular DNA molecule, 2 x 10⁶ r.m.m., with variable gap in one strand

DNA-dependent DNA polymerase enzyme

|←—27nm—→|
|←———42nm———→|

Fig. 19.9   Hepatitis B virus; its components and a diagrammatic representation of its structure. The intact virion has a diameter of 42 nm, often referred to as a Dane particle, from which the smaller core particle is derived. The surface antigen often occurs in a highly associated form visible under the electron microscope as cylindrical or rod-like structures. (Figure kindly provided by Dr Patricia MacKay.)

It comprises the surface, core and 'e' antigens, a viral DNA-dependent DNA polymerase, and a circular DNA molecule, relative molecular mass (r.m.m.) 2×10^6; the latter contains a nick in one strand and a large gap in the other through which it can be labelled *in vitro* by the endogenous polymerase on incubation with radioactive deoxynucleoside triphosphates. To obtain expression of the HBV genes in *E. coli*, fragments of DNA isolated from Dane particles can be produced by digestion with various restriction enzymes and cloned in the plasmid pBR322 at the *Pst* site by the 3' dC, dG tailing method. The plasmid contains a single target for the restriction enzyme R.*Pst*I, which is located within the gene that determines resistance to ampicillin (*Amp*^{r}) and through which it is converted into a linear molecule. Oligo(dG) sequences are then synthesized at the 3' termini by the action of polynucleotide terminal transferase. DNA fragments to be joined to these molecules are prepared either by digestion of a DNA preparation with restriction enzymes (or by shearing) or by the action of reverse transcriptase upon an RNA preparation, and oligo(dC) sequences are synthesized at their 3' termini as before.

Recombinant plasmids formed by annealing the DNA molecules with complementary 3' single-stranded projections effectively transform competent cells, the necessary exonucleolytic, synthetic and ligation reactions being completed *in vivo* to give the covalently linked recombinant. (Note that the recognition target for R.*Pst*I is regenerated at the junctions of the two components.) Transcription from the β-lactamase promoter gives an mRNA that is translated to a polypeptide comprising the first 182 residues of β-lactamase (penicillinase) and, when the number of connecting G residues is such as to maintain the correct translational phase, this is linked by a few glycine residues to the polypeptide sequence from the cloned gene or DNA fragment of interest. Although this does not give a functional penicillinase, the gene for tetracyclin resistance (*Tet*^{r}) remains intact and provides a means for selecting transformants. The crucial point of this method is that the population of recombinant DNA molecules produced in the biochemical reactions comprises a number of molecules that have the same plasmid and donor DNA fragments, but which differ in the number of G residues by which their two components are joined. There is thus a one in three probability (actually one in six depending upon relative orientations of the two fragments) that the nucleotide sequence of the donor DNA fragment will be in the correct translational phase to give a gene product consisting of the first 182 residues of β-lactamase fused by a number of glycine residues to the sequence corresponding to the cloned gene or its fragments (Fig. 19.10). Cells carrying the recombinant plasmids were screened for synthesis of the virus core and surface antigens by the disc radioimmunoassay method (Fig. 19.11) with ^{125}I-labelled antibodies; of some 400 colonies screened, 13 gave positive reactions for the core antigen. When injected into rabbits, extracts from these bacteria elicited the formation of antibodies that, in the Ouchterlony immunodiffusion assay,

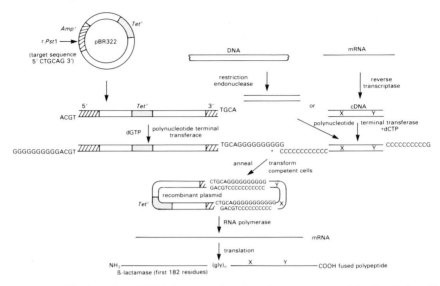

Fig. 19.10   The insertion of DNA fragments into the β-lactamase gene of the *E. coli* plasmid pBR322 for expression of the inserted genes as polypeptides fused to β-lactamase.

cross-reacted specifically with HBV core antigen from a human hepatoma (Fig. 19.12).

Several of the cloned viral DNA segments have been analysed in detail and the complete HBV DNA sequence determined. The nucleotide sequence corresponding to the core antigen has been identified and by suitable subcloning under other promoters the yield of the core antigen in *E. coli* can be increased greatly. This provides a source of the antigen for use in diagnosis, and this is now a commercial product.

Nucleotide sequences that correspond with the parts of the amino acid sequence that have been determined for the surface antigen were also located and appropriate fragments sub-cloned for expression of the surface antigen as a polypeptide fused to β-lactamase; yields were not high, but the product was immunogenic in rabbits.

More promising results for the expression of the surface antigen have been obtained from experiments where the surface antigen gene was transferred to a vector comprising part of the *E. coli* plasmid pBR322 and a selectable marker and origin for replication in yeast. *Saccharomyces cereviseae* transformed with this recombinant plasmid produces a particulate form of the antigen, and although yields may not be spectacular, the product will cross-react well with human antibodies to the HBV surface antigen and appears to be a good immunogen.

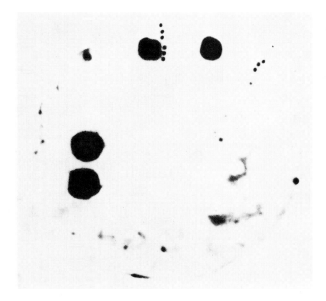

Fig. 19.11 The radioimmunoassay for detection of bacterial colonies producing an antigenic polypeptide. The example shows that 4 out of the 52 colonies imprinted upon the plastic disc coated with anti-HBc gave positive reactions when the disc was subsequently treated with ^{125}I-labelled HBc antibodies.

Many viruses have RNA genomes and in some picorna viruses, such as polio virus or foot and mouth disease virus (FMDV), the RNA is translated to give a very large polypeptide which is subsequently processed to yield the various viral capsid proteins. One of these (VP1) is of particular interest since it stimulates synthesis of neutralizing antibodies and may thus contribute to the high degree of antigenic variation encountered with this virus. Cloning and analysis of the corresponding segment of the viral genome was therefore of academic interest and also of applied value as a potential alternative source of vaccine.

Double-stranded cDNA preparations were made from the single-stranded viral genome and cloned in plasmid pBR322 for propagation in *E. coli*. Analysis of these recombinants gave a restriction map of the cDNA which was aligned with the known biochemical genetic map of the RNA genome. The major part of the coding sequence for VP1 was then dissected and fused in the correct translational phase to a vector carrying the information for the first 99 amino acid residues of the coliphage MS2 polymerase under the control of a thermosensitive phage λ repressor. Heat induction of cells transformed with this reconstructed plasmid gave good yields of

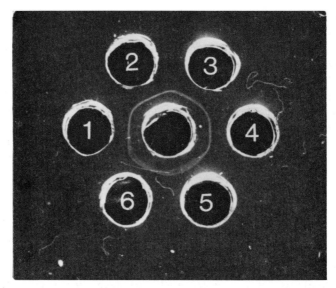

Fig. 19.12   Comparison, by the immunodiffusion method, of human antibodies to HBcAg
with those from rabbits injected with HBcAg made in *E. coli*. The centre well
of the gel contained partially purified HBcAg derived from human liver. Wells 1
and 4 contained human anti-HBc at a dilution of 1 in 30; wells 2 and 3 contained
undiluted serum from rabbits injected with the unfractionated extract of bacteria
carrying clone pHBV139; wells 5 and 6 contained undiluted rabbit serum raised
against HBcAg from the same bacterial extract, but fractionated on a column
of Sephadex G50.

a fused polypeptide of 396 amino acids (284 from the FMDV VP1 sequence)
which was detected initially by the disc radioimmunoassay method. The
product was subsequently purified and shown to be immunogenic when
inoculated into animals.

The experiments with hepatitis B virus DNA and FMDV RNA not only
illustrate the way in which the new genetic engineering methods can be
used for detailed molecular biology studies of a viral system that is otherwise
very difficult to work with, but they also show the power of these methods
and suggest that they may provide the preferred way for working with
pathogenic organisms. They already provide new and sensitive diagnostic
reagents and they promise new methods for vaccine production.

19.3.4   Peptide hormones

Somatostatin was the first animal hormone to be synthesized in *E. coli*,
by insertion of a synthetic oligonucleotide (of sequence corresponding to

the known amino acid sequence of the hormone) into an appropriate plasmid vector. Insulin and chorionic somatomammotropin, a polypeptide hormone resembling β-gonadotropin, have also been produced in *E. coli*, as, more recently, have enkephalins. However, the expression in *E. coli* of human growth hormone, a polypeptide of 191 amino acid residues, is a particularly interesting example, for it illustrates the powerful range of operations that can now be brought to bear upon this type of problem (Goeddel *et al.*, 1979). In these experiments, a gene for human growth hormone (HGH) expression in *E. coli* was assembled by a combination of chemical synthesis and biochemical reactions with mRNA isolated from pituitary glands.

A number of successes have been recorded with cDNA for the synthesis in *E. coli* of polypeptides fused to a prokaryotic peptide such as β-lactamase but these products cannot easily be processed to the native protein. Use of a cDNA preparation for HGH would give rise to the pre-hormone, which has 26 additional *N*-terminal amino acid residues and which would not be readily convertible into the hormone itself. The strategy chosen, therefore, made use of the known nucleotide sequence of the cDNA for HGH, which revealed a restriction enzyme (*Hae* III) target some 70 nucleotides into the coding sequence for the native hormone, and a synthetic oligonucleotide with a sequence corresponding to the *N*-terminus of the hormone and extending to this *Hae* III site.

Preparations of mRNA isolated from human pituitaries were used to synthesize cDNA which was digested with *Hae* III and the products fractionated electrophoretically so that molecules of the appropriate size (about 550 base pairs) could be selected for cloning in pBR322 at its *Pst* site. Here the 3' oligo dG:dC tailing procedure (Fig. 19.10) was used in order to simplify subsequent recovery of the cloned fragment (by digestion with *Pst* since the target sequence is regenerated). The cDNAs of human chorionic somatomammotropin and human growth hormone are nearly identical so that the former, when labelled with ^{32}P, provided an excellent hybridization probe for human growth hormone coding sequences. Transformants that hybridized with this probe thus provided the major segment (equivalent to amino acid residues 25–191) of the growth hormone gene which was isolated and ligated to the synthetic oligonucleotide carrying the information for the *N*-terminal peptide sequence. This coding sequence was then placed in a plasmid so that its control was regulated by the *E. coli lac* promoter and synthesis of the gene product was detected by radioimmunoassay methods. A number of derivatives were obtained with differing distances between the ribosome binding sequence and the initiating ATG. The best of these gave a high level of synthesis of human growth hormone, estimated to be nearly 200 000 molecules per bacterial cell. Some instability of the hormone was encountered when cells were grown to stationary phase, but this did not appear to be excessive in cells harvested in late log phase.

19.3.5 *Interferons*

One of the most exciting developments arising from the new gene technology has been the demonstration of the synthesis in *E. coli* of polypeptides with human interferon activity. This was first achieved with leukocyte (now often termed α) interferon, but fibroblast (or β) and immune (or γ) interferons have also been produced in a similar way.

Interferons are glycoproteins produced by cells of most vertebrates in response to invasion by viruses, or some other inducers. They induce a virus-resistant state in the infected or induced cell which is accompanied by the *de novo* synthesis of a number of proteins, and they may also regulate the immune response and killer lymphocyte activity. The different classes of interferons are encoded by different genes; they differ immunologically and in their target cell specificity. All the interferons are produced in minute quantities, and this presented a major obstacle to their purification and hence to both biochemical and biological studies. They were therefore prime candidates for preparation through molecular cloning methods and although identification and selection of their coding sequences at either mRNA or gene level was a particularly obtuse problem, the availability of a very sensitive bioassay made approaches through the detection of gene expression worth attempting.

In order to obtain interferon mRNA, leukocytes were primed with a little leukocyte interferon, induced with Sendai virus, and then used for preparation of total mRNA (i.e. polyA-RNA). That the preparation contained interferon mRNA was established by injection into oocytes and measurement of the interferon produced by a plaque reduction assay. This assay was also used to monitor fractionation of the mRNA to give roughly a hundred-fold enrichment before treatment with reverse transcriptase to produce cDNA which was then inserted into the plasmid pBR322 as in the instances described earlier. A large population of transformants was then screened for the presence of interferon coding sequences by a complex assay that combined the classical sibling selection procedure of bacterial genetics with a nucleic acid hybridization and translation assay.

Briefly, fractions of the total population of the recombinants were grown as mixtures in order to amplify the quantity of plasmids which, as a mixed population, were then linearized and hybridized with leukocyte mRNA and the hybrid nucleic acid was removed on a millipore filter. The hybridized RNA was then eluted and injected into oocytes which were subsequently assayed for interferon as before. Groups of recombinants giving a (weakly) positive indication of a capacity to direct interferon synthesis were then subdivided and the process repeated until individual clones giving a positive result were obtained. The first contained a DNA fragment smaller than anticipated, but they served as hybridization probes for the screening of other recombinants from which the full interferon cDNA sequence was recovered and transferred to other plasmids for expression via the β-lactamase promoter. A range of activities was observed when extracts from a number of transformants were assayed for interferon. Extracts from cells showing the highest activity were used to compare the interferon synthesized in *E. coli* with that produced by leukocytes. The interferon activity was destroyed by digestion with proteolytic enzymes, but resembled the leukocyte preparation in its stability to low pH, its cell specificity, its approximate size, and its reaction with specific antisera.

The cloned interferon coding sequence has subsequently been put into other plasmid constructions that give greatly enhanced levels of expression and the product is now being produced in large quantities by fermentation on a commercial scale. Some twenty-five years after the discovery of interferon it will now be possible to assess properly the value of interferons, both as antiviral agents, where α interferon already shows appreciable promise, and as antitumour agents, where much more information from extensive trials is needed for critical evaluation. Of no less importance has been the use of the cloned interferon coding sequence for further basic biological studies. Its use as a hybridization probe on genomic libraries has revealed the presence of at least fourteen human α interferon genes which diverge significantly in parts of their sequence and which do not contain introns. These genes have been cloned in systems for their expression so that the biological properties of the gene products can be studied. The β and γ interferons are being produced and studied in an analogous fashion, but they do not exhibit the same degree of polymorphism as the α interferons.

19.4 Conclusion

The purpose of this chapter has been to provide a description of the basic genetic engineering procedures and the principles behind them, and to illustrate through a few selected examples the range of applications to which this new technology is particularly suited. The examples are intended to show both the problems involved and the vast opportunity for future applications for basic biological studies as well as commercial enterprises. The principles and work described have centred very largely upon the microorganism *E. coli*, but this again is an example, albeit a widely and highly profitably used one. Increasing use is being made of *Bacillus subtilis*, various *Streptomyces* strains and the yeast *Saccharomyces cereviseae*, as well as some higher cells. The eukaryotic systems are of particular interest for the direct expression of higher organism genomic sequences that contain introns and for the synthesis of glycopeptides. In plant cells, the tumour-inducing (Ti) plasmid of *Agrobacter tumorfaciens* has been shown to be an effective vector, and with animal cell cultures direct transformation with DNA precipitated upon calcium phosphate is being used widely in addition to transformation with vector molecules based upon viral genomes such as simian virus 40, polyoma, and bovine papilloma virus. A particularly interesting vector for animal cells (a mutant Chinese hamster ovary cell line) is based upon the dihydrofolate reductase gene, the presence of which permits the mutant host line to grow, hence giving a selection for trans-formed cells. In the presence of methatrexate this gene becomes amplified many times and genes linked to it may be similarly amplified.

Altogether, a great deal of interest and opportunity lie ahead.

Reading list

Goeddel, D. V., Heyneker, H. L., Huzumi, T., Areutzen, R., Itakura, K., Yansura, D. G., Ross, M. J., Miozzari, G., Crea, R. and Seeburg, P. H. (1979). *Nature, Lond.* **281,** 544–548.

Hayes, W. H. (1968). 'The Genetics of Bacteria and their Viruses', 2nd Edn. Blackwell: Oxford.

Hinnen, A. and Meyhack, B. (1982). *In* 'Gene Cloning in Organisms other than *E. coli*' (Eds P. H. Hofschneider and W. Goebel), pp. 101–118. Springer-Verlag: Berlin.

Kehoe, M., Sellwood, R. S., Shipley, P. and Dougan, G. (1981). *Nature, Lond.* **291,** 122–126.

Kornberg, A. (1980). 'DNA Replication'. W. H. Freeman and Co.: San Francisco.

Murray, K. (1978). *In* 'Genetic Engineering' (Ed. A. M. Chakrabarty), pp. 113–122. CRC Press Inc.: West Palm Beach, Florida.

Murray, N. E. (1983). *In* 'The Bacteriophage λ II' (Eds R. Hendrix, R. A. Weisberg, F. W. Stahl and J. Roberts). Cold Spring Harbor Laboratories: Long Island, New York.

Nagata, S. Taira, H., Hall, A., Johnsnid, L., Strenli, H., Ecsodi, J., Boll, W., Cantell, K. and Weissman, C. (1980). *Nature, Lond.* **284,** 316–320.

Schell, J. and Van Montagu, M. C. (1982). *In* 'Gene Cloning in Organisms other than *E. coli*' (Eds P. H. Hofschneider and W. Goebel), pp. 237–254. Springer-Verlag: Berlin.

20 | Processes and Products Dependent on Cultured Animal Cells

R. E. SPIER

20.1 Historical

Although many investigators had previously studied the behaviour of animal cells *in vitro*, the first application of such cells which led to a useful product was in 1949 when J. F. Enders demonstrated that polio virus could grow in cultured animal cells from neural and nonneural tissues (kidney) derived from primates.

The pioneering work which led up to this event may be summarized briefly. It began as early as 1880 with the demonstration that leucocytes would divide outside the body (Arnold) which was followed by observations on the behaviour of excised pieces of animal tissues immersed in serum, lymph or ascites fluid. The hanging drop method of R. Harrison (1907), whereby a piece of tissue (tadpole spinal chord) was held in lymph on the underside of a coverslip which was sealed on to a hollowed-out microscope slide, is regarded as a turning point. This work was extended by Carrel (1913) who developed a complicated methodology for maintaining cultures free of extraneous contaminants (particularly bacteria). Few other people were so capable. However, media were developed to promote animal cell growth so that, by 1928, the Maitlands were able to grow virus in minced mouse or chick embryos maintained *in vitro*. This methodology was used by Enders for his polio work where he and his followers were considerably assisted by the use of antibiotics which had become available during the previous 10 years. The polio vaccines which were produced by Salk in the early 1950s were produced in roller tube cultures of monkey kidney or testicle. Once this technology became established other vaccines were produced from chick or primate embryo cells grown in culture [measles (1958); mumps (1951); adenovirus (1958)].

20.2 Types of products which can be obtained from cultured animal cells

Animal viruses were, and still are, the most commercial product derived from cultured animal cells. At present, about 1.5×10^9 doses of foot and mouth disease vaccine are produced annually, a figure which is approached by the poultry vaccines for Newcastle ($\sim 1 \times 10^9$) and Marek's Disease (0.5×10^9). Human viral vaccines are administered at the rate of 10^8 doses per annum or below. Processes for the production of interferon are also under development at scales approaching or exceeding those used for the veterinary vaccines, while the relatively unexploited area of the immunobiologicals, derived from specifically synthesized hybridoma cells, will undoubtedly become an area of significance during the next decade. In Table 20.1 the major products which either already are, or are likely to be, gener-

Table 20.1   Products generated from cultured animal cells

Virus vaccines	Immunobiologicals
Foot and Mouth	Monoclonal antibodies
Polio	diagnosis
Mumps	preparative
Measles	drug targeting
Rubella	passive vaccines or therapeutic agents
Rabies	investigative science
Tick Borne Encephalitis	
Newcastle	Hormones
Mareks	Growth Hormone
Rinderpest	Prolactin
Fish Rhabdovirus	ACTH
Cellular chemicals	Virus predators
α-Interferon (lymphoblastoid)	Insecticides
β-Interferon (fibroblastic)	
Interleukin-2	
Thymosin	
Plasminogen activator (Urokinase)	

ated from cultured animal cells are presented. This is not an exclusive listing but it does represent the major product areas, and the types of materials which can be generated within those areas.

20.3   Overview of methodology for product generation

The basic outline of the generation of a product from an animal cell culture system is depicted in Figs 20.1–20.3. It consists of three phases. The first

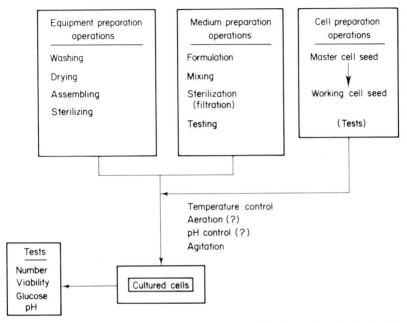

Fig. 20.1   Outline of the steps involved in generating a product from cultured animal cells.

phase is preparative, the second involves the cultivation of the animal cells, while the third phase is that of product generation. This last phase is not separated from the cell production phase in some systems (hormone and immunobiological production), while in others it involves the infection of a cell culture with a virus (for the production of a viral vaccine) or with an inducing agent(s) (for the production of the interferons: Fig. 20.3).

   Much of the effort involved in such production operations goes into quality control of all components, and the on-line and off-line monitoring of as many biological parameters as is feasible. The quality control tests are designed to ensure (a) that the materials used will promote the generation of cells or product, and (b) that they are free of exogenous organisms, i.e. contaminants.

   It is clear from Fig. 20.2 that preparation of the growth medium is a laborious and skilled process. While the defined materials of the medium are relatively easy to control, the two components which provide most difficulty are the water and the serum.

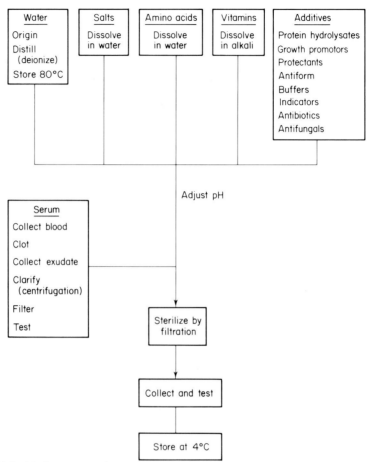

Fig. 20.2   Medium preparation.

Water can be obtained from a wide variety of sources and even after distillation and deionization it may contain materials detrimental to cell growth. For this reason, melted ice from deep within a Greenland glacier is a useful reference material! Storage of prepared water can lead to difficulties, as bacteria can reach a concentration of 10^5 cells ml^{-1} in distilled water held at ambient temperatures and the products of such bacteria, apart from constituting a variable, can be toxic in very small quantities. This is normally overcome by holding stored water at 80°C, at which temperature it is ostensibly sterile.

Medium can be prepared without serum and some, but by no means all, cells can be induced to replicate in it. However, such cell–medium combinations are not generally used for the production of saleable materials.

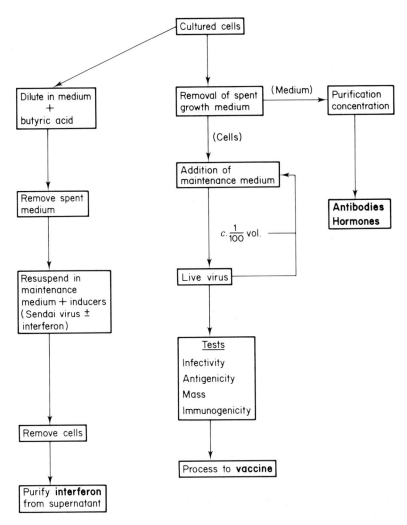

Fig. 20.3 Production generation.

Serum which has been treated to remove the immunoglobulin components can be most useful in virus production systems where locally-produced serum contains antibodies to the production virus. Although bought and tested serum is expensive, it is possible to produce it 'in house'. The thorough quality control of sera is essential for reliable reproducible operations. It is often advisable to maintain a reserve of pretested serum at −20°C, so as to cover for those times of the year (normally the early summer) when the quality of the fresh serum wanes. Fetal calf serum is often used for cell growth, yet this is expensive and scarce. It is also most liable to carry with it contaminants such as phage and mycoplasma. Batch-to-batch variations in the growth promoting properties of fetal calf serum are also observed. With cells derived from Baby Hamster Kidneys (BHK 21 Clone 13) grown in adult ox (bovine) serum a yield of 90% of that of a fetal calf control serum can be obtained.

Animal cells can be made to grow in two basically different modes. Most animal cells can be induced to divide if they are first allowed to attach to a surface or solid substratum, while some can be induced to grow freely in suspension. The former are called monolayer cells because they generally form a layer one cell thick on the substratum, though under conditions of continuous medium renewal tissue several layers thick can form. Cells which can grow independently of a substratum are referred to as suspension cells or anchorage-independent cells.

Cells which can only grow in monolayers are regarded as possessing (or expressing) fewer oncogenes (the genetic sequences which enable a cell to be carcinogenic) than suspension cells and, therefore, such cells are used to produce materials for direct human consumption (virus vaccines). It has also been shown many times that such cells are capable of producing a wider range of virus types than cells which have been selected to grow in suspension. However, if a virus or product can be produced from a cell which grows in suspension, then the scale-up of such a process is relatively easy. Products for veterinary use are made in this way (e.g. foot and mouth disease virus) as well as the α-interferons and the immuno-biologicals derived from hybridomas. Although the cells used for the latter two products can be shown to be carcinogenic in selected test systems, they do not pose a severe hazard either in the production phase or when the product has been freed of cellular contaminants prior to use. In Table 20.2 there is a summary of the advantages and disadvantages of using mono-layer (substrate or anchorage-dependent) cells. It can be concluded that it is necessary to be in a position to produce commercial quantities of both cell types and that two necessarily different technologies are required. Descriptions of these two technologies are presented in the next two sections.

20.4   Monolayer cell growth systems

Systems used to produce cells which require a substratum may be divided into two types. There are those in which one increases the *number* of units of equipment when one scales-up a process. These are the *multiple* processes. Alternative systems are now available whereby scale-up is achieved by increasing the *size* of the equipment—*unit* process systems.

Table 20.2   Advantages and disadvantages of monolayer cells

Advantages	*Disadvantages*
Fewer oncogenes expressed	Need a solid surface for growth (an extra component which adds to the cost)
More susceptible to a wider range of virus types	
Produce β-interferons	Scale-up more difficult
Easy to scale down (<1 ml)	More difficult to control (not homogeneous)
More cell types can grow this way (primary cells, limited life-span cells and some continuous line cells)	Surface parameters have to be precisely described
Easier to standardize the karyology (chromosome profile)	Difficult to obtain a fully continuous process
	Need a trypsinization step between successive cultures and scales of operation
	Cannot produce α-interferon
	Tend to need higher concentrations of better quality serum

20.4.1   Multiple processes for the production of monolayer cells

Traditionally, monolayer cells have been produced on the inner surface of stationary glass bottles or Petri dishes. Such bottles have had a variety of configurations and have gone under the names Brockway (USA), Roux, Carrel, Baxter or Medical Flats. Soda glass may be used, yet borosilicate could be preferred because although it is more expensive it is tougher and easier to repair. The bottles may be used individually, but are more usually stacked in customized racks for ease of transportation (Fig. 20.4).

A development of the stationary bottle culture system which enables cells to grow on all of the inner surface of a bottle rather than the base is the rolling bottle system. For this system cylindrical bottles are generally used, although bottles with rectangular cross-sections held in special jigs can also be rotated. There are many variations possible, with some bottles reaching lengths of 180 cm while others (Burghler bottles) are discarded 2 l Winchester bottles. The bottles are placed on rollers in mechanized racks. A standard piece of equipment can rotate 72 bottles and 6–10 such racks make a conveniently sized production unit. The rolling bottle system has been developed to its limit at the Istituto Zooprofilattico Sperimentale at Brescia, Italy, where 7000 bottles can be rotated simultaneously in each of four incubators (Fig. 20.5). The addition of medium, cells and virus to such a large number of bottles while maintaining freedom from bacterial contamination presents a formidable, but not insuperable, problem and the high degree of automation of the Brescia unit promotes the economic viability of this methodology.

Fig. 20.4   Roux bottle and racks of 10 Roux bottles.

20.4.2   Unit processes for the production of monolayer cells

The development of unit processes for the cultivation of monolayer cells began in earnest in the mid 1960s and has now become an accepted and preferred practical technology. Many of the details of this development have been reviewed recently by the author elsewhere (see Reading list) so that only the broad outlines of the different systems, along with the most recent developments, will be described below.

Following the reliable and successful operation of the multiple processes, the unit processes which were designed to supersede them followed the basic pattern of the multiple system. Thus, unit processes based on enclosing a stack of plates in a vessel were developed, reminiscent of a rack of bottles. Such systems were made from disposable polystyrene (multitray system) or polycarbonate as well as soda glass. Later developments took a stack of plates and rotated them either with the plane of the plates in the vertical configuration or, more recently, with the plates in the horizontal configuration. This latter system has been scaled-up and is at present under intensive exploitation as a means of producing the β- or fibroblastic interferon. The scale-up of such systems presents engineering difficulties as well as

Fig. 20.5 Seven hundred Roller bottles in one of the four such incubators of the Istituto Zooprofilattico, Brescia, Italy (by kind permission of Dr G. F. Panina).

manipulation problems during the cleaning and cell planting operations. The amount of surface which can be packed into a container in this way is limited, which has two consequences:

(a) both sides of the discs have to be used (a methodological problem); and

(b) the product is produced in a dilute form as all the plates have to be irrigated by the growth medium which must either fill, or at least half-fill, the total outer container volume.

The development of the roller bottle system can be seen in the wide variety of multitube cell propagators which have been developed. Such systems have reached a development scale of 200 l volume. In a recent piece of equipment, the Gyrogen, a large surface area for cell growth has been generated from an array of parallel tubes which are rotated by an external drive within an outer cylindrical shell with a volume of 100 l. This expensive piece of equipment can be made to work, but whether it is a cost-effective way of conveniently producing useful materials has yet to be demonstrated.

The contraction of tubes into hollow fibres has led to a number of developments which have unique features. At the small scale it is possible to simulate the structure of an animal tissue with cells growing on the outside of fibres and the nutrient medium percolating through the lumen of the fibre. This system could have advantages in two senses, as both waste materials (which could be cytotoxic) and product materials could be separated from the cells which produced them by the semipermeable membrane of the fibre. Systems of this nature have recently been developed up to the 0.2 l scale and are used commercially in well developed computer controlled and monitored systems.

Spirally wound films or bags have also been used and, although such systems have been shown to be practicable on the small scale, commercial-scale versions have not yet been launched.

Many groups are currently using packed beds for the production of cells and products such as β-interferon, foot and mouth virus and swine vesicular disease virus. The packings have varied from stainless steel springs through polystyrene jacks to glass spheres (Fig. 20.6). Yet the basic system of a packed bed through which medium is made to circulate, with most of the control activity (pH, DO_2) conveniently concentrated in the ancillary medium reservoir, is common to most such systems. They are relatively easy to operate and to scale up, as the mechanical problems of handling uniform column packings are easily solved. As glass is often used as a substratum, the reliability of the operations is improved and the relationship between the cell and the glass surface it has 'learned' to grow on in the laboratory is unchanged. Furthermore, the agencies which control the dissemination of new bioproducts raise fewer questions when presented with products made from cells grown on such a familiar substratum.

Fig. 20.6   One litre and 100 l propagators based on packed beds of 3 mm glass spheres.

Clearly, static packed beds are not without their own problems. Gradients across the bed are not uncommon and there is some difficulty in monitoring a culture one cannot put under the microscope. Although this latter problem may be solved with ease by monitoring components of the circulating medium (glucose concentration for cell growth; lactic acid dehydrogenase activity for cell breakdown and hence virus liberation), the problem of the inhomogeneity of a polyphasic system remains.

Quasi-homogeneous systems for the production of animal cells in mono-layers have been developed by A. J. van Wezel. In 1967 he showed that cells would grow on 0.2 mm diameter particles of DEAE Sephadex A50 coated with collodion. Such particles (relative density 1.05) can be held in suspension with very minimal agitation, and cells would attach to such microcarriers while the system was in its agitated mode (Fig. 20.7). Since

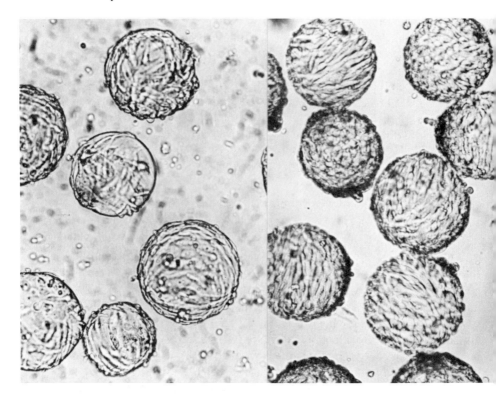

Fig. 20.7   Microcarriers supporting the growth of Baby Hamster Kidney Cell Line at 24 h and 96 h post planting.

these initial observations, many investigators have shown that it is possible to generate products from cells grown on such microcarriers at scales of up to 500 l. The latest such report is the production of polio virus from a green monkey kidney cell line called Vero, which was grown on a modern microcarrier called Cytodex. Other microcarriers have been produced from polystyrene (Biosilon), polypropylene and even glass microspheres. However, it would seem that for most purposes the Sephadex-based microcarriers provide the most suitable substratum.

While many investigators have been able to operate successfully and reliably with microcarrier systems, others have experienced difficulty. Some of these difficulties relate to the particular cell the investigator wishes to grow, which in itself is a product of its tissue of origin and the way it has been grown and subcultured. Other problems stem from the improper use of suitable or unsuitable equipment or the inappropriateness of the medium chosen, either for the period when the cells make the first and critical contacts with the beads or during the later replication of the attached cells.

In conclusion then, for the production of substratum-requiring cells, the microcarrier systems is the preferred methodology under those circumstances when it can be made to work well and reliably. However, packed beds of glass spheres can provide a reliable back-up system. Other systems also have a place, for the provision of that particular environment wherein a unique cell type can grow and generate product.

20.4.3 Processes for the production of suspension cells

The processes for the production of suspensions of animal cells are little different from those used to produce bacterial cultures. A recent review (by the author) describes many of the parameters which have to be monitored and controlled. The equipment used is a fairly standard fermenter modified to permit lower rates of agitation and aeration. Such a system may be defined as the conventional methodology and has been scaled up to 8000 l. At such scales it is used for the commercial production of foot and mouth disease virus from either Baby Hamster Kidney Cells or a different cell line, also derived from hamsters, called IFFA-3. Also, α-interferon produced from the Namalwa lymphoblastoid cell line is currently in production in similar equipment at about the same scale of operation. A basic modification of this agitator-driven system is that equivalent to the air-lift fermenter. Bulk mixing is provided by introducing large air bubbles at the bottom of the fermenter, as these are efficient at moving liquid yet do not damage the cells. Conventional sparging, which attempts to increase the air bubble surface area by creating masses of small bubbles, can cause severe damage to animal cells. The dissolved oxygen level *per se* is not deleterious until it drops below 15% of the amount of oxygen dissolved in the medium at NTP or rises above the amount of oxygen dissolved at NTP, but the interactions of the cell at the air–liquid interface can lead to loss of cell viability. Surprisingly, physical agitation resulting from rapid impeller speed seems to be much less damaging; however, an unambiguous quantitative study of the relative effects of these parameters taken singly and together remains to be made.

While the conventional stainless steel tank system has its place as the work horse of the technology, the system can be modified for use in technologically less developed countries. For this purpose useful quantities of vaccine (5×10^6 doses per annum) can be made in 10 l bottles agitated by an externally driven bar magnet. Such systems have been installed at

Surabaya, Indonesia, where the plant has been operated at about 250 000 doses a month, and at Polgolla, Sri Lanka, where the installation of basic equipment is nearing completion.

The continuous cultivation of animal cells is a practicable and useful method either for generating cells or for gathering information about those cells. Recent work has demonstrated the advantages of on-line computer monitoring and control of such a system, and the necessary computer hardware can be purchased relatively easily.

However, the exploitation of the continuous production of animal cells and the generation of product from such cells at levels which make it commercially attractive has not yet been achieved.

20.5   Downstream processing

While many of the downstream processing operations applied to the raw materials generated by animal cells have their equivalents in other areas of the chemical technology of proteins, there are some operations which are unique and will be considered in more detail below. Operations which are common to other process technologies, such as filtration, centrifugation (moderate speed), precipitation, chromatographic purification, and concentration (reverse osmosis) drying, will not be dealt with. The inactivation of living viral products, the generation of active subunits from whole virus particles and the formulation of materials into useful products are areas which may be considered unique to processes based on the use of animal cells.

20.5.1   Inactivation

Although many vaccines are made from infectious virus particles, others are made from killed or inactivated virus. In the latter case it is important to ensure that there is almost no risk of the recipient of the vaccine contracting the disease as a result of a residue of infective virus. Agents used for such inactivations are commonly formaldehyde (Waldman–Schmidt process for foot and mouth disease vaccine, killed polio vaccine, influenza), whereas β-propiolactone is used to inactivate Rabies Virus. More recently, the imines (acetylethyleneimine and ethyleneimine) have been used to inactivate foot and mouth disease virus. While the imines contribute to a slight destabilization of the foot and mouth disease virus structure, the formaldehyde treatment cross-links the capsomeres, holding the basic structure together more tightly. Other methods of inactivation such as ultraviolet irradiation or glycidaldehyde can be used, yet in all such cases it is important to ensure that the viral preparation to be inactivated is monodisperse so that the inactivant is not prevented from reaching virus embedded in the middle of a clump.

20.5.2   Subunit formation

Viral vaccines which constitute a danger to the recipient of the vaccine if improperly inactivated, can be made into a subunit vaccine whereby whole virus particles are disrupted using a detergent (Tween 80, Triton X, tri-(*n*-butyl) phosphate) and the final vaccine is made from an immunogenic fraction of the dissociated virus. Prime candidates for such subunit vaccines are rabies, hepatitis B, herpes simplex, cytomegalovirus and influenza. To date, trials with the isolated haemagglutinin glycoprotein immunogen from the influenza virus have shown that such a vaccine can be effective, although not quite so effective as the native virus. The studies which have been done so far with glycoproteins isolated from viral preparations show that such materials on a weight for weight basis are less efficient than native virus, but when sufficient material is administered adequate protective responses may be obtained.

20.5.3   Product formulation

Live virus vaccines generally contain from 1 to 4 different virus types, each of which is represented by about 10^3 live virus particles per dose. Such materials are freeze dried in the presence of bioprotectants such as sucrose, human or bovine serum or albumin glutamine and phosphate ions. Killed virus vaccines, on the other hand, are formulated at about 10^{11} particles per dose and are mixed with materials called adjuvants which potentiate the effect of the injected immunogens. The adjuvants are a heterogeneous group of chemicals such as the cardiac glycoside saponin (or its refined component Quil A), various aluminium salts (hydroxide, sulphate, phosphate) and oleaginous formulations where the aqueous suspension of virus is homogenized with a biocompatible oil, forming a water-in-oil emulsion which is then transformed into an oil-in-water emulsion by a second emulsification with an aqueous salt solution containing a detergent. Such materials may have many coincident effects as they provoke the immune system physically and chemically and make antigen available to the system at a rate which generates the maximum response.

20.6   Genetically engineered animal cells and bacteria

The ability to insert defined sequences of nucleotides into both prokaryotic and eukaryotic cells and then to collect, as a product, the proteins which are defined by the inserted nucleic acid, has far-reaching consequences for both bacterial and animal cell biotechnologies. While it is unlikely that animal cell based processes for the generation of live virus vaccines will be superseded by processes based on genetically engineered bacteria, it is clear that some animal cell products such as insulin and some of the α-interferons will be produced commercially from prokaryotes. Other materials such as the killed virus vaccines for foot and mouth disease virus and polio and the various interferons may be made commercially in either cell type. However, genetically engineered animal cells, which are capable of making glycosylated proteins could be the preferred substratum for the production of human growth hormone and the blood clot dissolving enzyme inducer, tissue plasminogen activator.

We are at present witnessing the beginning of a new era of augmented biotechnological capability. It is too early to predict prospective developments with confidence, yet it is certain that these multipotential methodologies will make a major impact on the ways in which we manufacture bioproducts in the future.

Reading list

Girard, H. C., Sütcü, M., Erden, H. and Gürhan, I. (1980). Monolayer cultures of animal cells with the Gyrogen equipped with tubes. *Biotechnol. Bioengng.* **22,** 477–493.

Horodniceanu, F. (1976). The problem of contaminants in the sera used for cell culture—a review. *In* Proceedings of the First General Meeting of the European Society for Animal Cell Technology, Amsterdam 1976 (Eds R. E. Spier and A. L. van Wezel), pp. 23–27. Rijks Institut voor de Volkgezondheid: Bilthoven.

Kurz, W. G. W. and Constabel, F. (1979). Plant cell suspension cultures and their biosynthetic potential. *In* 'Microbial Technology', Vol. 2 (Eds H. J. Peppler and D. Perlman), pp. 389–417. Academic Press: New York and London.

Spier, R. E. (1980). Recent developments in the large scale cultivation of animal cells in monolayers. *In* 'Advances in Biochemical Engineering', Vol. 14 (Ed. A. Fiechter). Springer-Verlag: Berlin.

Spier, R. E. (1982). Animal cell technology—an overview. *J. Chem. Technol. Biotechnol.* **32,** 304–312.

Waymouth, C. (1972). Construction of tissue culture media. *In* 'Growth Nutrition and Metabolism of Cells in Culture', Vol. 1 (Eds G. H. Rothblat and V. J. Cristofalo). Academic Press: London and New York.

21 | Products from Plant Cells

M. W. FOWLER

21.1 Introduction

21.1.1 Perspectives

Within the plant kingdom are to be found a vast array of chemical structures. Some are small and comparatively simple molecules, such as sugars and amino acids, while others are large and complex, for example starch and cellulose. Between these extremes are to be found a whole range of structures of varying degrees of complexity, many of which are often placed under the general heading of 'secondary products'; many of these 'secondary metabolites' undergo active metabolism within the plant and in many cases they play a significant part in the protection of the plant from attack by other organisms.

Over the centuries man has made extensive use of plant secondary products, for instance as perfumes, flavours, spices and particularly as medicines. Not all of these uses are for discrete single substances; some, such as cocoa butter fat or 'Attar of roses', are complex mixtures and blends, whose precise formulation and product quality control have long been a concern of industry. Where products consist of a single substance, attempts have often been made to produce these through chemical synthesis. Such an approach has however often been constrained by low yield, high costs, difficult chemical conversions, or the need for high purity of a particular isomer from a complex mixture, and in many cases the plant itself has continued to be the more effective means of synthesis.

Plants used as a source of fine or speciality chemicals have traditionally been grown in large plantations, more often than not located in the tropics or sub-tropics. During recent years techniques have however been developed which may in time come to rival and possibly replace traditional plantation systems as a source of plant products. Such techniques are generally encompassed by the terms 'plant cell biotechnology' or 'plant cell culture'. The technology involves the large-scale culture of isolated plant cells under conditions which induce them to synthesize commercially or socially desirable substances characteristic of the parent plant from which they were obtained. Plant cell culture offers many advantages over traditional plantation methods as a route to natural product synthesis. These advantages include:

- independence from environmental factors, including climate, pests, geographical and seasonal constraints;
- a defined production system with greater process control, and production as and when required;
- more consistent product quality and yield.

In consequence more and more research laboratories are investigating plant cell culture as an alternative route for natural product synthesis, and as an enabling tool to allow the further development of the plant kingdom as a major chemicals resource.

21.1.2   Historical background

Although cell culture technology has only come to the fore in the last five to ten years, the beginnings of the subject may be traced back to the late nineteenth century. The earliest documented work is that of Haberlandt who, in 1898, attempted to culture single cells isolated from various parts of different plants. Although Haberlandt's cell cultures apparently remained viable for quite some time, they showed no sign of cell division. Undeterred, other workers repeated his experiments and by the late 1930s White and Gautheret in particular were able to report growth and cell division in cultures established from a number of different species. As the requirements for successful culture became understood, more species were established in culture. An important development was the regeneration of whole plants from cultured cells. Many plant cells have the property of totipotency, that is, each cell carries in the genome the complete information required to give rise to an exact copy of the parent plant. In consequence it is possible to regenerate whole plants from cells taken from roots, leaves or stems. This ability to regenerate plants from culture is of key importance in horticulture and agriculture, providing a means of rapid propagation, in large numbers, of 'standard' plants. While plant propagation is not the subject of this chapter, it is none the less an interesting and increasingly important aspect of plant tissue culture. Readers who would like further details are referred to Murashige (1978).

In the early days the whole approach to cell culture was very 'hit and miss'. The composition of media was relatively unsophisticated and often ill-defined. Typical media consisted of mineral salts together with organic supplements such as glucose, thiamine, glutathione, cystein and indole-acetic acid (a key plant growth regulator). In many cases coconut milk, the liquid endosperm of the coconut and notorious for its variable composition, was an absolute requirement to sustain growth. During the period 1940–1960 steady progress was made in our understanding of plant cell culture systems; detailed information on the physiology and biochemistry of growth in suspension culture began to be published and studies of natural product synthesis became more extensive. In the mid 1950s the first major proposals for the industrial application of cell cultures to synthesize plant products were openly discussed, and Pfizer Corporation (USA) issued a patent in 1958 encompassing such a possibility. Prospects of achieving such an aim must none the less at that time have seemed limited; cell growth rates were still low, cultures were generally difficult to establish, media were in the majority of cases still undefined, and worse still, products characteristic of the parent plant were either absent in cell cultures or only present in vanishingly low concentrations.

The following decade from 1960–1970 saw major progress in many of these problem areas. Defined culture media were established and growth rates improved. Cultures were identified which would synthesize substances characteristic of the parent plant, and progress also occurred in the large-scale culture of plant cells and our knowledge of the biochemistry and physiology of cell cultures markedly improved.

21.1.3   The plant kingdom as a chemicals resource

The number of chemical structures reported from higher plants runs into hundreds of thousands and is constantly being added to. Something like 1500 new chemical structures from higher plants are reported each year, of which quite a number have some degree of biological activity. Quite obviously not all plants have the same constituents and it is normal to find secondary metabolites confined to a particular plant family, or unique to a particular species. Unfortunately our rudimentary knowledge of plant chemistry and biochemistry still restricts our ability to exploit plant natural product chemistry fully. On the other hand this provides all the more incentive to explore systematically the plant kingdom as a source of novel structures and activities.

Traditionally the plant kingdom is thought of as being a source of ill-defined herbal remedies and potions, many of doubtful efficacy. However, some 25% of prescribed and highly purified drugs are derived from the plant kingdom, including such key therapeutic agents as digoxin (a cardiatonic glycoside from *Digitalis lanata*, the foxglove); quinine (an antimalarial alkaloid from the bark of the *Cinchona tree*) and codeine (an analgesic alkaloid from the opium poppy). Not only do these three have different pharmacological properties, but they also serve to illustrate the diversity of chemical structure to be found in plant natural products (Fig. 21.1).

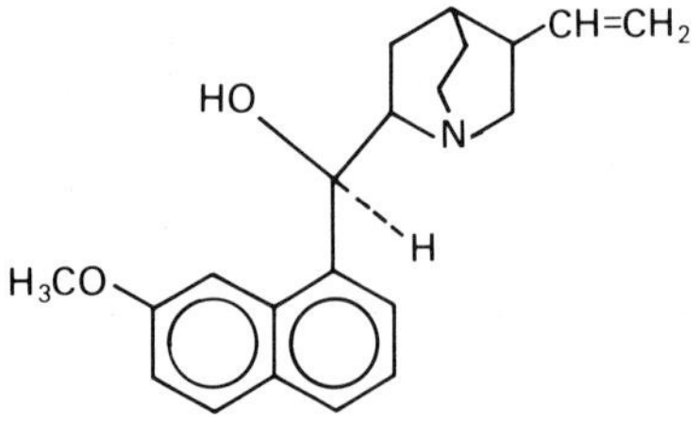

Quinine (antimalarial)

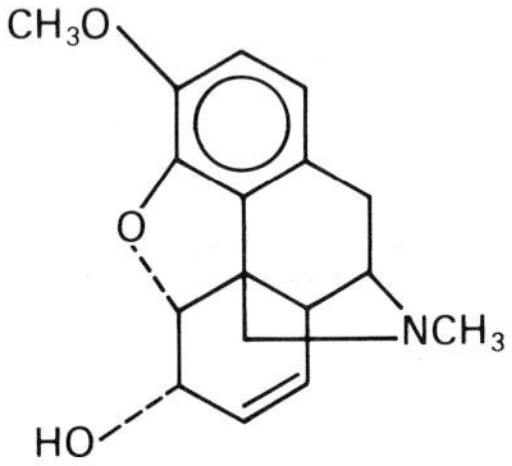

Codeine (analgesic)

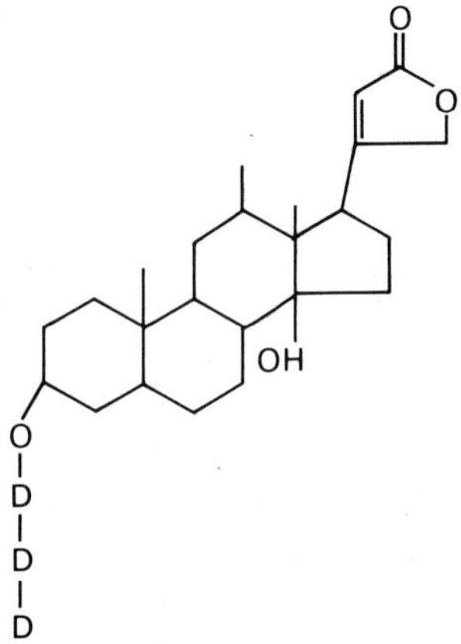

Digoxin (cardiatonic)

Fig. 21.1 Plant natural products: some examples of their diversity of structure and activity.

The range of therapeutic activity of plant-derived drugs is very wide and includes such properties as anticholinergics, antihypertensives and anti-

Table 21.1   Higher plants and drugs derived from them

Species	Drug	Activity
Atropa belladona	Atropine	Anticholinergic
Catharanthus roseus	Vincrystine ⎱ Vinblastine ⎰	Antileukaemic
Chondodendron tomentosum	Tubocurarine	Muscle relaxant
Cinchona ledgeriana	Quinine	Antimalarial
Colchicum autumnale	Colchicine	Anti-inflammatory
Datura metel	Scopalamine	Anticholinergic
Digitalis lanata	Digoxin	Cardiatonic
Dioscorea deltoidea	Diosgenin	Antifertility
Papaver somniferum	Codeine	Analgesic
Uragoga ipecacunha	Emetine	Amoebicide

leukaemics (Table 21.1). While perhaps having their major social and financial impacts in pharmacology, plant chemicals contribute to a wide range of other products, their properties being utilized in applications as diverse as human food additives, perfumes and agrochemicals. The degree of diversity is well illustrated by the examples listed in Table 21.2.

Table 21.2   Plant products and the chemical industry

Industry sector	Plant	Product
Pharmaceuticals	*Cinchona ledgeriana* *Digitalis purpurea* *Pilocarpus jabonandi* *Rauwolfia serpentina*	Quinine Digitoxin Pilocarpine Reserpine
Food and drink	*Cinchona ledgeriana* *Thaumatococcus danielli*	Quinine Thaumatin
Cosmetics	*Jasminum* sp. *Rosa* sp.	Jasmin 'Attar of roses.'
Agrochemicals	*Chrysanthemum* sp.	Pyrethroids

There are four ways in which cell cultures may contribute to the further exploitation of the plant kingdom. The first we have already touched on; that is as an alternative route to natural product synthesis. The second occurs where a novel substance with desirable properties has been isolated from a plant, which, for a variety of reasons may be difficult to grow. If that substance is also difficult to synthesize chemically then an alternative synthetic route through cell culture may well provide the answer. The third approach is in many ways a very exciting one. The plant cell genome contains large numbers of 'silent' genes, i.e. genes present and potentially active but which are not normally expressed under the environmental conditions prevailing in the field today. Cell culture is in itself a form of selection pressure which may allow the expression of these genes. As possible evidence of this there are instances where products of potential commercial interest have been obtained from cultures but which have not been identified in the plant from which the culture was initiated. Given the large amount of DNA present in most plant cells we may speculate on the possibility of a whole host of new substances and perhaps new enzymic mechanisms which may be uncovered from studies with cell cultures. The final approach is through 'biotransformations'. Plants possess a wide range of enzyme systems capable of effecting transformations of molecules which through synthetic organic chemistry are either not possible or are too complicated and costly to contemplate as an industrial application. Such biotransformation systems linked perhaps to immobilized enzyme technology could be the most important area for the development of cell culture technology in the future.

21.2 Nature of cell cultures

Although plant cell culture owes much to standard microbiological practice, there are a number of points of difference which need to be discussed before we move on to review mass growth and product synthesis.

21.2.1 Culture initiation

Plant cell cultures are initiated through the formation of a 'callus' which is essentially a mass of non-developed or non-differentiated cells. A callus is obtained by excising a piece of tissue, called an explant, from the parent plant and placing this onto a nutrient base solidified with agar. This nutrient base contains inorganic macro- and micro-nutrients, carbon and nitrogen sources and various plant growth regulators. The whole process must be carried out under aseptic conditions, and before the piece of explant tissue is transferred to the agar it must be carefully surface sterilized (the surface tissues of plants conceal an abundant microflora). The nutrient medium on which the tissue is placed is particularly rich and provides a good substratum for microorganisms which, unless checked, rapidly outgrow the plant cells, and the need for good aseptic technique cannot be overstressed. Surface sterilization is usually achieved by washing the tissue in 5% sodium hypochlorite, 2% mercuric chloride or 80% ethanol for varying lengths of time, after which the sterilizing agent is removed by repeated washing of the tissue in sterile distilled water. To enhance the effectiveness of the sterilizing agent a wetting agent such as Triton X-100 or Tween is sometimes added.

Caution must equally be exercised not to over-expose the tissues to the sterilizing agent, which may result in deleterious effects, particularly to the surface cells and tissues, and in turn in low cell viability and difficulty in establishing a culture. An alternative approach, with varying degrees of success, is to incorporate antibiotics such as streptomycin and nystatin into the nutrient medium. In some cases such substances not only prevent microbial infection but also affect the growth and metabolism of the cell culture. Instead of taking explants from the plant there is an increasing trend of initiating cultures from seeds. These are generally easier to surface sterilize, without quite the same problems of over-exposure to the sterilizing agent, and in many cases seeds will give rise to callus just as readily as tissue explants.

Once surface sterilized the tissue explant or seed is placed onto the solidified nutrient in a flask covered with material permeable to gases (oxygen, carbon dioxide) but which does not allow the entry of bacterial or fungal spores. Various types of flask closure are in use, ranging from cotton wool bungs, to aluminium foil and sheet polypropylene. Once closed the flask containing the explant is placed in an incubator, generally at 25°C, sometimes with light, sometimes in the dark. It should be noted at this point that although many plant cell cultures are green, there are relatively few reports of cultures which are genuinely autotrophic, a point which is discussed further below. After between 1 and 2 weeks, depending upon a variety of factors including the species, tissue origin, nutrient regime, etc., the explant begins to proliferate cells in the form of a general mass of cell material or callus. The initial sites of proliferation are often aligned with those parts of the explant where cell division had occurred in the parent plant. When this callus mass has reached a reasonable size, maybe 2–4 cm across, it is separated from the parent explant and placed on fresh nutrient media. It then continues to proliferate, typically into a large frizzy mass (Fig. 21.2).

Callus, although an almost obligatory start to the culture process, is far from being an ideal system with which to work. Slow growing and heterogeneous, it is not suitable for scale-up experiments or as a production system, and it is equally unsatisfactory as a tool for biochemical and physiological experiments. Many people who work with cell cultures therefore move quickly on to the next stage, that is into liquid culture.

Successful liquid culture of plant cells depends to a great extent on a 'friable' callus, that is a callus which when placed in liquid in a flask supported on a rotating platform rapidly breaks up under the swirling motion

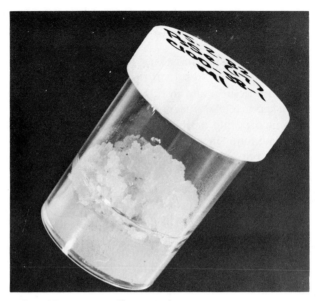

Fig. 21.2   A callus of *Papaver somniferum*—opium poppy.

to give a culture composed of a mixture of free cells and cell clumps. Such a system is much more amenable to biochemical investigation and process development studies than callus; growth is generally much more rapid, more cells are in direct contact with the nutrient and the degree of culture heterogeneity is much reduced.

Suspension or liquid cultures of plant cells are grown in much the same way as microbial cells. They are maintained on orbital shakers, at rotational speeds of about 120 r.p.m. (rather lower than with microbial cells) and the temperature is maintained at between 25 and 27°C (compared with 37°C typically used for bacterial cultures).

21.2.2   Growth regimes

Some aspects of the regimes under which plant cells are grown have been alluded to above, including temperature, rotational speeds on orbital shakers and general composition of nutrient media. Let us return to this latter topic in more detail.

Early growth media for plant cell cultures were largely undefined, and relied heavily on the addition of complex mixtures such as coconut milk to achieve cell growth and division. Such procedures have been largely superseded by the development of defined media and the use of specific natural and synthetic growth regulators. Examples of media used for plant

Table 21.3   Examples of media formulations for the growth of plant cells

Component	Formulation and component concentration $(mg\,l^{-1})$	
	Murashige & Skoog (1962)	Gamborg *et al.* (1968)
$(NH_4)_2SO_4$	—	134.0
$CaCl_2 \cdot 2H_2O$	440.0	150.0
$NaH_2PO_4 \cdot 2H_2O$	—	169.6
KH_2PO_4	170.0	—
NH_4NO_3	1650.0	—
KNO_3	1900·0	3000·0
$MgSO_4 \cdot 7H_2O$	370·0	250·0
$CoCl_2 \cdot 6H_2O$	0.025	0.025
$NaMoO_4 \cdot 2H_2O$	0.25	0.25
$CuSO_4 \cdot 5H_2O$	0.025	0.025
KI	0.83	0.75
H_3BO_3	6.20	10.0
$MnSO_4 \cdot 7H_2O$	22.30	13.20
FeNaEDTA	36.70	40.00
$ZnSO_4 \cdot 7H_2O$	8.6	2.0
meso inositol	100.0	100.0
Nicotinic acid	0.5	1.0
thiamine HCl	0.1	10.0
pyridoxine HCl	0.5	1.0
glycine	2.0	—
sucrose	20×10^3	20×10^3
pH	5.8	5.8

cell cultures are shown in Table 21.3. Although such media are 'defined', in that their constituents are precisely known, none have been properly optimized. Some attempts have been made to optimize the levels of the various inorganic constituents but only with a very restricted range of media and cell lines.

The nitrogen source is typically nitrate or ammonia, although successful growth has been achieved with nitrogen sources as diverse as urea, single amino acids such as glutamate and mixtures such as casein amino acids. The different nitrogen sources may have an effect on cell morphology. Organic sources tend to give cell cultures characterized by the presence of long thin sausage-like cells, while cells from cultures with inorganic nitrogen sources are generally more rounded.

It has already been mentioned that the majority of cell cultures are unable to maintain themselves autotrophically in spite of their producing chlorophyll. The reason for this is largely unknown although there are indications that a lesion in the fatty acid composition of the chloroplast membranes may much reduce the efficiency of electron trapping and transfer. To maintain cell growth and division it is therefore necessary to provide an exogenous carbon source. The major mobile carbohydrate of plants is sucrose and this has been the carbohydrate source of choice for many cell cultures. However in the last few years attempts have been made to grow cell cultures on a variety of carbon sources, some refined, some waste. Good cell growth has been achieved on glucose, maltose and galactose, and some degree of success has been achieved with more diverse substrates, but biomass yields and productivity levels are generally lower than with sucrose or glucose. Of the non-refined or waste substrates, starch produces variable results, depending upon the source, while molasses and milk whey result in little growth or none at all. With whey the low level of growth is probably due to a very low activity or absence of β-galactosidase needed to hydrolyse the lactose in the milk whey to glucose and galactose, both of which the cell cultures are able to utilize.

The utilization of sucrose by plant cells has interesting features. Plant cells possess an invertase in the cell wall which plays some part in hydrolysing the incoming sucrose to its two monomers, glucose and fructose, prior to entry into the cell through the plasmalemma. The degree of hydrolysis varies depending upon the origin of the cell line, the growth rate and the general growth conditions. So far total hydrolysis of the sucrose has only been found necessary for carbohydrate uptake and subsequent cell growth in one cell line, which was isolated from sugar cane (well known for its high rate of sucrose synthesis and level of sucrose accumulation). There is little understanding of the wide variation in the degree of cell wall sucrose hydrolysis which occurs in different cell lines and under different environmental conditions.

Comparatively little information is available on the need by plant cells for the range of macro- and micro-inorganic nutrients used in culture media. Many can be identified as being important cofactors in enzyme reactions, e.g. molybdenum for nitrate reductase, or magnesium for many of the kinases; others are important in electron transport processes, for instance iron. Variation in the level of phosphate has a marked effect on culture growth and productivity, but again little information is available on the precise interactions.

Plant growth regulators (auxin, kinetin, giberellin, etc.) have major effects on cell cultures, both quantitative and qualitative. The range of effects varies dramatically, but again our understanding of the biochemistry and physiology of these substances is extremely limited and their use is essentially empirical.

Temperature and pH also exert a major influence. In general, cell culture media are not purposely buffered, the media composition itself ensuring a limited degree of buffering capacity. The nutrient medium is initially adjusted to a pH in the range 5.2–6.5. Soon after inoculation with cells (2–12 h) the pH often declines by as much as 0.5 unit, to rise again as cell growth proceeds. In later stages of growth the pH generally stabilizes between 6.5 and 7.2. In those few instances where attempts have been made to maintain a stable and defined pH level from culture initiation onwards, reduced biomass yields and growth rates have been noted.

Plant cells show a low tolerance to high temperatures. Above 32°C culture viability is much reduced and productivity declines markedly. The optimal temperature for growth appears to be at about 27°C, although cultures are usually grown at a 'standard' 25°C. On a larger scale, unlike microbial processes which are often highly endothermic and require specialized cooling systems, with plant cells there may even be a need to provide heat to maintain an optimal operating temperature.

21.2.3   Growth kinetics

One of the greatest contrasts between the growth of cultured plant cells and microorganisms is in their respective rates of growth. While the pattern of growth may be the same (Fig. 21.3), plant cells have doubling times

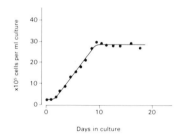

Fig. 21.3   Growth characteristics of a cell suspension culture of *Catharanthus roseus*.

or division rates measured in hours and days while many microorganisms have doubling times of the order of minutes or hours.

Figure 21.3 illustrates growth in batch culture of *Catharanthus roseus* cells, which tend to have faster doubling times than many plants but none the less are much slower than microorganisms. One of the fastest (and quite exceptional) recorded doubling times for a plant cell culture is 15 h, for tobacco cells.

Attempts have been made to study the kinetics of plant cell growth using chemostat continuous culture systems. Initial studies suggested that, at least in relation to carbohydrate substrates, uptake into plant cells obeyed classical Michaelis kinetics (Chapter 4). However more recent studies have questioned this and other kinetic models are now being explored.

21.2.4 Culture modes and productivity

Although predominantly grown in batch culture, plant cells have also been grown in semicontinuous, fed-batch or continuous (chemostat and turbidostat) culture. Biomass (dry wt) yields in excess of $25\,g\,l^{-1}$ have been achieved in batch culture, with carbon conversions greater than 50%. While these figures bear comparison with those from microbial systems, it must be remembered that the data for plant cells are accumulated over days (10–14 days being a fairly typical run time) compared again with hours for microbial cultures. In continuous culture run times of 2–3 months are not uncommon, and productivity figures in excess of $6\,g\,l^{-1}\,d^{-1}$ have been quoted.

The general approach used to continuous and semicontinuous culture with plant cells is however exactly the same as that for microbial systems and will not be discussed here.

21.3 Mass cell growth and production systems

21.3.1 Mass cell growth and properties of plant cells

The provision of sufficient enzymic machinery as viable biomass underpins all biotechnological processes, whether they be mass cell growth systems, single or multistage biosyntheses, immobilized systems or fluidized beds. This need has in turn led to the development of mass culture systems for plant cells. The approaches to this development are very much dependent on the properties of the cell systems for which they are required. Plant cells have a number of characteristics which not only set them apart from microbial cells and traditional fermenter designs, but also place particular constraints on the approach to vessel design and scale-up. The key features of plant cells in relation to mass growth are as follows:

- they have low growth rates, which necessitate long vessel residence times;
- the cells are large ($100\,\mu m$ in diameter) and dense;
- cell volume may change by a factor up to 10^5 during batch culture;
- cell clumps containing between 2 and 200 cells occur to varying degrees;
- older cells possess a large central vacuole which often contains toxic substances;
- metabolic and physiological activity is generally low in comparison with microorganisms (e.g. respiration rates of the order of $1\,\mu mol$ $O_2\,h^{-1}\,(10^6\ cells)^{-1}$;

- the plasmalemma is surrounded by a thick cellulose-based cell wall, which has a high tensile strength but low shear resistance.

Some of these features may be seen in Fig. 21.4.

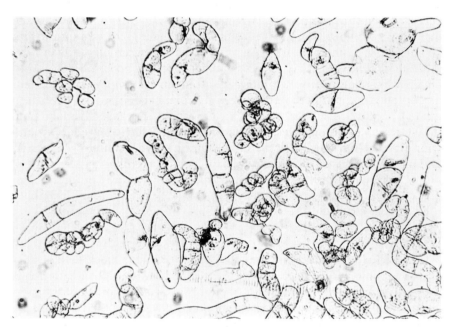

Fig. 21.4   Eight day old cells of *Catharanthus roseus*—periwinkle—grown in suspension culture. Note the varied nature of cell shape, the surrounding cell wall and the many cells grouped together in clumps.

Plant cells have now been grown in a whole range of vessel shapes and sizes. In Japan, tobacco cells have been successfully grown in conventional turbine-stirred bioreactors up to $20\,m^3$ in size. A wide range of cells from other species have been grown in vessels up to about $1.5\,m^3$. In general, however, conventional paddle stirred vessels have not been particularly successful. Those designed for handling microorganisms are generally designed to achieve high gas transfer rates, to cope with a very active metabolism and physiology, and therefore with a high level of shear (Chapter 5). Unfortunately such levels of shear are deleterious to plant cells and may rapidly result in cell lysis and death. To maintain plant cells in conventional bioreactors, internal baffles are normally removed, and the rotation speed of the turbine is reduced to 300 r.p.m. or less. Unfortunately while lowering the turbine rotation speed may reduce shear and cell lysis, problems of maintaining good mixing may then arise, particularly as the nutrient broth of plant cells tends to have a fairly high viscosity which increases during culture growth. One alternative is to modify the usual flatbladed impeller to a marine impeller to decrease shear but aid bulk mixing. Another approach finding increasing favour is to dispense with the impeller system and utilize the incoming gas stream not only to aerate the culture but also to provide mixing, using the air-lift approach with either an internal draught tube or an external loop (Chapter 5). Both systems have the major advantage of low shear characteristics. Bulk mixing in draught tube and loop reactors can however be poor, particularly at high biomass levels (over 15–20 g 1^{-1}), though scale-up to some extent ameliorates this problem. The much lower metabolic and physiological activity of plants results in a much lower respiratory oxygen demand, and in consequence venting rates need not be so high, which also reduces foaming problems. On the whole plant cell cultures do not tend to suffer from foaming problems to the degree observed with microbial systems, but a rather different problem arises. Many plant cell cultures in later stages of growth excrete large amounts of polysaccharide and protein which tends to accumulate around the foam bubbles at the top of the culture. Here it traps cells which are carried up into the foam, and a crust of cell material gradually forms in the head space above the culture. Accretion of cell material may also occur on the underside of the crust, which then 'grows' down into the culture broth aggregating more and more cell material as it does so. This soon results in a breakdown of mixing within the vessel and a highly heterogeneous culture. Though plant cells are not as sticky as many animal cells (Chapter 19), and do not adhere to the general surfaces of the vessel, they will often aggregate around probes, causing problems with sensing devices such as pO_2 and pH electrodes, as well as blocking orifices making the use of recycling systems or weirs in continuous cultures difficult. While the excretion of the polysaccharide may be regulated to some degree through the nature and manner of carbon supply to the culture, no really satisfactory way, either physical or chemical, has yet been found of controlling this phenomenon.

Another important aspect of plant cell mass growth involves the gas regime and the venting rate. As already stated, plant cells have a much lower oxygen demand than microbial cells. It is in fact easy to 'overgas' a plant cell culture, with deleterious effects on both growth rate and final biomass yield. At air flow rates of much more than about 0.5 v.v.m. or k_La values of $20 h^{-1}$, there is a marked extension of lag phase in batch culture, with a following reduction in μ_{max} and an eventual reduction in biomass yield, which can be up to 40% less than in cultures operated at k_La values of less than $20 h^{-1}$. The precise nature of this effect is not understood. One possibility is that high k_La and dissolved oxygen values may in some way reduce citric acid cycle activity (there are precedents for this from work with intact plant tissues). Another explanation is that high air flow rates lead to the stripping off of key volatiles. With nutrient broth pH values in the range of 5.5–6.5, loss of carbon dioxide is a strong possibility, and in experiments where a carbon dioxide bleed has been administered to cultures at high $k_L a$ values, some recovery of biomass yield has been noted.

21.3.2 Production systems

Only in exceptional cases can plant cell cultures, with their slow growth rates, be envisaged as a biomass source. The principle exception is for tobacco biomass, and possibly tea, coffee and ginseng powder. The major thrust of plant cell culture is therefore directed towards production systems for single, typically high added-value, speciality chemicals (see Section 21.4). Secondary product synthesis by plant cell cultures follows, in general, a pattern very similar to that observed in microbial systems; major synthesis typically occurs in the late log or stationary phase of batch culture when cell division has ceased and growth (as increase in biomass) has begun to decline. There are occasional exceptions to this, a good one being the alkaloid serpentine which is synthesized by cultures of *Catharanthus roseus* during active growth. Because of the apparent lack of coupling between cell division, active growth and secondary product synthesis, the general approach to product synthesis with plant cell cultures has either been to use two-stage systems, the first stage being for biomass production and the second for natural product synthesis, or more recently to move to 'zero' growth conditions, such as with immobilized cells or fluidized beds.

A number of nutrient regimes have now been developed which allow for effective two-stage operation. The first stage may be some form of continuous culture where conditions are adjusted for maximum biomass productivity, but which result in little or no product formation. Biomass is then transferred to a second vessel system and the nutrient regime geared to high natural product synthesis. This latter system is usually a batch culture and is characterized by low cell viability if the cells are subsequently subcultured. This two-stage approach has been particularly useful in the development of a process for cardiac glycoside formation.

During the last two or three years a number of laboratories have begun to study immobilized cells or fluidized beds; this is also a form of two-stage process. A wide range of plant cells have now been successfully immobilized and on a wide variety of supports, including starch, agarose and polyacrylamide. Cell viability has been retained for periods in excess of 150 days and continued product synthesis has been observed. A number of problems still remain: first, it is difficult to prevent cell growth and division; second, gas transfer in the immobilizing support may be critically restricted. A third problem, not unique to immobilized systems, concerns the 'leakage/excretion' of product from the cells. With immobilized systems release in some way of the desired product into the bathing nutrient is of key importance for product recovery. In general, it is still uncertain whether products appearing in the medium result from active release, or through cell lysis subsequent to cell death.

21.4   Products from cell cultures

21.4.1   The development of productive cell lines

The years 1973–1974 appear to have marked a major watershed in the development of plant cell culture technology. Prior to this, although many species had been established in culture, few had been shown to synthesize substances characteristic of the parent plant, and none at concentrations observed in high-yielding plant tissues. Even where desirable substances were synthesized it appeared that their production was linked to the development of shoots, roots and other tissue and organ systems. From a biochemical engineering standpoint, the development of tissues, organs and large slow-growing complex masses of cell material is far from desirable in large-scale bioreactors.

The period 1973–1974 was characterized by three major steps forward; the discovery of increasing numbers of cell lines which synthesized products characteristic of the parent plant, the observation of levels of desirable products equivalent to those found in the parent plant, and an increasing number of examples where product synthesis occurred in cell cultures and did not appear to be obligatorily linked to tissue or organ development. Examples which fulfil all of these criteria are listed in Table 21.4; note the wide range of plant species and chemical structures involved. The more general list, Table 21.5, illustrates the range of structures and activities which have been reported. A degree of caution must be exercised when appraising such a list; only a relatively small proportion of the total number will ever be economically viable targets for commercialization through plant cell culture.

Without a detailed discussion of the various products, it is worth noting the more important ones which are obvious targets. Under the heading medicinal agents there are a variety of substances including such important agents as the antileukaemic drugs vincrystine and vinblastine from the Madagascan periwinkle, *Catharanthus roseus*. These are complex dimeric alkaloids of very high added value and a major target for a number of

Table 21.4 Natural product yields from cell cultures and whole plants

Natural product	Species	Cell culture yield	Whole plant
Anthraquinones	*Morinda citrifolia*	900 nmol $(g\ dry\ wt)^{-1}$	Root, 110 nmol $(g\ dry\ wt)^{-1}$
Anthraquinones	*Cassia tora*	0.334% fr wt	0.209% seed, dry wt
Ajmalicine and serpentine	*Catharanthus roseus*	1.3% dry wt	0.26% dry wt
Diosgenin	*Dioscorea deltoidea*	26 mg (g dry wt)$^{-1}$	20 mg $(g\ drywt)^{-1}$ tuber
Ginseng saponins	*Panax ginseng*	0.38% fr wt	0.3–3.3% fr wt
Nicotine	*Nicotiana tabacum*	3–4% dry wt	2–5% dry wt
Thebaine	*Papaver bracteatum*	130 mg $(g\ dry\ wt)^{-1}$	1400 g $(g\ dry\ wt)^{-1}$ leaf and 3000 mg $(g\ dry\ wt)^{-1}$ root
Ubiquinone	*N. tabacum*	0.5 mg $(g\ dry\ wt)^{-1}$	16 mg $(g\ dry\ wt)^{-1}$ leaf

Table 21.5 Substances reported from plant cell cultures

Alkaloids	Latex
Allergens	Lipids
Anthroquinones	Naphthoquinones
Antileukaemic agents	Nucleic acids
Antitumour agents	Nucleotides
Antiviral agents	Oils
Aromas	Opiates
Benzoquinones	Organic acids
Carbohydrates (including polysaccharides)	Proteins
Cardiac glycosides	Peptides
Chalcones	Perfumes
Diathrones	Pigments
Enzymes	Phenols
Enzyme inhibitors	Plant growth regulators
Flavanoids, flavones	Steroids and derivatives
Flavours (including sweeteners)	Sugars
Furanocoumarins	Tannins
Hormones	Terpenes and terpenoids
Insecticides	Vitamins

research groups, but not one has yet succeeded in their synthesis through plant cell culture. The related monomeric alkaloids serpentine and ajmalicine, used as arrhythmic agents and also obtained from periwinkle, have likewise been intensely investigated and in this case cell lines have been isolated which synthesized both alkaloids at levels well above 1% of the dry weight, equivalent to or better than the parent plant. This could provide one of the first major plant cell culture processes. The opiate alkaloids, morphine and codeine, extensively used as painkillers, have been demonstrated in cultures of opium poppy cells (*Papaver somniferum*) and are perhaps an obvious target, not only from the viewpoint of drug supply but also in terms of control of narcotics abuse.

Plants are increasingly being turned to as a potential source of antimicrobials, particularly antifungal agents. Many cell cultures have some level of antimicrobial activity, but few have undergone any degree of rigorous screening. Undoubtedly this could be a most fruitful area for further study.

One of the classical plant products used in medicine is the cardiac glycoside digoxin. Not only have cell cultures of *Digitalis lanata* (the foxglove) been established which will synthesize this cardiatonic, but also cell lines have been isolated which will carry out the chemically difficult single-step conversion of the low-value digitoxin to high-value digoxin. This particular biotransformation was perhaps the first to find a true process application.

In addition to the more obvious pharmaceutical targets, there are substances of importance in foods and agrochemicals. Cell cultures of *Cinchona ledgeriana* have been isolated which will produce the alkaloid quinine, which is a bittering agent and ingredient of many soft drinks, and is still used as a key agent against malaria. Although the alkaloids tend to predominate in secondary product applications, other chemical structures do make an important contribution, for instance terpenoids and the aromatic oils in perfumes and aromas. Cultures have now been established which possess very pleasant aromas, but in perfumery the precise aroma is often a product of a complex mix of chemical structures, difficult to reproduce or mimic. Steroids also have been isolated from cell cultures, in particular diosgenin, which is a key precursor in the steroidal component of the oral contraceptive steroids. However in this case the raw material from traditional plantations has such a low market value that it is doubtful if plant cell culture will ever compete commercially.

Natural colours provide an interesting target. With the increasing problems caused by the various regulatory agencies in response to public concern over synthetic colours in foods, the plant kingdom is an obvious search area for reds, blues, yellows, etc. A variety of reds have now been produced, and recent work has shown that both blue and yellow may be feasible. However in this case there are likely to be problems with the stability of the natural product. Many foods are either near neutrality or slightly alkaline in pH, whereas most natural colours are more stable in acid conditions. Whether it will prove possible to stabilize these products in an acceptable way is difficult to say.

Some of the plant neurotoxins match anything that some groups of the animal kingdom, such as reptiles and arachnids, are able to synthesize. Ricin and curare are particularly good examples. Whether they will ever be targets for plant cell culture one hesitates to contemplate.

21.4.2 Selection and screening

A key aspect of the development of a plant cell culture process is to select cells with a high yield of the desired product. The importance of this cannot be overemphasized. Equally this has so far proved to be one of the most problematical areas of plant cell culture. It is exceedingly difficult to isolate and properly clone single plant cells which may be high yielding. At the same time techniques for the assay of desired substances are often insufficiently sensitive to measure the minute traces of substances present in a single cell. Over the last two or three years developments in radioimmunoassay and enzyme immunoassay have helped overcome this particular problem. Coupled with classical microbial approaches to application of selection pressures this should bring major progress in the years ahead. Yield increases in the region of 10-fold would dramatically change the economic viability of a number of potential products.

21.4.3 Future horizons

In the long term it is likely that the most important applications of plant cell culture will be in biotransformations and in novel products. The potential for a wide range of biotransformations using either immobilized cells or extracted plant enzymes undoubtedly exists, the major restriction to development being our lack of fundamental knowledge of the general biochemistry and enzymology of natural product synthesis. Novel products raise all sorts of possibilities.

The availability of plant cell culture technology brings with it a possible field of applications for gene manipulation techniques, and where a desirable plant product is an enzyme or a simpler peptide then undoubtedly this will be a target for transfer to a more amenable host/production system for process development. The non-nutritive protein sweetener thaumatin is a good example. However where the product is non-proteinaceous, of low molecular weight and at the end of a long multienzyme sequence it is extremely unlikely that gene transfer to another host will be feasible. In this case gene manipulation within the plant cell itself may be of assistance in de-controlling a pathway to allow greater throughput.

21.5 Conclusion

Our knowledge of both plant natural product biochemistry and of plant cell culture lags far behind that for microbial systems. In consequence our time horizons for plant cells must be extended that much further, possibly into the next century, while the fundamental science base is established. The prize is worth working for, but the level of patience required may be very high.

Reading list

Barz, W., Reinhard, E. and Zenk, M. H. (Eds) (1977). 'Plant Tissue Culture and its Biotechnological Application', 419 pp. Springer-Verlag: Berlin.

Fowler, M. W. (1981). The large scale cultivation of plant cells. *Prog. Ind. Microbiol.* **16,** 207–230.

Fowler, M. W. (1982). Commercial applications and economic aspects of mass plant cell culture. *In* 'Plant Cell Biotechnology', (Eds H. Smith and S. Mantell). Cambridge University Press: Cambridge.

Murashige, T. (1978). The impact of plant tissue culture on agriculture. *In* 'Frontiers of Plant Tissue Culture 1978' (Ed. T. A. Thorpe), pp. 15–26. The Bookstore, University of Calgary: Calgary.

Staba, E. J. (Ed.) (1980). 'Plant Tissue Culture as a Source of Biochemicals'. CRC Press: Boca Raton, Fl.

Street, H. E. (Ed.) (1974). 'Tissue Culture and Plant Science 1974', 502 pp. Academic Press: London and New York.

Street, H. E. (Ed.) (1977). 'Plant Tissue and Cell Culture', 614 pp. Blackwell Scientific Publications: Oxford.

Thorpe, T. A. (Ed.) (1978). 'Frontiers of Plant Tissue Culture 1978', 556 pp. The Bookstore, University of Calgary: Calgary.

Index